Oceanography Challenges to Future Earth

Wave breaking against the reef (Copyright Yasuyuki Koike)

Matsushima Bay (Japan) (Copyright Yutaka Okumura)

Spat collection in Matsuchima Bay (Copyright Yasuyuki Koike)

Link between the two cultures: *Crassostrea* and *Ostrea* (Copyright Yasuyuki Koike)

Teruhisa Komatsu · Hubert-Jean Ceccaldi ·
Jiro Yoshida · Patrick Prouzet ·
Yves Henocque
Editors

Oceanography Challenges to Future Earth

Human and Natural Impacts on our Seas

Société franco-japonaise d'Océanographie

La mer

Springer

Editors
Teruhisa Komatsu
SFJO Japan
Masion franco-japonaise de Tokyo
Yokohama College of Commerce
Atmosphere and Ocean Research Institute
The University of Tokyo
Tokyo, Japan

Jiro Yoshida
SFJO Japan
Masion franco-japonaise de Tokyo
Department of Ocean Sciences, Laboratory of Physics and Environmental Modeling
Tokyo University of Marine Science and Technology
Tokyo, Japan

Yves Henocque
Politique maritime et gouvernance
Correspondant Asie-Pacifique (JAMSTEC)
Institut Français pour l'Exploitation de la Mer (Ifremer)
Issy-les-Moulineaux, France

Hubert-Jean Ceccaldi
SFJO France
Académie des Sciences, Lettres et Arts de Marseille
Marseille, France

Patrick Prouzet
SFJO France
Paris, France

ISBN 978-3-030-00137-7 ISBN 978-3-030-00138-4 (eBook)
https://doi.org/10.1007/978-3-030-00138-4

Library of Congress Control Number: 2018954611

This Springer imprint is published by the registered company Springer Nature Switzerland AG
The registered company address is: Gewerbestrasse 11, 6330 Cham, Switzerland

French and Japanese scientists taking part in the SFJO Symposium at Shiogama. Copyright SFJO Japan

French and Japanese scientists taking part in the synthesis meeting at the "Maison franco-japonaise de Tokyo". Copyright SFJO Japan

Organization Committee

Prof. Teruhisa Komatsu (President SFJO Japan)
Prof. Hubert-Jean Ceccaldi (President SFJO France)
Prof. Shiro Imawaki (JAMSTEC/SFJO Japan)
Dr. Yves Hénocque (Ifremer/JAMSTEC/SFJO France)
Dr. Patrick Prouzet (Vice-president SFJO France)
Prof. Tsutomu Morinaga (SFJO Japan)
Prof. Jiro Yoshida (TUMSAT)
Prof. Yasuyuki Koike (SFJO Japan/ex TUMSAT)
Prof. Jota Kanda (TUMSAT)
Prof. Yuji Tanaka (TUMSAT)
Prof. Hisayuki Arakawa (TUMSAT)
Prof. Yujiro Kitade (TUMSAT)
Prof. Takeyoshi Nagai (TUMSAT)
Prof. Toshiki Nakano (Tohoku University)
Dr. Hiroshi Uchida (JAMSTEC)
Prof. Kaoru Ichikawa (Kyushu University)
Prof. Mitsuru Hayashi (Kobe University)
Prof. Tomoharu Senjyu (Kyushu University)
Dr. Yutaka Okumura (FRA)
Dr. Hideki Takami (FRA)
Dr. Tetsuo Seki (SFJO Japan)
Prof. Kenji Ookoshi (Toho University)

Scientific Committee

Prof. Teruhisa Komatsu (President SFJO Japan)
Prof. Hubert-Jean Ceccaldi (President SFJO France)
Prof. Shiro Imawaki (JAMSTEC/SFJO Japan)
Dr. Yves Henocque (Ifremer/JAMSTEC/SFJO France)

Editorial Board

List of Chairmen and Chairwomen

Prof. T. Komatsu
Prof. H.-J. Ceccadi
Dr. Y. Kitade
Dr. S. Ruitton
Prof. H. Arakawa
Dr. Y. Hénocque
Prof. R. Kimura
Dr. S. Vaz
Prof. S. Imawaki
Dr. H. Thébault
Prof. Y. Koike
Prof. J.-C. Dauvin
Dr. T. Seki
Dr. F. Galgani

Foreword

Message de S.E. Monsieur l'Ambassadeur Thierry Dana—Ambassadeur de France au Japon

C'est un grand plaisir pour moi de préfacer cet ouvrage qui regroupe les actes du 16ème colloque franco-japonais d'océanographie qui s'est tenu à Shiogama en novembre 2015.

Le nombre et la régularité impressionnante de ces colloques tenus alternativement en France et au Japon attestent de la vivacité des échanges dans ce domaine. Ce dynamisme doit beaucoup aux deux Sociétés franco-japonaises d'océanographie, basées respectivement en France et au Japon, qui, depuis maintenant plus de trente ans, nourrissent les relations académiques, professionnelles et non moins amicales entre scientifiques et gestionnaires de la mer de nos deux pays.

Le lieu de ce 16ème colloque n'est bien sûr pas un hasard puisqu'il se situe au cœur de la région du Tohoku frappée par un puissant tremblement de terre et le terrible tsunami qui s'en est suivi en mars 2011. Alors que les efforts de reconstruction continuent, je sais combien les deux Sociétés franco-japonaises d'océanographie ont œuvré ensemble au redémarrage des activités, plus particulièrement d'ostréiculture, une activité majeure de la région qui, dans les années 70, avait généreusement sauvé l'ostréiculture française.

Domaine historique de notre coopération scientifique bilatérale, l'océanographie continuera à jouer un rôle essentiel dans l'avenir. J'ai la conviction que la Mer et les Océans représentent un enjeu majeur de nos sociétés modernes, que ce soit sur les plans environnemental, économique ou stratégique, qui prendra une place de plus en plus importante dans les échanges entre les deux grandes puissances maritimes que sont la France et le Japon. Nul doute que les deux Sociétés franco-japonaises d'océanographie continueront à alimenter activement cette relation, notamment avec le 17ème colloque déjà en préparation en 2017 à Bordeaux.

Je rends donc hommage aux présidents et aux membres des deux Sociétés franco-japonaises d'océanographie ainsi qu'à tous les participants qui ont fait de ce 16ème colloque un succès comme l'atteste la richesse du contenu de cet ouvrage.

Tokyo, Japan

Thierry Dana
Ambassadeur de France au Japon

Preface

The Sea Under Human and Natural Impacts: Challenge of Oceanography to the Future Earth

The sea is an indispensable component of the Earth. Human societies obtain a great benefit from the sea.

Global warming threatens the sea through seawater temperature and sea level rises, acidification, and huge depressions.

The tsunami repeats destroying coastal zones. Increase in population and industry activities along the coast, especially in Asia, cause degradation of coastal ecosystems through direct and indirect manners such as reclamation, overexploitation of bioresources, and pollution.

We need to understand physical, chemical, and biological mechanisms of the sea and threats of human and natural impacts on the sea and its ecosystems at the present day.

We base ourselves on a global understanding of the sea, including the impacts that affect it, in order to predict the future importance of the sea to the planet under several global warming scenarios.

Our common goal is to maintain the Oceans and the Earth in the future in a sustainable way. We are working and will continue to work together in scientific areas that explore the sustainability of natural processes. We need to invest efforts to educate people for maintaining marine environments the best possible for present and future generations from points of view of not only a scientific aspect but also cultural and social ones which constitute a relationship between Mankind and Nature.

The 16th France-Japan Symposium entitled as "Oceanography Challenges to Future Earth" was organized by the two SFJOs and held in Shiogama, Tokyo and Hinase, Japan from 17 to 23 November 2015, on marine sciences focusing on progress in oceanography to understand the present sea from physical, chemical, biological, ecological aspects including fishery sciences and social aspects. We discuss the sea in the future Earth under different scenarios and how to establish sustainable ocean through marine science.

Finally, we acknowledge here the supports and cooperations of Maison franco-japonaise de Tokyo, Fondation franco-japonaise de Sasakawa, Fisheries Agency of Japan, French Embassy in Tokyo, Hinase Fisheries Cooperative, and various funding agencies and individuals that permit us to realize the 16th France-Japan Symposium on marine sciences in Japan.

Tokyo, Japan — Prof. Teruhisa Komatsu
President of SFJO Japan

Marseille, France — Prof. Hubert-Jean Ceccaldi
President of SFJO France

Contents

Part I Introduction

1 **Evolution and Progress Accomplished During Previous French-Japanese Symposiums of Oceanography** 3
Hubert-Jean Ceccaldi

2 **Challenge to Resolve Problems in the Ocean and Coastal Waters in Future Earth with Stronger Cooperation Between the Two Societies *Franco-Japonaise* of Oceanography** 15
Teruhisa Komatsu

Part II Natural and Anthropogenic Impacts

3 **Slower Decrease in Radioactive Concentrations in Some Fish Species After the Fukushima Daiichi Nuclear Power Plant Disaster** 21
Hisayuki Arakawa

4 **Influence of Behavioral Patterns of Several Fish Species on Their Radioactive Cesium Concentrations Revealed with a Biotelemetry System After the Nuclear Accident Caused by the 2011 Great East Japan Earthquake** 29
Keiichi Uchida, Kohei Hasegawa, Yoshinori Miyamoto, Hisayuki Arakawa, Seiji Akiyama and Naoto Hirakawa

5 **Estimate of Water Quality Change in Osaka Bay Caused by the Suspension of Marine Sediment with Mega Tsunami** 45
Mitsuru Hayashi, Satoshi Nakada, Shunich Koshimura and Eiichi Kobayashi

6 **Litter in the Mediterranean Sea** 55
François Galgani

7 The European eel (*Anguilla anguilla*) in France: An Example of Close Cooperation Among Researchers and Fishers to Study and Manage an Endangered Species 69
Patrick Prouzet, Elsa Amilhat, Catherine Boisneau, Philippe Boisneau, Eric Feunteun and Nicolas Michelet

8 Trophic Cascade in Seaweed Beds in Sanriku Coast Hit by the Huge Tsunami on 11 March 2011: Sea Urchin Fishery as a Satoumi Activity Serving for Increase in Marine Productivity and Biodiversity 95
Teruhisa Komatsu, Shuji Sasa, Hiroki Murata, Shuhei Sawayama, Masahiro Hamana, Minami Asada, Ryo Tsujimoto, Genki Terauchi and Tetsuo Yanagi

9 The English Channel: Becoming like the Seas Around Japan 105
Jean-Claude Dauvin, Jean-Philippe Pezy and Alexandrine Baffreau

Part III Physical Oceanography

10 Recent Research Results and Future Project in the Antarctic Ocean by Umitaka-Maru Research Group for Physical Oceanography 123
Yujiro Kitade, Keishi Shimada, Shigeru Aoki and Kay I. Ohshima

11 Response of Near-Inertial Internal Waves to Various Typhoon-Tracks Around the Tango Peninsula, Japan 137
Keiichi Yamazaki, Yujiro Kitade, Yosuke Igeta, Yutaka Kumaki and Tatsuro Watanabe

12 A High-Resolution Unstructured Grid Finite Volume Model for Currents Around Narrow Straits of Matsushima Bay 161
Hidekazu Shirai, Ritsuki Kunisato, Shinya Magome, Teruhisa Hattori, Takamasa Takagi, Katsuaki Okabe, Kazufumi Takayanagi and Shigeho Kakehi

13 Observation of Near-Bottom Current on the Continental Shelf Off Sanriku 171
Daigo Yanagimoto, Kiyoshi Tanaka, Shinzou Fujio, Hajime Nishigaki and Miho Ishizu

Part IV Innovative Research

14 Control of Pressure-Driven Microdroplet Formation and Optimum Encapsulation in Microfluidic System 181
Mathias Girault, Akihiro Hattori, Hyonchol Kim, Kenji Matsuura, Masao Odaka, Hideyuki Terazono and Kenji Yasuda

15 Development of a De-oiling System for Seabed Sediments 195
Yoshichika Ikeda, Motohiro Miki, Hisayuki Arakawa and Mitsuru Izumi

16 Development of an Optical Detection System of Fuel Oil on Seabed Sediments 203
Akira Matsumoto, Kazuki Toguchi, Yoshichika Ikeda and Hisayuki Arakawa

17 Retinomotor and Stress Responses of Marbled Sole *Pseudopleuronectes Yokohamae* Under the LEDs 217
Rena Shibata, Yasuyuki Uto, Kenichi Ishibashi and Takashi Yada

18 Metabolome Profiling of Growth Hormone Transgenic Coho Salmon by Capillary Electrophoresis Time-of-Flight Mass Spectrometry 223
Toshiki Nakano, Hitoshi Shirakawa, Giles Yeo, Robert H. Devlin and Tomoyoshi Soga

19 Estimating the Diets of Fish Using Stomach Contents Analysis and a Bayesian Stable Isotope Mixing Models in Sendai Bay 235
Hiroyuki Togashi, Yukinori Nakane, Yosuke Amano and Yutaka Kurita

Part V Coastal Ecosystem and Management

20 Ecological Status of Atlantic Salmon (*Salmo Salar* L.) in France: Need for an Ecosystemic Approach 249
Patrick Prouzet and Nicolas Michelet

21 Challenges to Harmonize Sustainable Fishery with Environmental Conservation in the Coastal Ecosystems Under Oligotrophication 277
Masakazu Hori, Masahito Hirota, Franck Lagarde, Sandrine Vaz, Masami Hamaguchi, Naoaki Tezuka, Mitsutaku Makino and Ryo Kimura

22 One-Year Colonization by Zoobenthic Species on an Eco-Friendly Artificial Reef in the English Channel Intertidal Zone 285
Jean-Claude Dauvin and Aurélie Foveau

23 New Installations of Artificial Reefs Along the Coast of the Landes (South–West Atlantic Coast of France) 295
Gérard Fourneau, Florence Dufour, Aurélie Penne, Nelly Ferrou, Thomas Scourzic, Aurore Laborde and Elodie Zaccari

24 Marine Ecosystem Services: Perception of Residents from Remote Islands, Taketomi Town 305
Kazumi Wakita, Keiyu Kohama, Takako Masuda, Katsumi Yoshida, Taro Oishi, Zhonghua Shen, Nobuyuki Yagi, Hisahi Kurokura, Ken Furuya and Yasuwo Fukuyo

25 Quantitative Mapping of Fish Habitat: From Knowledge to Spatialised Fishery Management 313
Sandrine Vaz and Olivier Le Pape

26 Do Our Ocean Policies Make Any Difference in the Wellbeing of Coastal Communities? 325
Yves Henocque

Part VI Aquaculture

27 Heterogeneity of Japanese Oyster (*Crassostrea Gigas*) Spat Collection in a Shellfish Farmed Mediterranean Lagoon 341
Franck Lagarde, Martin Ubertini, Serge Mortreux, Adeline Perignon, Axel Leurion, Patrik Le Gall, Claude Chiantella, Slem Meddah, Jean-Louis Guillou, Gregory Messiaen, Béatrice Bec, Cécile Roques, Delphine Bonnet, Hélène Cochet, Ismaël Bernard, Erika Gervasoni, Marion Richard, Gilles Miron, Annie Fiandrino, Stephane Pouvreau and Emmanuelle Roque D'orbcastel

28 Suitable Oyster Culture Density in Oginohama Bay, Miyagi, Japan 351
Yutaka Okumura, Akatsuki Nawata, Hiroshi Ito, Akio Oshino and Motoyuki Hara

29 Population Dynamics of the Manila Clam *Ruditapes Philippinarum* and Implications of the 2011 Tsunami Impact in Two Shallow, Semi-enclosed Bays in Northeastern Japan 365
Hirokazu Abe, Masami Hamaguchi, Naoto Kajihara, Yuichi Taniai, Akio Oshino, Akihiro Moriyama and Takashi Kamiyama

30 Feed and Feeding in Certification Schemes of Sustainable Aquaculture 387
Catherine Mariojouls, Raphaëla Le Gouvello and François Simard

Part VII Short and Preliminary Communications

31 French Bluefin Tuna Longline Fishery Bycatch Programme 401
François Poisson, Sophie Arnaud-Haond, Hervé Demarcq, Luisa Métral, Blandine Brisset, Delphine Cornella and Bertrand Wendling

32 ^{137}Cs and Tritium Concentrations in Seawater off the Fukushima Prefecture: Results from the SOSO 5 Rivers Cruise (October 2014) 407
Michio Aoyama, Hervé Thébault, Y. Hamajima, Sabine Charmasson, Mireille Arnaud and Céline Duffa

Author Index 411

Part I
Introduction

Chapter 1
Evolution and Progress Accomplished During Previous French-Japanese Symposiums of Oceanography

Hubert-Jean Ceccaldi

Abstract France and Japan are partners in marine science and technology for a long time. Léonce Verny and Louis-Emile Bertin created and organized the dockyards and the Navy of Japan in the late nineteenth century. This country built many warships on its plans and this new Japanese fleet defeated the Russian fleet in May 1905 at Tsushima. Continuous exchanges took place between the two countries in the marine and maritime fields. Paul Claudel, Ambassador of France in Japan, founded the « Maison Franco-Japonaise » in Tokyo, which is the seat of 26 specialized Franco-Japanese Societies including Franco-Japanese Society of Oceanography of Japan. In marine science and technology, a friendly and effective cooperation extended exchanges with the French bathyscaphes dives during the 1960s, in the depths of Japanese waters and the establishment, at that time, of the French-Japanese Society of Oceanography of Japan. Many other areas of cooperation focused on shellfish, aquaculture of various marine organisms, artificial reefs, coastal development, Law of the Sea, etc. Since 1983, Franco-Japanese symposiums took place between Japanese and French experts, alternately in each country. They all brought new ideas to the participants. These differences always enrich cultural approaches by which selected topics are discussed and addressed in each country. The consideration of the issues in previous conferences shows that participants express similar interests in each country, on the functioning of marine ecosystems, especially coastal ecosystems, exploitation of these ecosystems, changes in operation or under effects of human activities. They provided a good highlight about the human factors involved in the use patterns of marine ecosystems. The necessities are now required to Mankind to make his primitive ecosystems natural abilities, which will require the study of new topics of multidisciplinary research.

H.-J. Ceccaldi (✉)
Président de la Société Franco-Japonaise d'Océanographie,
Académie des Sciences, Lettres et Arts de Marseille, Marseille, France
e-mail: ceccaldi.hubert@orange.fr

T. Komatsu et al. (eds.), *Oceanography Challenges to Future Earth*,
https://doi.org/10.1007/978-3-030-00138-4_1

Keywords Cooperations · Symposiums · Paul claudel · Shibuzawa Eiichi · Warships · Bathyscaphs · Maison Franco-Japonaise · Aquaculture · Littoral · Marine productivity · Perturbations · Drugs · Résilience · Sato-umi · Governance

1.1 Some Historical Reminders

Each country suffers attacks of nature, and certainly Japan more than France. In economic development, which followed the industrial revolution, each country has largely used its ports and coastal areas. These were modified, polluted, disturbed under the influence, often harmful, of human activities. On these issues, France and Japan have been almost permanent exchanges, which are based on a long common history in the marine and maritime fields.

In 1864, French engineer Léonce Verny, performs at the request of the Shogun, the great naval arsenal of Yokosuka, which was the first modern shipyard of Japan and he led from 1866 to 1875. He created a school of naval architecture. He was also the source of a large lighthouses building program along the Japanese coasts.

After him, Louis-Emile Bertin was the creator of the Japanese Navy during the Meiji Era. He spent four years in Japan, from 1886 to 1890 and was special adviser to the emperor Mutsuhito. He founded the arsenals of Sasebo, near Nagasaki and Kure, near Hiroshima.

No less than 68 major warships were built according to his plans and its guidelines. Thus, Japan put an end to his vis-à-vis inferiority of the Chinese fleet at the time, which had built its warships in Germany. It is this new Japanese fleet destroyed in May 1905 off the Tsushima Islands, a Russian war fleet coming from the Baltic.

Another major step in exchanges between the two countries was the creation by Paul Claudel, then ambassador of France in Japan, the « Maison Franco-Japonaise » (Nichi-Futsu Kaikan) in Tokyo, with the support of a major Japanese private foundation, with the active support of the Viscount Shibusawa Eiichi.

French-Japanese Society of Oceanography (Nichi-Futsu Kaiyo Gakkai) still has its headquarters in this place, as well as 26 other French-Japanese scientific Societies specialized in various disciplines: literature, economics, sociology, laws studies, art, etc., and in many other areas.

This friendly and efficient cooperation has been strengthened with the French bathyscaphes dives during the 1960s, in the depths of Japanese ocean trenches and the establishment in Japan at that time, of the French-Japanese Society of Oceanography of Japan. For example, Professors of Marine Biology Jean-Marie Pérès, of the Faculty of Sciences of Marseilles and Pierre Drach, of the University of Paris, were received by the Emperor Hirohito.

Henri-Germain Delauze, who later founded the International Diving Company COMEX specialized in deep diving, was at that time in charge of bathyscaphes technology. He often went to Japan later for technical cooperations.

Furthermore, a rich flow of exchanges took place between the marine science experts from Japan and France as well as in formal settings thanks to researchers from the Associations activities.

Themes of joint research were chosen in close cooperation by the two Societies, and information about research, then in progress, were exchanged in several specialities. Thus, Kaïko I and II campaigns, who studied some marine deep sea bottoms of Japan, under the direction of Xavier Le Pichon and Y. Kobayashi, were a great success and achieved very good results (Le Pichon et al. 1987). These deep dives, conducted by the submersible "Nautile"—which dipped for the first time up to 6,000 m deep—and its support ship the "Nadir" led to new discoveries (Le Pichon 1986). Another great success was also achieved by the world tour of the Ifremer research vessel, the "Jean Charcot".

Specialized technical exchanges took place in the area of polymetallic nodules, marine technology, offshore structures or the environment and, more recently, marine energy and biotechnology.

Many other examples of successful cooperation have been conducted between the two Franco-Japanese Societies of Oceanography, often with exchanges of researchers and doctoral students could be given in the field of shellfish farming, farming of eels, aquaculture of penaeid shrimps, lobsters and fish, algae, physiology and biochemistry of marine, artificial reefs, coastal development, the Law of the Sea, etc.

It was the same for artificial reefs area in which technicians and industrial of Japan have extensive experience (Yamane 1985). But for their part, the French teams specialized in this field have given results of researches extremely useful in the design of new types of reefs and habitats for marine organisms.

It became clear that we need to clearly differentiate the "productive reefs " designed to accommodate fish that are then harvested, and reefs for maintaining and enhancing biodiversity, so that artificial habitats are home to species that are part of local ecosystems.

A group of French-Japanese studies dedicated to coastal developments was also established in 1988 as part of the « Maison Franco-Japonaise », a group whose activities were harmonized by François Simard. This group has been active for four years and has published several interesting pamphlets. Other books on tourism development of the Japanese coast, on the legal regime of coastal development, and about general conflicts of coastline, taking French examples and Japanese problems.

1.2 General Themes of Recent French-Japanese Symposiums

Since 1983, the French-Japanese Symposiums are held fairly regularly between Japanese and French experts in marine sciences who are members of both French-Japanese Societies of Oceanography, alternately in Japan and France. They all brought new ideas to the participants of both countries. They always enrich the differences between scientific, technical and cultural approaches by which selected topics are discussed and addressed in each country. The description of the first 12 Symposiums was made in short summaries (Ceccaldi 2011; Koike 2015). Also worth mentioning is that the French-Japanese Symposiums of Oceanography of Japan is publishing since its foundation, a scientific journal whose title is French « La Mer » (« The Sea »).

This Communication aims to highlight some of the most interesting general themes that were most often defined and treated together between the French-Japanese Societies, and due to the common interest was expressed in each the two countries at the time of the holding of conferences.

Thus, the general theme of the 13th Symposium was: "*Global Change: Mankind-Marine Environment Interactions*".

The 14th Conference—in which the Fiftieth Anniversary of the Creation of the French-Japanese Society of Japan was celebrated—was titled: *« Towards sustainable use and management of the oceans »* .

The theme of the 15th Symposium: "*Marine Productivity: Disturbance and Resilience of Socio-Ecosystems*".

A review of these titles, it appears that participants express similar interest on the following major topics:

- How the marine ecosystems are functioning and working, especially coastal ecosystems;
- How can man exploit them without much change, even without destroying them;
- How their functioning has been altered, either by natural phenomena (storms, typhoons, tsunamis) or by the effects of human activities;
- What are their possibilities of returning to an initial or similar state;
- What are the human factors involved in the use of the wealth of marine ecosystems.

Also worth remembering that these exchanges involve numerous higher education and research institutions. The 15th French-Japanese symposium of Oceanography has contacted over 30 research organizations in France and Japan.

1.3 Experiments Carried Out Thanks to the Seminars and Lessons Learned

Many mutually beneficial cooperation examples can be cited. We did not here the opportunity to mention the numerous papers that were presented in previous conferences. We just want to mention a few that are particularly characteristic even among the oldest. The reader may refer to the content of previous conferences on the website of French-Japanese Societies of Oceanography.[1]

Among the topics of common interest, we can mention the breeding of marine shellfish, especially oysters (Grizel and Héral 1991, Mariojouls and Prou 2015), abalone (Stott et al. 2004), mussels, lobsters (Kittaka 1984), cephalopods (Bourguignon 1992), the Saint-Jacques shells (Lucien-Brun 1983; Lucien-Brun and Lachaux 1983; Matsumoto et al. 2015), clams, and other species (Le Borgne and Le Pennec 1983). The techniques of artificial breeding (Hénocque 1983; Kakimoto 1985), hatcheries (Chiba 1982), producing larvae (Uno 1984), post-larvae (Yagi 1984), fisheries (Bailly et al. 1992; Bourguignon 1992), spat exchanges (Koike 2015) resulted in numerous co-operations for the benefit of both countries.

In the field of aquaculture, the exchanges of experience were numerous. This is the case of sea bream (Châtain 1982), bar, turbot, eel (Chiba 1982), salmon (Nomura 1984), penaeid shrimp (Pastoureaud 1972; Guary 1973; Guary et al. 1976; Van Wormhoudt 1980). In almost all cases, basic researches were more developed in France (Ceccaldi 1982), but the technical achievements and practical achievements to medium and large scale were made in Japan (Yamane 1985).

We learned together, not only during conferences, but during the field visits, examining the consequences of the storms, large earthquakes, tsunamis, as well on the material, scientific, ecological, technical, professional, economical and social.

Regarding the perturbations, it also appeared that we have to clearly differentiate disturbances due to natural phenomena—against which one can do nothing—,and perturbations due to human activities against which we must act, even for a long term period. In almost all cases, our exchanges are based on quantitative data obtained using technology more and more sophisticated and original.

Thus, new techniques using e.g. ultraviolet radiations and lasers have been used to study deep ecosystems characteristics near hydrothermal vents (Sasano et al. 2015). The use of small cameras attached to the fish helped to understand certain characteristics of their behaviour (Tanoue et al. 2013).

It is the same in algal farming, in developing and amenities of coastal areas, in the Law of the Sea, in the flow of radioactive elements in food webs, in the biology of polychaetes in the distribution of planktonic species, etc.

Several other major problems were found during our common works: the invasion of plastics in the marine environment, both on the seabed and in the surface, the fact that marine protected areas are not really protected; and even

---

[1]http://www.socfjp.com.

the presence of molecules produced to modify and to treat human physiology and health of populations, hormones, antibiotics and drugs that act directly on the species in the ecosystems. These themes will be necessarily taken in the future.

1.4 Governance

In our previous conferences, it became apparent that the management, between nations, of marine natural resources must be unavoidable, because the oceans have no borders. That management obviously still spends a lot of time before a regulation of the exploitation of marine resources is established globally (de Montbrial 2015).

However, it begins to do locally, regionally and internationally. But the exploitation of resources can only be done in good conditions, that when we know precisely what is the part being removed without disturbing significantly the ecosystem of which it is part. Or ecosystems vary over time and especially with the physical and chemical characteristics of the environment in which they exist (Girault et al. 2011).

Forging links between the different countries of specialists, either as part of informal networks, either as part of formal twinning for example, is real consistent progress, and can serve as examples. So we need to achieve to create, implement and operate new effective governance in particular areas.

The content of the papers presented in previous conferences shows that it will be essential to deal in near future contemporary issues related to the existence and the depletion of natural resources, their management, and human factors that play a major role—but usually inconspicuous—in their operations. One of the major concepts was highlighted: that of an indicator based on ecosystems to enable rational exploitation of fisheries (Trenkel et al. 2015; Lehuta et al. 2013; Dauvin 2015).

So far, the decisions of politicians are always to prioritize economic, industrial and social considerations, and put into the background ecological imperatives nevertheless the basis of the existence of wealth upon which the economy and structures social, so the survival of the Mankind himself, in each country. These fundamental ideas began to be expressed quite clearly during our recent meetings.

1.5 Resilience

The theme of the resilience of disturbed ecosystems was treated during our previous symposia, too modestly compared to the importance of the disturbances created by man, as the latter will not stop overnight. Industrial machineries, economic and financial, major sources of ecosystem disturbances, are operating today at their full capacity.

This is notoriously the case of China, India and many emerging countries. It will be years before these products are modified significantly to become less polluting, during which they will continue to exert their effects. We will keep these themes in priority objectives to consider not only now but later.

Many facts show that today significantly disturbed ecosystems almost never return to their original state. Parts of their original characters are recovered, but most of the time, leaving some features are lost permanently. We can assert that it never returns to its original condition. A destroyed primary forest for timber harvesting, can not return to its original state. Coastal funds that were often the place dredging can not return to their original state. Seabed in which were collected large amounts of sand and gravel will centuries to regain their equilibrium departure profile.

In our scientific meetings, we realized that many human societies, who live directly or indirectly from the use of marine natural resources, already had to evolve to continue to exploit these modified resources and disturbed environments (Simard 1990).

The resilience thus affects human populations themselves, forcing them to adapt to new situations: this area of research has not led to detailed studies to date.

1.6 Importance of the Role of Cultural Factors in the Exploitation of Resources

The man almost always played a negative role on the exploitation of marine natural resources. This exploitation has often resulted in various marine cultures and specific behaviors in both ports in coastal areas around the world. Overfishing and destruction of many coastal species habitats nurseries are often the consequences characteristics.

In the relations established between French researchers and Japanese researchers in marine science, we were able to compare the behavior vis-à-vis the marine fishermen and the very different role of fishing cooperatives in each country. The historical and cultural factors largely explain the differences.

1.6.1 Sato-Yama and Satoumi

It is important to remember that Japan was virtually closed off from any relationship with foreign countries for 260 years, until 1868. During this long period of each village in the countryside or in the mountains had to stand on himself and practically lived in autarky from the management of natural resources or of crops used in common are located in their immediate vicinity (Komatsu and Yanagi 2015). This attitude wise use of the natural environment has been named "Sato-Yama" or village of the mountain. This positive attitude and thrifty found itself greatly facilitated by the culture and traditions of the country. Indeed, the Japanese population has received in its history, on the bases of Confucianism sometimes overestimated (Ansart 2015), the major influence of Buddhism, which protects life, and Shintoism, which emphasizes the relationship with nature. These cultural traits have strongly influenced the mentality of the Japanese people until today.

When extrapolating on coastal environments and these fundamentals, and we consider the operation of fishing villages, the same basic attitude persists and fishing cooperatives operate the marine environment so as to not destroy the ecosystems in which they will collect them and taken their own subsistence. The establishment of aquaculture cages or long ropes on which to grow their oysters, or when installing aquaculture cages, or when they implant on the seabed artificial reefs, all these activities are conducted with the desire not disrupt the functioning of marine natural environments.

The quasi-permanent implantation of aquaculture cages in a determined coastal area, and immersion and facilities on the seabed of artificial reefs to be operated thereafter, permit the inference that the marine environments, although they formally belong to the state, is in fact operated almost exclusively, as would private property by members of the fishing cooperative which is responsible for such equipment.

As in mountain villages or countryside, the seaside community activities take place. This attitude also exists in the fishing villages: it ensures the cohesion of the group, of the cooperative and of the village, and induce a common approach to the frugal use of natural resources, animal and plant, of the surrounding environment.

In this perspective, the man—the fisherman in this case—is easier to feel part of the ecosystem in which it is built and on which it feeds.

In France and in many western countries, the individualistic behavior often prevails, as well as the culture of immediate profit. The changes in the economic and social organization of oyster farmers have done based on long traditions and spat successive imports of foreign origin, including Japan (Dupont 1992; Ikematsu 1992).

It is this kind of ideas that comparisons and cooperation between members of the two French-Japanese Societies of Oceanography became the most successful.

1.6.2 Human Integration in Natural Ecosystems

One of the major conclusions of our previous conferences is to view conclusively that man, according to the geographical area where he lives, is part of specific ecosystems and that all its activities must now be oriented according to its harmonious integration natural ecosystems or roughly artificialised which it draws its sustenance.

In fact, the relationships established between people and animals living beings should be deeply reviewed not only in economy or sociology, but to a view of anthropology, because we have a vision of the world ethnocentric and anthropocentric, we take an obvious first. These concepts have been well developed by outstanding anthropologists (Descola 2005, 2016) some of which have highlighted the importance of relations between popular and traditional cultures on the one hand and the relationship between human and natural resources.

This shows that these cultural factors are extremely important in how natural resources were and are exploited.

Now man must make to ecosystems in one form or another, he took it, or rather, what he borrowed. We began to understand the concept of "socio-ecosystem", which deserves further development.

The ideas of preservation of natural environments are binding on all sectors of society, and more or less rapidly according to the different countries, as many of them have no tradition and no culture of preserving the naturel environment.

Mostly outside of Asia, the monotheistic religions have a prominent place in one God to explain the creation of all things on the planet, especially the creation of living beings. They also give a prominent place to the human species compared to other animal species: Man on Earth is to use the wealth that nature—and God—put at his disposal.

Surprising as it may seem at first, Pope François has published 18 June 2015 an Encyclical « climate » very detailed, 191 pages, a great novelty, entitled « Laudato Si » (Praised be you) « on safeguarding the common house », which emphasizes the excesses of consumption, social injustice especially reach the lowest of human populations, on water shortage will reach way large areas of the planet, on the excessive consumption of natural resources, biodiversity loss, on weak political decisions facing this enormous problem that will affect future generations. He warned those who seek to minimize the effects of global warming, and its dramatic consequences. Referring to St. François of Assisi, the Pope recommends a return to simplicity, which in no way diminishes the intensity of life. This attitude represents a significant shift in one of the great monotheistic religions of the world. It gives priority to protecting the natural environment in order to protect humans.

These recommendations join the traditionally frugal lifestyle of the Japanese, often imbued with Buddhism, Shintoism and Confucianism even, that most French researchers who have stayed in Japan have discovered in this country. Japan remained closed to foreign imports for 260 years: he retained its frugality until today.

1.7 Conclusion

The above remarks have highlighted the interest of exchanges between marine science specialists from both France and Japan Societies of Oceanography.

Knowledge of the evolution of technical measures and marine observation helps take new steps in the understanding of how function the seas and the oceans. These natural areas are subject to disruption due to natural phenomena against which can do nothing, and other disturbances due to human activities. Natural ecosystems see the features of their operation vary all the effects of these disturbances. Then, when the disturbance ceases, these ecosystems back more or less to a state close to their usual natural state. But human populations living from the production of these ecosystems must also evolve during these stages, with varying degrees of success following their resilience.

We see and differentiate more clearly the concept of socio-ecosystem showing the very close relationship between man and the natural environment.

The great lesson we have learned from our French-Japanese meetings is that Mankind must make the natural marine environment sufficient evidence that these ecosystems are functioning normally again. The choice is simple: either we favor human and economic mechanisms he devised, destroying irrevocably the natural environment, or we preferred normal functioning of ecosystems which will enable man to survive.

References

Ansart O (2015) L'étrange voyage de Confucius au Japon. *Anima Viva*, 137 p

Bailly D, Bourguignon G, Kase K (1992) An introduction to Japanese and French fisheries. Oceanis 18(4): 371–395

Bourguignon G (1992) Japan's cephalopod industry. Océanis 18(4):467–477

Ceccaldi HJ (1982) Contribution of physiology and biochemistry to progress in aquaculture. Bull Japan Soc Sci Fish 48:1011–1028

Ceccaldi HJ (2011) Introduction: a brief history of the activities between the two Sociétés Franco-Japonaises d'Océanographie, pp xxxv-xli. In: H-J Ceccaldi et al (eds) Proceedings of 13th French-Japanese oceanography symposium. *Global change: mankind-marine environment interactions*, Springer Science, 447 p

Châtain B (1982) Contribution à l'étude de l'élevage larvaire de la daurade Japonaise (*Pagrus major*). Thèse Université Aix Marseille II, 121 p

Chiba K (1982) Water quality and eel production in Japanese greenhouse ponds using: 33–63. In: First symposium on France-Japon, Aquaculture, Montpellier, 16 Déc 1982, 152 p

Dauvin J-C (2015) The estuary-bay of seine continuum: the need for an ecosystem-based management: 197–204. In: H-J Ceccaldi et al (eds) Proceedings of 15th French-Japanese oceanography symposium: *Marine productivity: perturbations and resilience of socio-ecosystems*, Springer Science, 391 p

Descola P (2005) Par delà nature et culture. Gallimard éd. 640 p

Descola P (2016) L'écologie des autres: L'ethnologie et la question de la nature. Quae édit, 112 p

Dupont P (1992) Dynamique économique et sociale en milieu ostréicole. Océanis 18(4):487–495

Flasch JP, Leborgne Y (1992) Introduction in Europe, from 1972 to 1980, of the Japanese Manila clam (*Tapes philippinarum*) and the effects on aquaculture production and natural settlement. In: Proceedings of international council for the exploration of the sea (ICES) marine science symposium, pp 12–13 Juin 1992, Halifax. ICES J Mar Sci 194:92–96

Girault M, Arakawa H, Kanda J (2011) Limiting factors of phytoplancton communities along the Ogasawara transect in North Pacific Ocean. La mer (Tokyo) 49(3–4):93–101

Grizel H, Héral M (1991) Introduction into France of the Japanese oyster *(Crassostrea gigas)*. J Cons Int Explor Mer 47(3):399–403

Guary J-C (1973) Contribution à l'étude du métabolisme des lipides chez le crustacé décapode *Penaeus japonicus* Bate. Thèse de Doctorat de spécialité, Université Aix Marseille II, 147 p

Guary J-C, Kayama M, Murakami Y, Ceccaldi H-J (1976) The effects of fat-free diet compound and compounded diets supplemented with various oils on moult, growth and fatty acid composition of prawn *Penaeus japonicus* Bate. Aquaculture 7(3):245–254

Hénocque Y (1983) Les plages artificielles: phase intermédiaire de repeuplement de la crevette Japonaise *Penaeus japonicus*. La pêche maritime, mai 1983: 263–268

Ikematsu M (1992) Cooperative fisheries management in Japan: the exemple of abalone and topshell fisheries in Southern Izu Peninsula. Océanis 18(4):437–444

Kakimoto H (1985) Distribution et comportement des poissons dans les récifs artificiels. In: *Actes du 3 ème colloque pluridisciplinaire Franco-Japonais d'Océanographie, Marseille,* pp 123–132, 16–21 Sept 1985, Tome 6, 167 p

Kikuchi S, Uki N (1974) Technical study on artificial spawning of abalone (genus *Haliotis*. VI—On sexual maturation of *Haliotis gigantic* (Gmelin) under artificial conditions. Bull Tohoku Reg Fish Lab 35:85–90

Kittaka J (1984) Transplantation of useful Atlantic crustacean into Japan: 67–80. In: Actes 2nd coll. Franco-Japonais océanographie, *Aquaculture*, Sendai, 102 p

Koike Y (2015) An example of friendship and cooperation between France and Japan: Oyster farming in Sanriku Area (Tohoku Region, Northern Japan) before and after Tsunami—restoration and technical adaptation of culture systems: 3–14. In: H-J Ceccaldi et al (eds) Proceedings of 15th French-Japanese oceanography symposium: *Marine productivity: perturbations and resilience of socio-ecosystems*, Springer Science, 391 p

Komatsu T, Yanagi, T (2015) Sato-umi: an integrated approach for sustainable use of coastal waters, lessons from human-nature interactions during the Edo period of eighteenth-century Japan: 283–290. In: Proceedings 15th French-Japanese oceanography symposium: *Marine productivity: perturbations and resilience of socio-ecosystems*. Springer Science, 391 p

Le Borgne Y, Le Pennec CM (1983) Elevage expérimental de l'huître asiatique *Ostrea denselamellosa* (Lischke). Vie Marine 5:23–28

Le Borgne Y (1996) Echanges internationaux de mollusques bivalves et situation actuelle en France et en Europe. Rev Sci Tech Off Int Epiz 15(2):491–498

Lehuta S, Mahévas S, Petitgas P, LeFloch P (2013) A simulation-based approach to assess sensitivity and robustness of fisheries management indicators for the pelagic fishery in the Bay of Biscay. Can J Fish Aquat Sci 70:1741–1756

Le Pichon X (1986) Kaiko, voyage aux extrémités de la mer. Odile Jacob - Le Seuil édit:255p

Le Pichon X, Kobayashi K, et al (1987) Introduction to KAIKO Special Issue. Earth Planet Sci Lett 83: 183–185

Lucien-Brun H (1983) Elevage de juvéniles d'ormeaux à fin de repeuplement au Japon. Publ Ass Dév Aquaculture (A.D.A.) 10:54 p

Lucien-Brun H, Lachaux A (1983) Evolution de la pectiniculture au Japon. La Pêche Maritime Juil 1983:388–396

Mariojouls C, Prou J (2015) Changes, adaptations and resilience: the case of French Oyster farming: 299–308. In: H-J Ceccaldi et al (eds) Proceedings of 15th French-Japanese oceanography symposium: *Marine productivity: perturbations and resilience of socio-ecosystems*, Springer Science, 391 p

Matsumoto Y, Noro T, Takami H, Fujinami Y, Kuji Y, Kawamura T (2015) A study on the new method for larval settlement, post larval growth and survival of abalone *Haliotis discus hannai* using a cultural benthic diatom *Cylindrotheca closterium* and a pre-grazed green alga *Ulvella lens*. Nippon Suisan Gakkaishi 81(6):995–997 (in Japanese)

de Montbrial T (2015) Une goutte d'eau et l'océan. Albin Michel édit 368:p

Nomura T (1984) Actual state of Pacific salmon culture in Japan. In: Matsuura et al (ed) Actes 2nd coll. Franco-Japonais océanographie, *Aquaculture*, Sendai, pp 23–32, 102 p

Pastoureaud A (1972) Contribution à l'étude de l'alimentation de *Penaeus kerathurus*, Crustacé Décapode. Thèse de doctorat de spécialité, Université Aix Marseille II, 56 p

Sasano M, Nakajima Y, Yamamoto J, Furushima Y (2015) Use of bio-fluorescent characteristics for ecosystem monitoring on hydrothermal deposits, pp 253–258. In: H-J Ceccaldi et al (eds) Proceedings of 15th French-Japanese oceanography symposium: *Marine productivity: perturbations and resilience of socio-ecosystems*, Springer Science, 391 p

Simard F (1990) S'affranchir du milieu naturel - pêche côtière et gestion des ressources -. France Japon Eco 44:15–21

Stott AE, Takeuchi T, Koike Y (2004) Performance of a new artificial abalone hatchery culture system in terms of settlement of larvae and growth and survival of postlarval *Haliotis discus discus* (Reeve). Fish Sci 70(6):1070–1081

Trenkel V, Brind'Amour A, Lehuta S, Lorance P, Mahévas S, Rochet M-J (2015) Indicators for ecosystem based management: methods and applications, pp 215–221. In: H-J Ceccaldi et al (eds) Proceedings of 15th French-Japanese oceanography symposium: *Marine productivity: perturbations and resilience of socio-ecosystems*. Springer Science, 391 p

Tanoue H, Komatsu T, Natheer A, Mitani I, Watanabe S, Watanabe Y, Hamano A, Miyazaki N (2013) Measurement of diurnal body tilt angle distributions of threeline grunt *Parapristipoma trilineatum* using micro-acceleration data loggers. J Mar Sci Eng 1(1):3–9

Uno Y (1984) Recent mariculture activities in Japan: 99. In: Actes 2nd Coll. Franco-Japonais océanographie: *Aquaculture*, Sendai, 102 p

Van Wormhoudt A (1980) Adaptation des activités digestives, de leurs cycles et de leur contrôle aux facteurs du milieu chez *Palaemon serratus* (Crustacea Natantia). Thèse Université Aix-Marseille II, 351 p

Yagi H (1984) Some data on the larval rearing of two crustacean species *Palaemon serratus* and *Penaeus japonicus*: 61–65. In: Actes 2nd coll Franco-Japonais océanographie, *Aquaculture*, Sendai, 102 p

Yamane T (1985) A national scale fishing ground development plan by means of artificial reefs placement in the sea and its socio-economic effects in Japan, pp 7–31. In: Actes du 3ème colloque pluridisciplinaire Franco-Japonais d'Océanographie, Marseille, 16–21 Sept 1985, Tome 6, 168 p

Chapter 2
Challenge to Resolve Problems in the Ocean and Coastal Waters in Future Earth with Stronger Cooperation Between the Two Societies *Franco-Japonaise* of Oceanography

Teruhisa Komatsu

Abstract In 1958, Bathyscaphe FNRS-III visited Japan to carry out deep-sea expeditions in Japan Trench and Kuril Trench with Japanese scientists. Inspired by this event, Prof. Tadayoshi Sasaki and his colleagues founded the "*Société franco-japonaise d'Océanographie de Japon*" in 1960 to stimulate scientific exchanges between Japanese and French scientists and research groups working on oceanography and fisheries. Based on active exchanges of fisheries scientists between France and Japan since 1970s, Prof. H. J. Ceccaldi founded the "*Société franco-japonaise d'Océanographie de France*" in 1984 with his French colleagues. Two societies have continued scientific exchanges through French-Japanese symposiums of oceanography and other related fields. In the 21st century, we are facing the limit of the planet which will be followed by abrupt and irreversible consequences for human communities and ecological systems. It is very important to challenge to resolve such problems in the ocean and coastal waters in future Earth with stronger cooperation between the two societies between Japan and France because the conception on relation between humankind and nature is different between the two countries and the differences in conception bring more prospective perspectives on their solutions.

Keyword Société franco-japonaise d'Océanographie · French-Japanese symposium of oceanography · Future Earth · Japan-France cooperation

T. Komatsu (✉)
Atmosphere and Ocean Research Institute, The University of Tokyo,
5-1-5 Kashiwanoha, Kashiwa 277-8564, Japan
e-mail: komatsu@aori.u-tokyo.ac.jp

T. Komatsu et al. (eds.), *Oceanography Challenges to Future Earth*,
https://doi.org/10.1007/978-3-030-00138-4_2

In 1950s, French scientists lead the deep-sea research using manned submersibles. In 1958, Bathyscaphe FNRS-III visited Japan to carry out deep-sea expeditions in Japan Trench and Kuril Trench with Japanese scientists. They dived to a bottom depth of 3000 m in Japan Trench on 20 June 1958. Inspired by this event, Prof. Tadayoshi Sasaki and his colleagues founded the "*Société franco-japonaise d'Océanographie de Japon*" (hereafter SFJO Japan) on 7 April 1960 to stimulate scientific exchanges between Japanese and French scientists and research groups working on oceanography and fisheries. After this foundation, bathyscaphe "Archimède" succeeded to dive at a bottom depth of 9,545 m with Prof. Sasaki and French team off Kuril Archipelago on 15 July 1962. Scientific cooperation, particularly in geology, has been increased after the first observations made by the bathyscaphe around Japan.A periodical publication of SFJO Japan, "La mer", which contains articles written in Japanese, French or English, was created in August 1963.

In 1960s, oyster production was abruptly decreased in France due to oyster disease. In 1966, Dr. P. Trochan, Director of La Tremblade Institute, approached Prof. Takeo Imai of Tohoku University on export of oyster seeds from Japan to France. Oyster seeds of 10 million were sent from Sanriku in Tohoku Region to France by air cargo in 1969. Japanese oyster seeds could contribute to recovery of oyster production in France. Since 1970, exchanges of fisheries scientists have been active between France and Japan. Based on these achievements, Prof. H. J. Ceccaldi founded the "*Société franco-japonaise d'Océanographie de France*" in 1984 with Y. Hénocque, C. Mariojouls, F. Simard, J. M. Thiérry, A. Nishikawa, N. Lucas and D. Bailly. Cooperation in marine sciences and oceanography under the two SFJOs started at that time. As Prof. Ceccaldi introduced the French-Japanese Symposiums of Oceanography, the two SFJO organized 17 symposiums in total in France and Japan. We can say "Continuer, c'est pouvoir" (Table 2.1).

Huge tsunami hit Pacific Coast of Japan on 11 March 2011. Sanriku Coast is the nearest to and in front of the epicenter of the East Japan Great Earthquake. Coastal fisheries including oyster culture were damaged by the tsunami. Immediately, SFJOs of France and Japan started active support to fishermen in Sanriku Coast with French Association for Promotion of Aquaculture, Rotary Club Marseille Saint-Jean and Fondation Air Liquide through donations of 17 binocular microscopes and 10 plankton nets to Miyagi and Iwate Prefectures. This story with donations by other French organizations and French oyster farmers is known as a requite for supply of Japanese oyster seeds by Sanriku fishermen to French oyster farmers in 1960s. This beautiful story highlights that the cooperation is indispensable to resolve problems.

In the 21st century, we are recognizing the limit of the planet, which is called as the planetary boundary. We can no longer exclude the possibility that our collective actions will trigger tipping points, risking abrupt and irreversible consequences for human communities and ecological systems. Sciences are requested to resolve environmental problems for future generations and future sustainable earth being

Table 2.1 Numbers, years, themes and places of the French-Japanese symposiums of oceanography held from 1983 to 2017

Num.	Year	Theme	Place
1	1983	Aquaculture	Montpellier
2	1984	Aquaculture	Sendai
3	1985	Coastal management and littoral planning	Marseille
4	1988	General oceanography	Shimizu
5	1989	Founding an algal park in Seto Inland Sea	Hiroshima and Higashino
6	1990	Coast line and conflicts	Tokyo
7	1990	Determination of biological recruitment at sea	Tokyo
8	1991	Determing factors of the growth in aquaculture economy and management of fisheries	Nantes
		Co-development of fisheries and leisure in coastal zones	
9	1991	Oceanic fluxes	Tokyo
10	1992	Biotechnology and environment	Tokyo
11	1997	Coastal zone observation and the medium and longterm forecasting	Paris
12	2005	Mutual new understanding for research in oceanography and fisherie in France and Japan	Tokyo
13	2008	Global change: interactions mankind/marine environments	Marseille and Paris
14	2010	Towards sustainable use and management of the oceans	Kobe and Tokyo
15	2013	Marine productivity: perturbation and resilience of coastal socio-economic systems	Boulogne sur Mer and Marseille
16	2015	The sea under human and natural impacts: challenge of oceanography to the future Earth	Sendai, Tokyo and Hinase
17	2017	Vulnerability to climate change, natural hazards and anthropogenic pressures	Bordeaux

with and for society. It is very interesting to compare approaches in coastal and marine environmental problems between Japan and France because the conception on relation between humankind and nature is different between the two countries like Ukiyo-e in Japan and paintings of impressionists in France in 19th century.

This symposium was designed to introduce researches concerning oceanography and marine sciences conducted in two countries for resolving problems in the ocean and coastal waters in future Earth. Readers can understand why cooperation between the two SFJOs must be enforced for future Earth.

Acknowledgments On behalf of SFJOs Japan and France, I thank to His Excellency Mr. Thierry Dana, former Ambassador of France in Japan, Dr. Jacques Maleval, former Science Counselor, Embassy of France in Tokyo, Mme Tomoko Ito, Director of Fondation franco-japonaise Sasakawa and Dr. Hideki Akiyama, former Director of Tohoku National Fisheries Research Institute for their supports to realize the 16th French-Japanese Symposium of Oceanography entitled "The sea under human and natural impacts: challenge of oceanography to the future Earth".

Part II
Natural and Anthropogenic Impacts

Chapter 3
Slower Decrease in Radioactive Concentrations in Some Fish Species After the Fukushima Daiichi Nuclear Power Plant Disaster

Hisayuki Arakawa

Abstract Large quantities of radioactive material were released into the ocean as a result of the Fukushima Daiichi Nuclear Power Plant disaster. Because of the movement of the highly radioactive water mass, coastal seawater, seabed sediments, and many marine organisms were contaminated by radioactive material. Although radioactive concentrations in many species decreased over time, concentrations in some fish species decreased very slowly (e.g., Japanese rockfish and skate). In this study, the changes in radioactive cesium (Cs) concentrations in marine organisms are reported over the 4 years since the accident and possible reasons for the slower decrease in radioactive concentrations observed in some fish species are discussed. Marine organisms and seawater and seabed sediment samples were collected from May 2012 to May 2014. Radioactive Cs concentrations in the collected organisms and samples were measured using a germanium semiconductor detector. An ultrasonic pinger was inserted into the bodies of some fish, and fish movements were recorded by biotelemetry for 1 year. Japanese rockfish were cultured in a large water tank for 1 year, and the biological half-life of radioactive Cs was estimated in the fish. Radioactive Cs concentrations in benthic and omnivorous fish, such as Japanese rockfish (367 days) and a skate species (560 days), had relatively long ecological half-lives. Japanese rockfish remained within the same rocky area (within a radius of 200 m) in open sea areas throughout the year. The ecological half-life of radioactive Cs was approximately 270 days longer in rockfish than that in other marine fish. These results suggest that relatively high radioactive Cs concentrations remained in coastal benthic fish owing to slow Cs excretion rates.

Keywords Biological half-life · Ecological half-life · Fish movement · Fukushima Daiichi Nuclear Power Plant disaster · Radioactive contaminantions · ^{137}Cs

H. Arakawa (✉)
Department of Marine Sciences, Tokyo University of Marine Science and Technology, Tokyo, Japan
e-mail: arakawa@kaiyodai.ac.jp

T. Komatsu et al. (eds.), *Oceanography Challenges to Future Earth*,
https://doi.org/10.1007/978-3-030-00138-4_3

3.1 Introduction

Large quantities of radioactive material were released into the ocean as a result of the Fukushima Daiichi Nuclear Power Plant (FNPP) disaster. The amount of cesium-137 (^{137}Cs) released was estimated to be 3.5 PBq (Tsumune et al. 2012), and the release of radioactive material occurred during a concentrated period of <1 month. Immediately following the accident, it is thought that a water mass containing high concentrations of radioactive material released from the FNPP flowed southward with the coastal current and was dispersed eastward when it reached the Kuroshio extension (Yoshida and Kanda 2012; Kawamura et al. 2011; Honda et al. 2012; Kumamoto et al. 2014). Because of the movement and diffusion of this water mass, coastal seawater, seabed sediments, and many marine organisms were contaminated by radioactive material. Fisheries activities were banned after marine organisms contaminated with high radioactive concentrations were caught in April 2011.

Seawater ^{137}Cs concentration immediately after the accident was very high, but decreased rapidly (Tsumune et al. 2012), and the concentration has declined slowly since April 2014 to the present concentration of about 10 mBq/L. No difference in radioactive concentration has been detected between near the FNPP and 35 km south of the FNPP (Fig. 3.1).

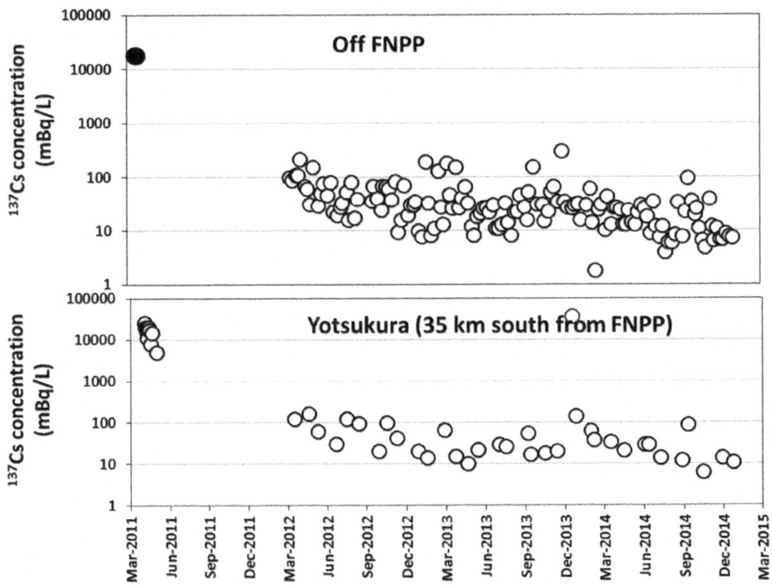

Fig. 3.1 Temporal changes in seawater ^{137}cesium (Cs) concentrations in the area near the Fukushima Daiichi Nuclear Power Plant (FNPP) and off the coast of Yotsukura. ● indicates the ^{137}Cs concentration in the FNPP immediately after the accident (Tsumune et al. 2012). The figures are created by TEPCO data, (http://www.tepco.co.jp/decommision/planaction/monitoring/index-j.html)

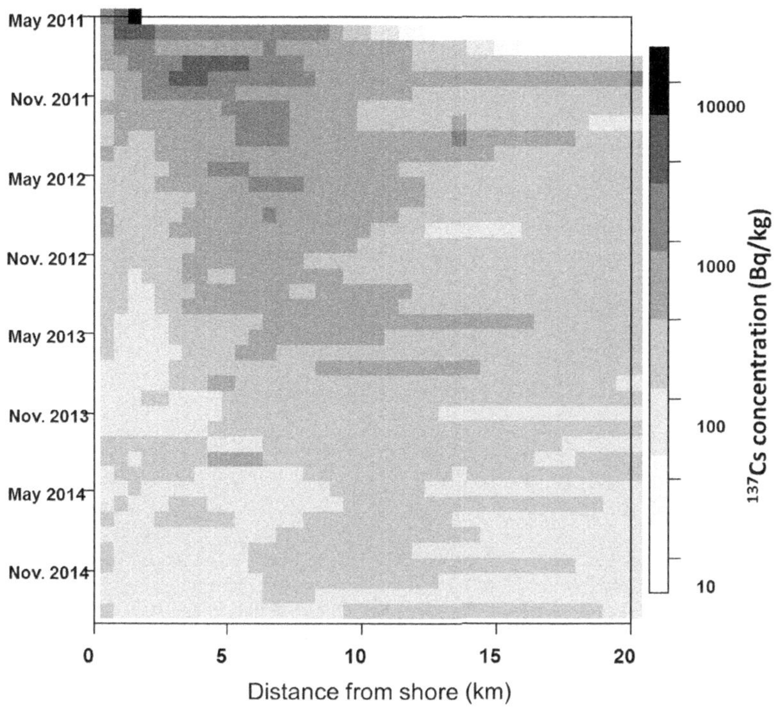

Fig. 3.2 Temporal and specific changes in seabed sediment ^{137}cesium (Cs) concentrations off the coast of Yotsukura 35 km south of the Fukushima Daiichi Nuclear Power Plant (FNPP). The figure is created by MEXT (Ministry of Education, Culture, Sports, Science & Technology in Japan) data. (http://radioactivity.nsr.go.jp/ja/list/275/list-1.html)

On the other hand, the ^{137}Cs concentration in seabed sediments was >9000 Bq/kg in May 2011, just after the accident. The concentration near the coast decreased with time, and the high sediment concentration indicated that the radioactivity had dispersed offshore. The ^{137}Cs concentration in sediments was <100 Bq/kg in end of 2014 (Fig. 3.2). Radioactive concentrations in marine organisms were very high immediately after the accident. However, the concentration in many species decreased rapidly over time (Buesseler 2012; Wada et al. 2013; Sohtome et al. 2014). In particular, ^{137}Cs concentrations in Cephalopoda decreased. The radioactive concentrations in pelagic fish also decreased rapidly, and those in many marine organisms decreased with time. However, the concentrations in some species of coastal and benthic fish, such as Japanese rockfish *Sebastes cheni* and skate *Okamejei kenojei*, decreased very slowly (Wada et al. 2013; Iwata et al. 2013) (Fig. 3.3).

In this paper, we discussed about the changes in ^{137}Cs concentrations in various marine organisms over the 4 years since the accident and the reasons for the slower decrease in radioactive concentrations observed in some fish species.

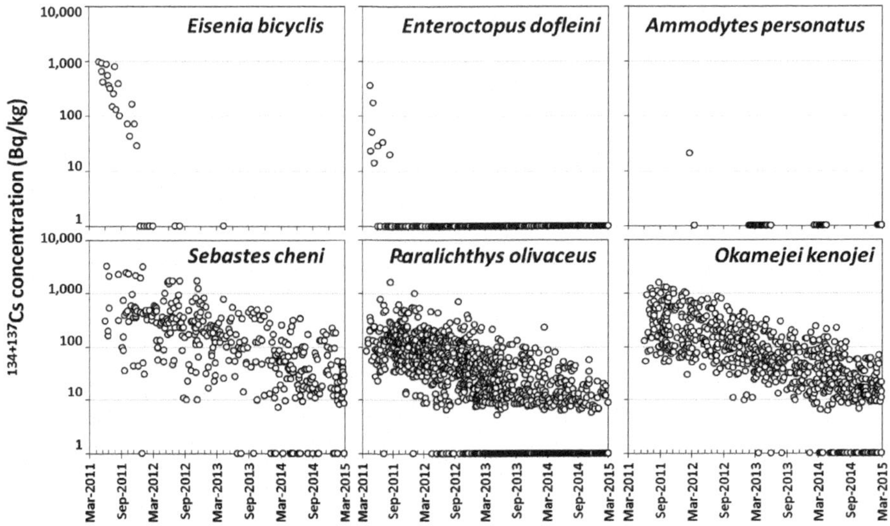

Fig. 3.3 Temporal changes in $^{134+137}$cesium (Cs) concentrations in marine organisms off the Fukushima coast. N.D. was assumed to be 1 for convenience. The figures are created by Japan Fisheries Agency data. (http://radioactivity.nsr.go.jp/ja/list/2751/list-1.html)

3.2 Materials and Methods

We examined (1) the distribution and changes in ^{137}Cs concentrations in the environment and individual marine organisms, (2) the dispersion of ^{137}Cs attributable to fish movement, and (3) the reasons for the slower decrease in radioactive concentrations in several fish species (Arakawa et al. 2015).

(1) Three sampling points were chosen off the coasts of Iwaki (Yotsukura and Ena) and Soma, Fukushima Prefecture. Marine organisms, seawater, and seabed sediment samples were collected at the three sampling points from May 2012 to May 2014. The ^{137}Cs concentrations in the samples were measured using a germanium semiconductor detector, and the ecological half-lives in the organisms were estimated from the results.
(2) Japanese rockfish, fat greenling *Hexagrammos otakii* and skate, which continue to have high ^{137}Cs concentrations, were selected as sample fish species. An ultrasonic pinger was inserted into the bodies of the rockfish and greenling, and fish movements were recorded using a stationary type biotelemetry system (Vemco Inc., Bedford, NS, Canada) off the coast of Iwaki city, Fukushima Prefecture, from November 2014 to February 2016. The skate has a broad area of movement and was surveyed using a tracking type system (Vemco Inc.). Six skate were tracked off the coast of Ena during 2014. The details of the measurement method are described in Uchida et al. (2018).

(3) Japanese rockfish were cultured in a large tank for 1 year. The seawater used for culture was collected off Iwaki city, Fukushima Prefecture. Antarctic krill, which had no detectable ^{137}Cs, was used as food for the fish. Cs concentrations were measured in each fish once or twice per month, and the biological half-life of ^{137}Cs in the fish was estimated from these data. The details of this experimental method are described by Matsumoto et al. (2015).

3.3 Results and Discussion

3.3.1 Ecological Half-Life of ^{137}Cs

Six species of brown algae, five species of red algae, and one species of sea grass were collected. The ^{137}Cs concentrations in annual algae *Undaria pinnatifida* during May 2012 were 1.6–6.5 Bq/kg and those in perennial algae species *Eisenia bicyclis* and *Phyllospadix iwatensis* were 6.0–7.8 Bq/kg. However, no species had >1 Bq/kg ^{137}Cs in May 2014.

The invertebrate species in which ^{137}Cs concentrations were analyzed differed by season and location. About 14–33 species were collected, and 1–20 species were analyzed. ^{137}Cs was detected in almost all invertebrates in September 2012. However, ^{137}Cs was almost not detectable in cephalopods and crustaceans in 2013. In contrast, ^{137}Cs was still detectable in the echinoderms sampled in May 2014.

About 20–28 fish species were caught and analyzed during the study. ^{137}Cs was detected in 18 fish species during September 2012, except in Japanese amberjack and black scraper, and the ^{137}Cs concentrations in Japanese rockfish and skate were >100 Bq/kg, which is the limit for food in Japan. The ^{137}Cs concentrations decreased in all fish species by May 2014. However, two rockfish species, such as *S. cheni* and *Sebastes vulpes*, skate, and slime flounder *Microstomus achne* registered ^{137}Cs concentrations >50 Bq/kg. The species in which ^{137}Cs concentrations decreased at the slowest rate were all coastal benthic fish (Table 3.1).

Obvious decreases in ^{137}Cs concentrations were detected in many invertebrates and fish. The ecological half-lives in most of the fish species were 36–281 days. However, ^{137}Cs in benthic and omnivorous fish species, such as Japanese rockfish (367 days) and one of the skate species (560 days), had relatively longer ecological half-lives. The ecological half-lives of ^{137}Cs in the Japanese rockfish and skate were the same as those reported in previous studies (Wada et al. 2013; Iwata et al. 2013).

3.3.2 Diffusion via *Fish Movement*

Diffusion *via* fish movement within the Iwaki study area was investigated using ultrasonic biotelemetry. The signals of seven of 20 Japanese rockfish were detected

Table 3.1 ^{137}Cesium (Cs) concentrations in fish species off the Fukushima coast during 2012 and 2014

Month and year	Site	^{137}Cs (Bq/kg)	Species
Sept. 2012	Ena	N.D.	*Seriolaquinque radiata, Thamnaconus modestus*
		≦10	*Upeneus japonicus*
		≦50	*Zeus faber, Ditrema temmincki, Takifugu snyderi, Mustelus manazo, Chelidonichthys spinosus, Kareius bicoloratus, Pleuronectes yokohamae, Paralichthys olivaceus*
		≦100	*Nibea mitsukurii*
		100<	*Sebastes cheni, Okamejei kenojei*
	Yotsukura	N.D.	—
		≦10	—
		≦50	*Dasyatis matsubarai, Takifugu snyderi, Mustelus manazo, Paralichthys olivaceus*
		≦100	*Pagrus major, Platycephalus* sp.2, *Paraplagusia japonica, Cynoglossus joyneri, Squatinae japonica, Chelidonichthys spinosus, Pleuronectes yokohamae*
		100<	*Okamejei kenojei*
May 2014	Ena	N.D.	*Pennahia argentata*
		≦10	*Squalus acanthias, Ditrema temmincki, Takifugu poecilonotus, Nibea mitsukurii, Takifugu snyderi, Scyliorhinus torazame, Trachurus japonicas, Pagrus major, Cynoglossus joyner, Platycephalus sp.*2
		≦50	*Hexagrammos otakii, Lateolabrax japonicas, MIicrostomus achne, Sebastes pachycephalus, Okamejei kenojei*
		≦100	*Sebastes cheni*
		100<	—
	Yotsukura	N.D.	*Takifugu poecilonotus, Pennahia argentata*
		≦10	*Hemitripterus villosus, Takifugu snyderi, Scyliorhinus torazame, Sqoalus acanthias, Paraplagusia japonica, Mustelus manazo*
		≦50	*Physiculus maximowiczi, Lateolabrax japonicas, Cynoglossus joyneri, Paralichthys olivaceus, Pleuronectes yokohamae, Gadus microcephalus, Eopsetta grigorjewi*
		≦100	*Sebastes vulpes, Sebastes cheni, Okamejei kenojei, MIicrostomus achne*
		100<	—

in the sea area during 1 year. The Japanese rockfish remained within the same rocky area (within a radius of 200 m) in the open sea throughout the year (refer to Uchida et al., 2018). Therefore, the reason for the slower decline in radioactivity in some species was not movement of polluted fish from the sea near the FNPP.

3.3.3 Biological Half-Life of ^{137}Cs in Fish

The culture experiment results showed that the biological half-life of ^{137}Cs in rockfish was approximately 275 days (Matsumoto et al. 2015) longer than the 50–100 days of other marine fish (Baptist and Price 1962; Kimura 1984; Zhao et al. 2001), and no significant difference was detected between sex, suggesting that metabolism of ^{137}Cs in adult Japanese rockfish is slower than that in other marine fish.

3.3.4 Reason for the Slower Decrease of ^{137}Cs in Coastal Fish

The decrease in radioactive Cs concentrations in the coastal benthic fish species was slow according to the results of several surveys (Wada et al. 2013; Iwata et al. 2013). Three reasons have been proposed for this phenomenon: (1) uptake from high radioactivity food by the food web; (2) slow fish metabolism; and (3) fish polluted with a high ^{137}Cs concentration moving away from the sea near the FNPP.

First, previous observations have indicated that ^{137}Cs concentrations decrease rapidly from food-like invertebrates, and food does not increase Cs concentrations in fish. Thus, the slower rate of ^{137}Cs elimination from some fish was unlikely to have been caused from pollution by uptake.

The coastal benthic fish (e.g., Japanese rockfish) with high Cs concentrations remained in a small territorial area of the coast during the entire year. Therefore, it is unlikely that fish with high Cs concentrations were recruited from the sea area near the FNPP.

Furthermore, the biological half-life of ^{137}Cs in Japanese rockfish (Matsumoto et al. 2015; Tagami and Uchida 2016) is much longer than the 50–100 days of other marine fish (Baptist and Price 1962; Kimura 1984; Zhao et al. 2001).

Matsumoto et al. (2015) used adult fish (body length, 26.2–32.6 cm) in their experiment. Isoyama et al. (2008) investigated the relationship between body length of marbled flounder and Cs concentration in the body and found that larger individuals have higher Cs concentrations. This finding indicates that adult flounder, which is a species with a high Cs concentration, excretes Cs very slowly.

These results indicate that the relatively higher ^{137}Cs concentrations in the coastal benthic fish were caused by relatively slower ^{137}Cs excretion rates, and that marine organisms off the Fukushima coast should be monitored for radioactivity in the future.

Acknowledgements This study was supported by the Environment Research and Technology Development Fund (No. 4ZD-1201) from the Ministry of the Environment, Japan.

References

Arakawa H, Tokai T, Miyamoto Y, Akiyama S, Uchida K, Matsumoto A, Narita M, Myose H, Agatsuma Y, Katayama S, Aoki M, Matsumoto I, Hirakawa N (2015) Distribution of radioactive material in marine ecosystems off the Fukushima coast. Radioactive Cesium levels in Fukushima marine organisms. In: Marine productivity: perturbations and resilience of socio-ecosystems. Springer, pp 71–78

Baptist JP, Price TJ (1962) Accumulation and retention of cesium 137 by marine fishes. In: Fish and wildlife service, superintendent of documents, US Government

Buesseler KO (2012) Fishing for answers off Fukushima. Science 338:480–482

Honda MC, Aono T, Aoyama M, Hamajima Y, Kawakami H, Kitamura M, Masumoto Y, Miyazawa Y, Takigawa M, Saino T (2012) Dispersion of artificial caesium-134 and -137 in the western North Pacific one month after the Fukushima accident. Geochem J 46(1):e1–e9

Isoyama N, Oikawa S, Misonoo J, Nakahara M, Nakamura R, Suzuki N, Yoshino M, Suzuki C, Sato H, Hara T (2008) Relationship between growth of Marbled flounder (*Pleuronectes yokohame*) and Cs concentration in their muscle determined by ICP-MS. Bunseki Kagaku 57:763–769

Iwata K, Tagami K, Uchida S (2013) Ecological half-lives of radiocesium in 16 species in marine biota after the TEPCO's Fukushima Daiichi Nuclear Power Plant accident. Environ Sci Technol 47:7696–7703

Kawamura H, Kobayashi T, Fruno A, In T, Ishikawa Y, Nakayama T, Shima S, Awaji T (2011) Preliminary numerical experiments on oceanic dispersion of ^{134}Cs and ^{137}Cs discharged into the ocean because of the Fukushima Daiichi Nuclear Power Plant disaster. J. Nuclear Sci Technol 48:1349–1356

Kimura K (1984) Accumulation and retention of cesium-137 by the common goby. Bull Jpn Soc Sci Fish 50:481–487

Kumamoto Y, Aoyama M, Hmajima Y, Aono T, Kouketsu S, Murata A, Kawano T (2014) Southward spreading of the Fukushima-derived radiocesium across the Kuroshio extension in the North Pacific. Sci Rep 4:4276

Matsumoto A, Shigeoka Y, Arakawa H, Hirakawa N, Morioka Y, Mizuno T (2015) Biological half-life of radioactive cesium in rockfish *Sebastes cheni* contaminated by the Fukushima Daiichi nuclear power plant accident. J Environ Radioact 150:68–74

Sohtome T, Wada T, Mizuno T, Nemoto Y, Igarashi S, Nishimune A, Aono T, Ito Y, Kanda J, Ishimaru T (2014) Radiological impact of TEPCO's Fukushima Dai-ichi Nuclear Power Plant accident on invertebrates in the coastal benthic food web. J Environ Radioact 138:106–115

Tagami K, Uchida S (2016) Consideration on the long ecological half-life component of ^{137}Cs in demersal fish based on field observation results obtained after the Fukushima accident. Environ Sci Technol 50:1804–1811

Tsumune D, Tsubono T, Aoyama M, Hirose K (2012) Distribution of oceanic ^{137}Cs from the Fukushima Dai-ichi Nuclear Power Plant simulated numerically by a regional ocean model. J Environ Radioact 111:100–108

Uchida K, Hasegawa K, Miyamoto Y, Arakawa H, Akitama S, Hirakawa N (2018) Influence of behavioral patterns of several fish species on their radioactive cesium concentrations revealed with a biotelemetry system after the nuclear accident caused by the 2011 Great East Japan Earthquake. In: Teruhisa K et al (eds) Oceanography challenges to future earth, pp (xx–xx)

Wada T, Nemoto Y, Shimamura S, Fujita T, Mizuno T, Sohtome T, Kamiyama K, Morita T, Igarashi S (2013) Effects of the nuclear disaster on marine products in Fukushima. J Environ Radioact 124:246–254

Yoshida N, Kanda J (2012) Traking the Fukushima radionuclides. Science 336:1115–1116

Zhao X, Wang W, Yu K, Lam P (2001) Biomagnification of radiocesium in a marine piscivorous fish. Mar Ecol Prog Ser 222:227–237

Chapter 4
Influence of Behavioral Patterns of Several Fish Species on Their Radioactive Cesium Concentrations Revealed with a Biotelemetry System After the Nuclear Accident Caused by the 2011 Great East Japan Earthquake

Keiichi Uchida, Kohei Hasegawa, Yoshinori Miyamoto, Hisayuki Arakawa, Seiji Akiyama and Naoto Hirakawa

Abstract The Great East Japan Earthquake and subsequent tsunami on March 2011 damaged Fukushima Daiichi Nuclear Power Plant (FDNPP), and a part of radioactive materials released from FDNPP entered in coastal waters of north–eastern Japan. Fisheries Cooperative of Fukushima Prefecture has autonomously banned coastal fisheries since 15 March 2011. Japanese Government restricts foods including fishes that contain radioactive cesium (Cs) above the maximum limit of 100 Bq kg^{-1}. Given this serious situation, we have studied whether fishes can transfer radioactive materials to other areas. For this purpose, we selected seven target fish species containing relatively high concentrations of radioactive Cs measured by authorities and monitored their behaviors around rocky beds in the open sea and in a port near the coast off Fukushima Prefecture. The monitoring was performed by an ultrasonic biotelemetry system consisting of fixed receivers and pingers inserted in fish. The results revealed two types of behaviours: site fidelity and migration. Fish species with high concentrations of radioactive Cs preferentially indicated the site fidelity over the migration. A decline of radioactive Cs

K. Uchida (✉) · Y. Miyamoto · H. Arakawa · S. Akiyama
Tokyo University of Marine Sciences and Technology,
4-5-7 Konan Minato-ku, Tokyo 108-8477, Japan
e-mail: kuchida@kaiyodai.ac.jp

K. Hasegawa
Hokkaido University, 3-1-1, Minato-cho, Hakodate, Hokkaido 041-8611, Japan

N. Hirakawa
Fukushima Fisheries Office, 18-2 Obama Oikawa, Soma-shi Fukushima 976-0022, Japan

T. Komatsu et al. (eds.), *Oceanography Challenges to Future Earth*,
https://doi.org/10.1007/978-3-030-00138-4_4

concentrations in fish may be attributed to a natural decrease overtime and a replacement of contaminated individuals by immigrant ones with a low concentration of radioactive Cs from less contaminated waters.

Keywords Biotelemetry system · Site fidelity · Migration an concentrations of radioactive

4.1 Introduction

Following the Great East Japan Earthquake and tsunami on 11 March 2011, radioactive materials from the Fukushima Daiichi Nuclear Power Plant (FDNPP) accident were found in the coastal sea area of north–eastern Japan. To prevent human contamination through an uptake of fish which inhabit around Japan, authorities such as Fisheries Agency of Japan and prefectural governments have been monitoring concentrations of radioactive materials in marine organisms just after the accident. They found that radioactive cesium was the most important radioactive substance and the radioactive cesium (Cs) concentrations in many fish species were 1000 Bq kg^{-1} or higher in waters around Fukushima Prefecture (Nemoto et al. 2013). Japanese Government decided that foods including fish containing radioactive Cs more than 100 Bq kg^{-1} couldn't be supplied in a market since 1 April 2012 (Director of Food Safety Division, Ministry of Health, Labour and Welfare of Japan 2012). Since commercial fish species caught off Fukushima Prefecture accumulated high radioactive Cs, Fisheries Cooperative of Fukushima Prefecture has autonomously banned coastal fisheries since 15 March 2011 just after the hydrogen explosions of three nuclear reactors (Omori 2014). It might be possible that radioactive materials reached other waters through fish migration (Arakawa et al. 2015). Although radioactive Cs concentrations have reduced in some species, they have not entirely diminished in other species (Matsumoto et al. 2015). In addition, radioactive Cs concentration widely varied among fish species (Matsumoto et al. 2015). It is important to know mechanisms that produce differences in decline tendencies of their concentrations among fish species for deciding to stop the voluntary ban of fisheries by Fisheries Cooperative of Fukushima Prefecture in the future. We hypothesize that differences in behavior among fish species may bring different decline tendencies of their concentrations among fish species. Therefore, we investigated whether fishes have transferred radioactive materials to other areas through migration and why radioactive Cs concentrations in fish decreased at different rates depending on species in waters off Fukushima Prefecture. To this end, we observed behaviors of several fish species by a biotelemetry method (Mitamura 2005a; Yokota et al. 2006) applied to fish species distributing in waters 2–4 km off the coast and in a port in Fukushima Prefecture.

4.2 Materials and Methods

We selected seven experimental fish species containing relatively high radioactive Cs concentrations according to Arakawa et al. (2015), and with sufficiently large bodies to insert an acoustic transmitter into the abdominal cavity. The area, period, experimental fish and equipment for the two study sites described below.

4.2.1 Fishing Ground Close to the Coast

We selected a typical and traditional fishing ground where fishermen have used before the accident as a study site, which was located off Iwaki City, Fukushima Prefecture and 2–4 km distant from the coast (Fig. 4.1). To monitor the long-term site fidelity of fish species, we employed an acoustic monitoring receiver (VR2W, VEMCO, Canada), which receives signals from pingers implanted in the experimental fish. The receiver records data of target fish that are currently within a detection range. Since the pinger has a depth sensor, we can know the current depth of the experimental fish within the detection range. However, this method cannot clarify a migration route or measure positions of the fish. We used two types of pressure-sensing acoustic coded pinger tags: V9P (φ = 9 mm, L = 47 mm) and V13P (φ = 13 mm, L = 45 mm) made by VEMCO. Battery lives of V9P and V13P are approximately in 180 d and 352 d, respectively. Experimental fish were two species containing high concentrations of radioactive Cs: rockfish *Sebastes cheni* and greenling *Hexagrammos otakii*. The experimental fish were caught by gill net or a line-fishing in the study site, chosen with a body size enough to insert a pinger in its abdominal cavity (Table 4.1) and were kept in a tank (1000 L) with running seawater supplied from the sea on the fishing boat. Experimental fish were maintained in the tank for 0.5–1 h till they were transported to shore for surgery. The fish were individually anesthetized in another tank containing 2-phenoxyethanol of 8 mg L^{-1} and seawater of 40 L. A pinger was inserted in the abdominal cavity of experimental fish after anesthetizing it and opening its cavity with a surgical knife. The incision was closed with two to three separate silk sutures (F25-10B2, Natsume Seisakusho Co., Ltd., Japan). Anesthesy, surgery, and recovery duration for each experimental fish are about 4 min, respectively. The time from selecting experimental fish in the tank to release in the sea was about 3 h.

Three sets of an acoustic monitoring receiver were set near the capture points of the experimental fish (Fig. 4.1). The first, second and last receiver sets were deployed at a height of 2 m above the bottom on 22 November 2012, 10 February 2013 and 22 May 2013, respectively. They were retrieved on 25 May 2014. Data were not obtained during storm and typhoon events because the receiver sets were moved on land to avoid their lost. To investigate a relationship between a season and a behavior of experimental fish, we released experimental fish in different seasons (Table 4.1).

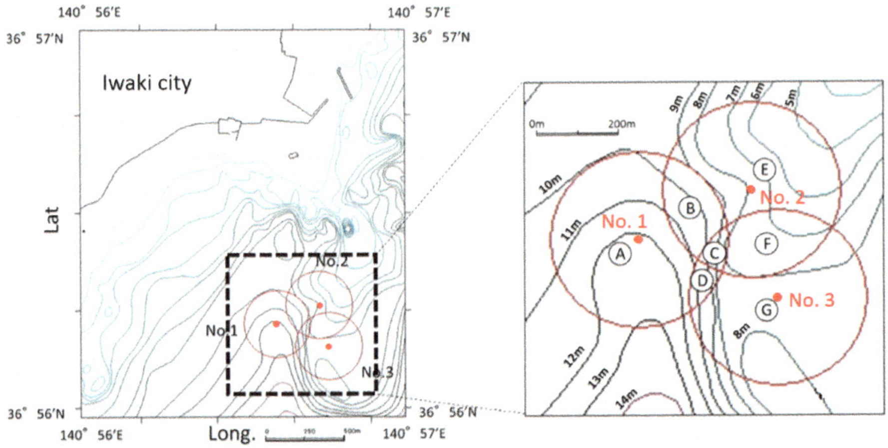

Fig. 4.1 Left map showing the study site off Iwaki City, Fukushima Prefecture indicated with a bold broken line, and right map showing positions and their ranges of three receiver sets indicated with red dots with the number and red circles, respectively. The areas labeled B, C, D and F, and those labeled A, E and G in the right-hand map are overlapped detection areas with neighbouring receivers and non-overlapped areas covered by only one receiver among three, respectively

4.2.2 Port Area Near FDNPP

We selected one of ports near FDNPP as another study site. The monitoring period was from 25 September 2013 to 29 September 2014. We prepared two sets of a receiver (FMR 1000, Fusion, Japan) and a water temperature logger (HOBO Water Temp Pro v2, ONSET, USA) attached to the upper part of the receiver, and deployed them at a depth of 1 m from the lowest low sea level from the pier at Stn. A in the head of the port and at Stn. B near the port entrance (Fig. 4.2). The port entrance was within the detection range of receiver at Stn. B, which was approximately 200 m. The experimental fish were captured by a trap or a line-fishing on 24 September and 9 December 2013, and on 19 February and 1 May 2014. The species containing high radioactive Cs were selected as experimental species. We chose individuals of these fish species with a sufficient body size to insert a pinger in the abdominal cavity. The experimental fish were eleven white spotted conger eels (*Conger myriaster*), seven spotbelly rockfishs (*Sebastes pachycephalus pachycephalus*), six brown hakelings (*Physiculus maximowiczi*), two rockfishs (*Sebastes cheni*), two black rockfish (*Sebastes schlegelii*) and one greenling (*Hexagrammos otakii*). Detail informations on the fish were listed in Table 4.2. After being caught, the experimental fish were kept in a small holding tank on land filled with 50 L sea water in which air bubbles were supplied by a pump. The method to insert a pinger into the abdominal cavity of fish was the same as

Table 4.1 Experimental fish data and their living areas off Iwaki City. Fish species (Sp.) are rockfish *Sebastes cheni* (S) and greenling *Hexagrammos otaki* (H). Period: number of days from the release day of an individual to the day when the last signal was received. Living area indicates most frequently detected area of fish during the survey period of individual among the areas of A, B, C, D, E, F and G in Fig. 4.1 during the daytime (d) and nighttime (n) defined as the times from 6 am to 6 pm and from 6 pm to 6 am, respectively. Average, description, min and max are average, minimum and maximum depths of individual during its whole survey period, respectively. N is the number of detected signals during the period defined above

Experimental fish							Recorded living area and depth of experimental fish											
Day of release	Sp.	ID	TL	W	Period (Days)	Type	Living area		Average		Dispersion		Min		Max		N	
			(cm)	(g)			d	n	d	n	d	n	d	n	d	n	d	n
22/11/12	S	1	27.0	353	200	V9P	A	A	14.83	15.19	0.62	1.94	11.6	7.3	18	18.7	679	2157
22/11/12	S	2	27.3	334	200	V9P	A	A	13.29	14.28	1.03	2.76	9.2	7.7	18	18.7	3040	7564
22/11/12	S	3	27.0	342	200	V9P	G	G	11.99	12.94	1.11	2.05	8.6	6.6	17.1	17.8	677	2760
22/11/12	S	4	31.5	535	200	V9P	A	A	13.15	14.14	1.56	3.06	6.6	4	17.1	18.2	1060	2933
22/05/13	S	5	32.3	665	195	V9P	G	G	8.42	8.52	1.33	1.14	0.2	0.2	13	15.2	4137	3378
22/05/13	S	6	28.7	464	144	V9P	G	G	9.31	9.59	0.54	0.71	7	6.6	16.5	14.3	2014	664
22/05/13	S	7	26.5	375	104	V9P	E	E	13.14	12.61	1.03	3.83	10.1	3.3	18	18.2	2852	1997
22/05/13	S	8	25.7	327	195	V9P	E	F	11.73	11.69	1.81	1.65	7.9	5.7	17.6	18	8964	9620
22/05/13	S	9	21.0	176	0	V9P	–	–	–	–	–	–	–	–	–	–	–	–
22/05/13	S	10	27.5	415	195	V9P	E	E	13.75	13.47	1.17	3.55	10.3	7.5	18.2	19.3	3390	4977
07/10/13	S	11	28.0	423	0	V13P	–	–	–	–	–	–	–	–	–	–	–	–
07/10/13	S	12	25.0	291	2	V13P	–	–	–	–	–	–	–	–	–	–	–	–
07/10/13	S	13	21.5	212	0	V9P	–	–	–	–	–	–	–	–	–	–	–	–
12/06/13	S	14	30.5	619	63	V13P	A	A	12.86	11.18	1.85	3.96	1.8	0.4	15.4	16.9	2304	4548
12/06/13	S	15	33.0	887	0	V9P	–	–	–	–	–	–	–	–	–	–	–	–
12/06/13	S	16	25.0	331	155	V9P	A	A	14.5	12.59	0.96	2.25	4.6	1.3	16.3	16.9	4117	8348
12/06/13	S	17	22.5	203	0	V9P	–	–	–	–	–	–	–	–	–	–	–	–
12/06/13	S	18	26.5	413	171	V9P	E	A	14.77	12.56	1.36	2.92	6.6	3.7	18.2	19.6	2711	9533

(continued)

Table 4.1 (continued)

Experimental fish							Recorded living area and depth of experimental fish											
Day of release	Sp.	ID	TL	W	Period (Days)	Type	Living area		Average		Dispersion		Min		Max		N	
			(cm)	(g)			d	n	d	n	d	n	d	n	d	n	d	n
12/06/13	S	19	26.5	327	2	V9P	–	–	–	–	–	–	–	–	–	–	–	–
12/06/13	S	20	24.5	312	97	V9P	A	A	13.79	13.82	2.57	3.26	7.9	2.4	16.5	17.8	257	2868
22/11/12	H	A	37.0	670	0	V13P	–	–	–	–	–	–	–	–	–	–	–	–
22/11/12	H	B	31.4	375	0	V13P	–	–	–	–	–	–	–	–	–	–	–	–
22/05/12	H	C	34.5	636	355	V13P	A	A	13.16	12.58-	2.03	2.59	1.7	1.1	19.5	19.1	30898	43872
22/05/12	H	D	37.0	595	33	V13P	G	G	11.97	12.42	0.51	0.47	10.3	11.2	18.0	18.2	4909	3767

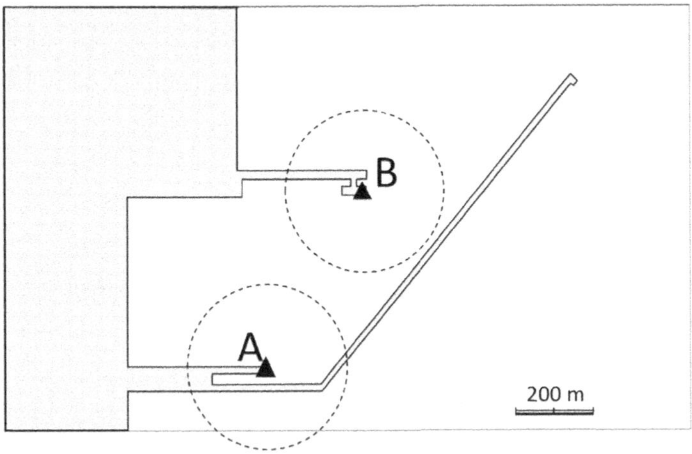

Fig. 4.2 Map showing two stations where a set of an acoustic receiver and a water temperature logger (marked with closed triangles). Dotted circles indicate the detection range of the receivers at Stns. A and B

above-mentioned. We used pingers of FPXG1030 (battery life of 70 d) and FPXG1040 (battery life of 126 d). The captured and released points are as shown in Table 4.2.

4.3 Results and Discussion

4.3.1 Fish Behaviors in the Fishing Ground Close to the Coast

4.3.1.1 Fish Behaviors

Among 20 rockfish, nine remained in the detection ranges of the receivers until the end of the monitoring period limited by the battery life, and two were recaptured: one about one year later after the end of the monitoring period and the other about a year and a half later. Four of the remaining 11 individuals eventually moved out of the detection ranges a couple of months later. These results show that rockfish strongly favor site fidelity over migration. However, seven rockfish individuals moved beyond the detection ranges within two days after release. Some went out the signal receiving ranges within several hours after release. Mitamura et al. (2005b) show that rockfish exhibit homing behavior to their home range in Osaka Bay. It is estimated that the seven individuals exited from the receiver ranges were caught when they came from their homes located outside the ranges near fishing points and returned to home in two days after the release.

Table 4.2 Experimental fish data and their tracking results in the port near FDNPP. ID is the identification number of experimental individuals. "out" in remarks means that an individual went outside the range of the receiver at Stn. B

Species and common name	ID	Capture and release point		Total length (cm)	Weight (g)	Pinger	Day of release	Last day of signal received	Monitoring period	Remark
Conger myriaster white spotted conger eel	1	A	A	58.5	362.5	FPXG1040	25/09/13	15/10/13	21	
	2	A	A	68.0	504.0	FPXG1040	25/09/13	04/10/13	10	Out
	3	A	A	82.5	1153	FPXG1040	25/09/13	16/11/13	53	Recapture
	4	B	A	77.0	946.0	FPXG1030	10/12/13	04/02/14	57	Recapture
	5	B	A	73.0	716.0	FPXG1040	10/12/13	30/12/13	21	Out
	6	A	A	73.0	600.0	FPXG1040	20/02/14	28/04/14	68	
	7	A	A	84.0	1066	FPXG1040	20/02/14	28/03/14	37	
	8	A	A	70.0	500.0	FPXG1030	02/05/14	15/06/14	45	Out
	9	A	A	77.5	827.7	FPXG1030	02/05/14	02/05/14	1	Out
	10	A	A	80.0	918.8	FPXG1030	26/06/14	03/09/13	70	
	11	A	A	68.0	584.3	FPXG1030	26/06/14	03/09/14	70	
Sebastes pachycephalus pachycephalus spotbelly rockfish	1	A	A	23.5	325.0	FPXG1040	25/09/13	31/01/14	129	
	2	A	A	18.0	124.0	FPXG1030	25/09/13	16/10/13	22	
	3	B	B	20.5	186.0	FPXG1040	25/09/13	19/10/13	25	Out
	4	B	A	25.8	423.0	FPXG1040	10/12/13	04/28/14	140	
	5	A	A	20.0	161.8	FPXG1030	26/06/14	01/09/14	68	
	6	A	A	19.0	156.2	FPXG1030	26/06/14	03/09/14	70	
	7	A	A	24.2	272.8	FPXG1030	26/06/14	26/06/14	1	

(continued)

Table 4.2 (continued)

Species and common name	ID	Capture and release point		Total length (cm)	Weight (g)	Pinger	Day of release	Last day of signal received	Monitoring period	Remark
Physiculus maximowiczi brown hakeling	1	A	A	26.8	188.0	FPXG1030	10/12/13	23/01/14	45	
	2	B	B	27.2	199.0	FPXG1030	20/02/14	28/02/14	9	Out
	3	B	B	27.0	185.1	FPXG1040	20/02/14	21/02/14	2	Out
	4	A	A	28.0	241.4	FPXG1030	02/05/14	24/05/14	23	Out
	5	A	A	28.5	225.0	FPXG1040	02/05/14	02/09/14	124	
	6	A	A	29.0	255.1	FPXG1030	26/06/14	03/09/14	70	
Sebastes cheni rockfish	1	B	B	27.5	364.3	FPXG1040	20/02/14	28/03/14	37	
	2	B	B	29.0	442.2	FPXG1040	20/02/14	28/05/14	97	Out
Sebastes schlegelii black rockfish	1	A	A	43.5	1372	FPXG1030	10/12/13	22/01/14	44	
Hexagrammos otaki greenling	1	A	A	38.8	699.1	FPXG1040	20/02/14	02/06/14	102	

Among the four greenling individuals, only ID C remained within the detection ranges of the receivers until the end of the monitoring period and ID-D did during about one month (Table 4.1). The numbers of dots in Fig. 4.3 indicate that ID C mainly inhabited Area A. On the other hand, IDs A and B moved beyond the detection ranges within one day and ID D did one month after the release. After the release, individuals IDs C and D settled more readily than individuals IDs A and B. The former and the latter were released in May 2012 and in November 2012, respectively. Since the two greenling individuals of IDs C and D have stayed in certain areas, greenling may be a resident type fish like black rockfish except spawning season. As greenling spawn in early winter when the water temperature begins falling (Izumi 1999), seasonal differences might have influenced the behavior of the greenling individuals released in November 2012.

4.3.1.2 Diurnal Patterns of Fish Behavior

The duration of day and night depend on the calendar. In this study, we regarded the lengths of day and night as fixed periods from 6 am to 6 pm and 6 pm to 6 am, respectively, to roughly compare activities of fish in a day. When a signal from an experimental fish was detected by more than one receiver, we assumed that the fish occupied one of areas indicated as Areas B, C, D and F that correspond to overlapped detection ranges between or among three receivers (Fig. 4.1). We first surveyed behaviors of 13 rockfish over three months. The night detection exceeded the daytime detection in 10 of the 13 individuals (Table 4.1). Excluding two individuals of IDs 5 and 8, depth ranges of the others during the night were wider than during the daytime. Greater depth range at nighttime than in the daytime likely reflects a circadian activity pattern of rockfish, suggesting that they stayed still in the daytime and forage at nighttime. As the detection range of each receiver was about 200 m, it is estimated that a diameter of one rockfish's habitat range regarded as a circle is about 400 m. Although the experimental fish were caught at the same place at nighttime, the habitat range of the rockfish differed among the individuals (Table 4.1).

Receiver 1 recorded a widely varying depth of ID C, and depth range at nighttime was wider than in the daytime (Fig. 4.3). Conversely, signals obtained by Receivers 2 and 3 indicated that the depth of ID C varied only from a depth of 11 to 13 m. Individual ID D demonstrated no difference in depth between the daytime and nighttime. Individual ID D remained at a fixed depth in Area G and occupied a narrower depth range. These results suggest that vertical diurnal movements of individuals may be caused by a difference in their environments such as bottom depths, tidal currents, prey distributions and so on between an area covered by Receiver 1 and areas covered by Receivers 2 and 3. For example, the former area was deeper than the latter area (Fig. 4.1).

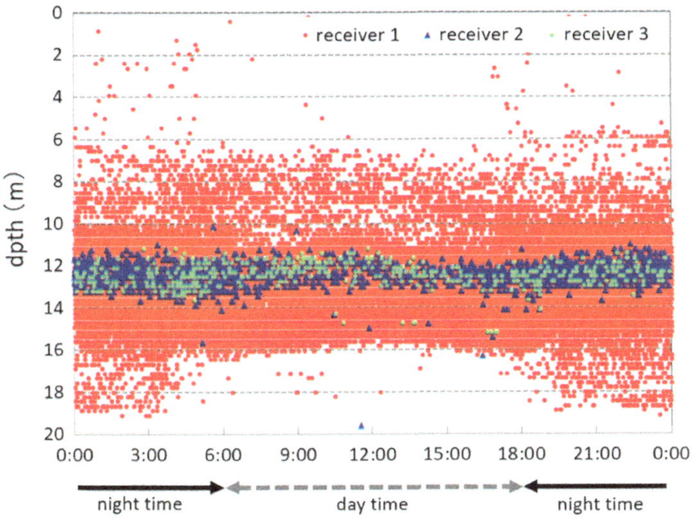

Fig. 4.3 Diurnal plots of depth of greenling ID C detected with Receiver 1 (red closed circle), Receiver 2 (blue closed triangle) and Receiver 3 (green closed circle) in the fishing ground off Iwaki City form May 2012 to May 2013

4.3.2 Fish Behaviors in the Port Near FDNPP

4.3.2.1 Water Temperature

Water temperature in the port showed seasonal variations (Fig. 4.4). Water temperature dropped from October 2012 to March 2013 when the water temperature was 6 °C, the minimum temperature in a year, and rised from April to September when the water temperature was about 23 °C, the maximum temperature in a year.

4.3.2.2 Tracking Duration

The tracking durations of the released experimental fishes (28 individuals of six species) are shown in Table 4.2. The signals of 8 individuals consisting of two white spotted conger eels, 4 spot belly rockfish, two brown hakelings were detected throughout the battery life of the pinger. Two white spotted conger eels were recaptured at the study site within the monitoring term.

We assume that if the last signal from an experimental fish was detected by the receiver at Site B near the port entrance, the experimental fish had moved outside the port or remained within the port but outside the detection ranges. Nine individuals were in that situation. The experimental fish remaining inside the port and those migrating beyond the receiver detection range were classified with the detected signal patterns. Schematic view of habitat range patterns of white spotted conger eel (IDs 1–5) are presented in Fig. 4.5. Individuals IDs 1, 2 and 3 of conger eel were released in September 2013. Individual ID 1 remained within the detection range of receiver at Stn. A for 21 days while it has never been detected by the receiver at Stn. B. It indicates that this individual remained inside the port until the end of battery life. Individual ID 2 which was initially released at Stn. A remained in the range of the receiver at Stn. A for 10 days, then moved to the range of the receiver at Stn. B. Then, it moved beyond the range of receiver at Stn. B. It suggests that ID 2 had migrated from the port to the outside of the port. Individual ID 3 released at Stn. A remained within the range of receiver at Stn. A and was finally recaptured with a trap in the port by sampling for a new experiment. Individuals IDs 4 and 5 were released on 10 December 2013. Individual ID 4 remained in the range of the receiver at Stn. A until the end of battery life and was recaptured there on 2 May 2014. Individual ID 5 remained in the range of the receiver at Stn. A for 10 d, then moved to the range of the receiver at Stn. B for 20 days and disapperared. It suggests that this individual migrated outside the port.

The apparent migrators were four individuals of IDs 2, 5, 8 and 9 among 11 white spotted conger eels, three individuals (IDs 2, 3 and 4) among six brown hakeling and one individual (ID3) among two spotbelly rockfish. Six among seven spotbelly rockfish, one rockfish, one black rockfish and one greenling didn't migrate from the ranges of two receivers. The *Sebastes* species exhibited especially strong site fidelity behavior.

4.3.2.3 Migration Outside the Port

The events that the experimental fish exited the port corresponded to the water temperature drops (Fig. 4.4). When the temperature dropped, five individuals two white spotted conger eels (IDs 2 and 5), two brown hakeling (IDs 2 and 3) and a spotbelly rockfish (ID 3) exited the port. We estimate that the falling water

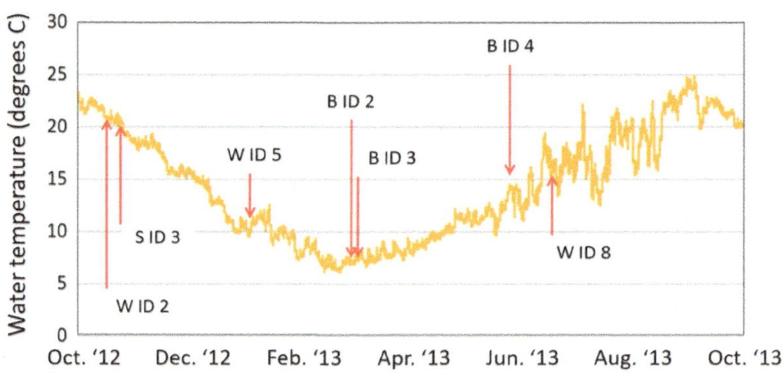

Fig. 4.4 Hourly water temperature profile in the port from October 2012 to October 2013. Red arrows indicate the events that the experimental fish might exit the port. W: white spotted conger eel, S: spotbelly rockfish and B: brown hakeling

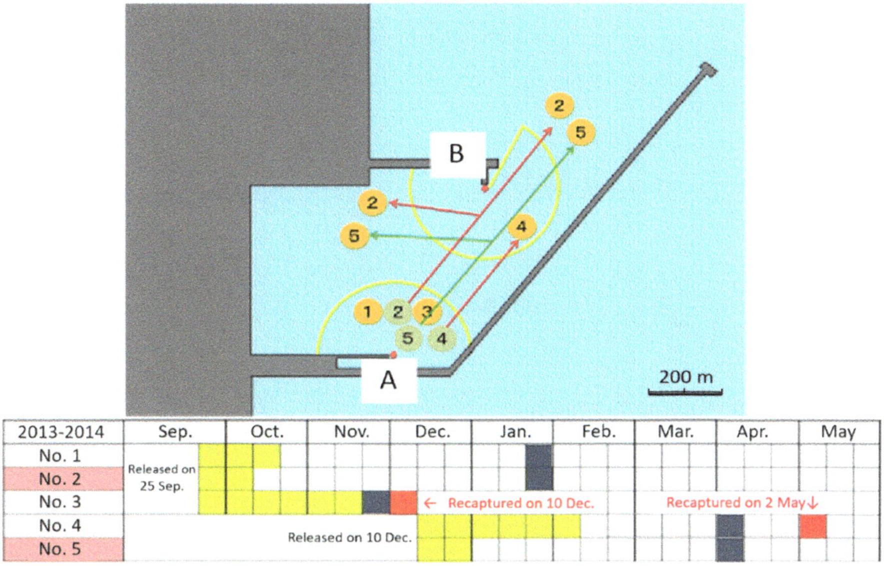

Fig. 4.5 Estimated behavior patterns of five white spotted conger eels. Schematic view of movements of IDs 1–5 released at Stn. A (upper panel). Individuals of IDs 1 and 3 remained in the range of receiver at Stn. A. Cells of ID 1–ID 5 from September 2013 to May 2014 corresponding to a term of a month are in yellow when one of two receivers received signals from these fish (lower panel). Red or black Cells are a recapture day and the end of battery life, respectively

temperature triggered the migratory behavior to deeper waters. Kitagawa et al. (1983) reported that brown hakeling off Sanriku Coast migrate to deeper waters for spawning in winter. The eggs and larvae were mainly distributed in the eastern part of the Izu Islands and along the Boso Peninsula, central Japan, which is situated in the inner region of Kuroshio Current (Kitagawa et al. 1985). Thus, spawning grounds are probably formed in offshore deeper waters around bottom depths of 130–330 m (Kitagawa et al. 1983). White spotted conger eel is one of the species that migrate their spawning area located in the western North Pacific along the Kyushu-Palau Ridge (Kurogi et al. 2012). Thus, it is considered that the events that exit of the white spotted conger eels and brown hakeling from the port may be related to a spawning migration behavior triggered by the water temperature drop.

4.4 Relation Between Fish Behavior and Radioactive Cs Concentration

To examine a relation between fish behaviors and their radioactive Cs concentrations, we referred to monitoring results of radioactive Cs concentrations in fish around Fukushima Prefecture announced by the Fisheries Agency (Fisheries

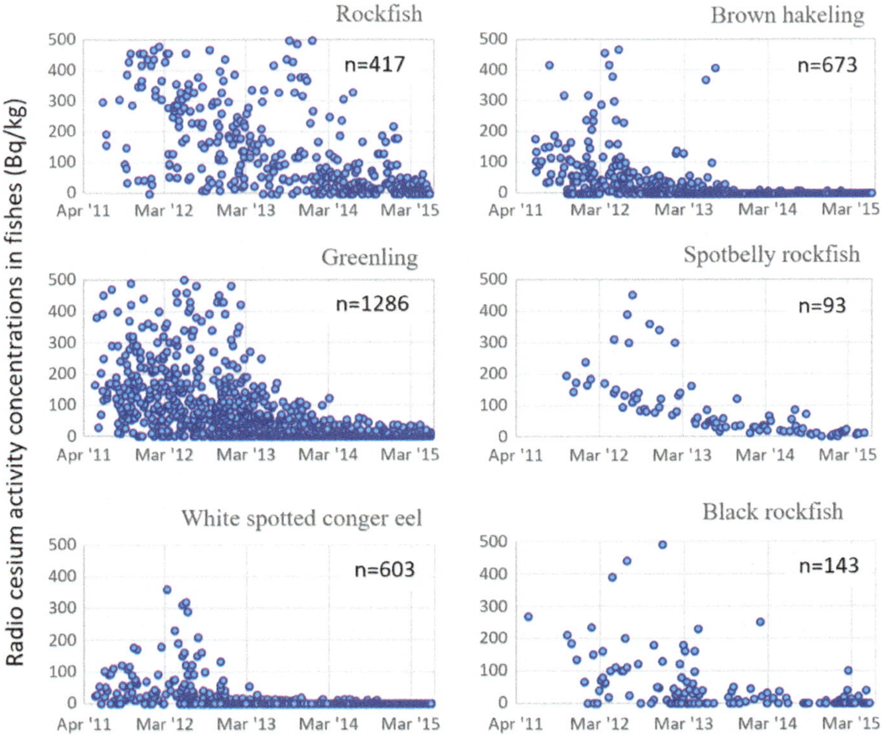

Fig. 4.6 Time series data of radioactive Cs concentrations of the experimental fish species around Fukushima Prefecture based on information released by the Fisheries Agency of Japan (2016). The original data can be downloaded from the Fisheries Agency website (http://www.jfa.maff.go.jp/index.html)

Agency of Japan 2016). Figure 4.6 shows temporal trends of radioactive Cs concentrations ($^{134}Cs + {}^{137}Cs$) in fish species inhabiting off Fukushima Prefecture from June 2011 to June 2015 (Fig. 4.5). It is noticed that the number of sampled individuals differed from species to species. The radioactive Cs concentrations in rockfish, greenling, spotbelly rockfish and black rockfish exceeded 100 Bq kg^{-1} at day 1000 after the accident of FDNPP on 11 March 2011. Rockfish recorded high concentrations of radioactive Cs. This species showed strong site fidelity both in the fishing ground near the coast in the Pacific Ocean and in the port near FDNPP. All of these species remained in the detection ranges of the receivers for various lengths of time without straying. Rockfish and spotbelly rockfish exhibited especially strong site fidelity and might spawn in waters near sampling points. These species may likely remain in the region of high radioactive Cs concentrations. Thus, they are unlikely to transfer radioactive Cs to other waters.

The radioactive Cs concentrations in white spotted conger eel and brown hakeling were under 50 Bq kg^{-1} at day 1000. Spawning grounds of white spotted conger eel and brown hakeling are far from their sampling areas (Kitagawa et al. 1983, 1985;

Kurogi et al. 2012). The radioactive Cs concentrations of these migrating species declined more rapidly than those of species with strong site fidelity. Therefore, these two species can potentially convey radioactive Cs to other waters.

Another trend in radioactive Cs concentrations (up to 500 Bq kg^{-1}) was a wide difference among individuals of the same species. Differences of radioactive Cs concentrations among individuals might be explained by diurnal vertical movements and migration. Our biotelemetry results indicate different site fidelities and migration behaviors of the experimental fishes. The radioactive Cs concentrations of resident species exhibiting site fidelity were higher than those of migrating species. The latter type shows the radioactive Cs concentration is likely to fall by two mechanisms: natural decline over time and a replacement of contaminated fish with less contaminated fish coming from waters with low radioactive Cs concentrations through migration. These results are useful information and contribute to decide to stop a ban of fisheries by Fisheries Cooperative of Fuksushima Prefecture in the future.

Acknowledgements We thank all fishers of the Iwaki-shi Fisheries Cooperative for maintaining the equipments and catching the experimental fish. We also thank T. Nakagawa, a graduate student at Tokyo University of Marine Science and Technology, for analyzing the data. This study was supported by an Environment Research and Technology Development Fund from Ministry of the Environment, Japan.

References

Arakawa H, Tokai T, Miyamoto Y, Akiyama S, Uchida K, Matsumoto A, Narita M, Myouse H, Agatsuma Y, Katayama S, Aoki M, Matsumoto I, Hirakawa N (2015) Distribution of radioactive material in marine ecosystems off the Fukushima coast: radioactive cesium levels in Fukushima marine organisms. In: Ceccaldi, HJ, Hénocque Y, Koike Y, Komatsu T, Stora G, Tusseau-Vuillemin M-H (eds) Marine productivity: perturbations and resilience of socio-ecosystems. Springer International Publishing Switzerland, Cham, pp 71–78

Director of Food Safty Division, Ministry of Health, Labour and Welfare of Japan (2012, in Japanese) Revision of ministerial ordinance on the part revision of the standards for food, additives etc. Circular of Food Safety Division, Ministry of Health, Labour and Welfare of Japan, 0315, No 1, 1–5

Fisheries Agency of Japan (2016, in Japanese) Information on radioactive materials and fishery products. Fisheries Agency of Japan, Tokyo, Japan. http://www.jfa.maff.go.jp/e/index.html

Izumi S (1999, in Japanese) Growth and Maturing of fat greenling *Hexagrammos otakii* in North Joban Sea. Bull Fukushima Prefect Fish Exp Stn 8:41–49

Kitagawa D, Nagahora S (1983, in Japanese) Estimation of the spawning season of the Morid fish Physiculus maximowiczi collected from the coastal waters of Iwate Prefecture, Japan. Bull Jpn Soc Sci Fish 49:1649–1654

Kitagawa D, Kuroda K, Tsuruta Y (1985, in Japanese) Description and distribution of eggs and larvae of the brown hakeling *Physiculus maximowiczi* in Japanese waters. Bull Jpn Soc Sci Fish 51:1627–1630

Kurogi H, Mochioka N, Okazaki M, Takahashi M, Miller JM, Tsukamoto K, Ambe D, Katayama S, Chow S (2012, in Japanese): Discovery of a spawning area of the common Japanese conger *Conger myriaster* along the Kyushu-Palau Ridge in the western North Pacific. Fish Sci 78:525–532

Matsumoto A, Shigeoka Y, Arakawa H, Hirakawa N, Morioka Y, Mizuno T (2015, in Japanese) Biological half-life of radioactive cesium in Japanese rockfish *Sebastes cheni* contaminated by the Fukushima Daiichi nuclear power plant accident. J Environ Radioact 150:68–74

Mitamura H, Arai N, Mitsunaga Y, Yokota T, Takeuchi H, Tsuzaki T, Itani M (2005a) Directed movements and diel burrow fidelity patterns of red tilefish *Branchiostegus japonicus* determined using ultrasonic telemetry. Fish Sci 71:491–498

Mitamura H, Arai N, Sakamoto W, Mitsunaga Y, Tanaka H, Mukai Y, Nakamura K, Sasaki M, Yoneda M (2005b) Role of olfaction and vision in homing behaviour of black rockfish Sebastes inermis. J Exp Mar Biol Ecol 322:123–134

Nemoto Y, Sohtome T, Sato M, Fujita T, Kamiyama K, Shimamura S (2013, in Japanese) Radioactive substance effects on marine products off fukushima prefecture. Bulltin Fukushima Prefect Fish Exp Stn 16:63–89

Omori T (2014, in Japanese) Current conditions and problems of marine products of Fukushima Prefecture and efforts to resolve the problems. National Fisheries Cooperative Association, Tokyo, 23 pp

Yokota T, Mitamura H, Arai N, Masuda R, Mitsunaga Y, Itani M, Takeuchi H, Tsuzaki T (2006) Comparison of behavioral characteristics of hatchery-reared and wild red tilefish *Branchiostegus japonicaus* released in Maizuru Bay by using acoustic biotelemetry. Fish Sci 72:520–529

Chapter 5
Estimate of Water Quality Change in Osaka Bay Caused by the Suspension of Marine Sediment with Mega Tsunami

Mitsuru Hayashi, Satoshi Nakada, Shunich Koshimura and Eiichi Kobayashi

Abstract Huge tsunami will be caused by the Nankai-Trough Earthquake, and will suspend marine sediment in Osaka Bay. Various materials contain in the marine sediment, and will be released to the water column by the tsunami. The spatial distributions of concentration of heavy metals, Cd, Pb, Cr, Hg and Zn, total nitrogen and phosphorus and suspended solid in the water column after the tsunami were estimated. The environmental standards of water quality may be exceeded in the inner part of Osaka Bay in the case of the worst conditions. And the suspend matters will drift during the tsunami.

Keywords Tsunami · Marine sediment · Osaka bay · Nankai-Trough

5.1 Introduction

The Nankai Trough Earthquake, with magnitude 8 or 9, will hit with a probability of 70% within 30 years, and a mega tsunami will be caused by the earthquake (Headquarters for Earthquake Research Promotion, Government of Japan 2013). It will attack Osaka Bay, and will suspend and transport the marine sediment in the water column. In fact, it is reported based on the bathymetric data before and after the 2004 Indian Ocean tsunami that the erosion, deposition of the bottom and the movement of shoreline were occurred inside of the port in Sri Lanka (Goto et al. 2011). Kondou et al. (2011) estimated the maximum erosion and deposition depths in the port of Osaka, and also gave that the dredging for marine transportation will

M. Hayashi (✉)
Research Center for Inland Seas, Kobe University, 5-5-1 Fukaeminami, Higashinada, Kobe 658-0022, Japan
e-mail: mitsuru@maritime.kobe-u.ac.jp

S. Nakada · E. Kobayashi
Graduate School of Maritime Sciences, Kobe University, Kobe, Japan

S. Koshimura
International Research Institute of Disaster Science, Tohoku University, Sendai, Japan

T. Komatsu et al. (eds.), *Oceanography Challenges to Future Earth*,
https://doi.org/10.1007/978-3-030-00138-4_5

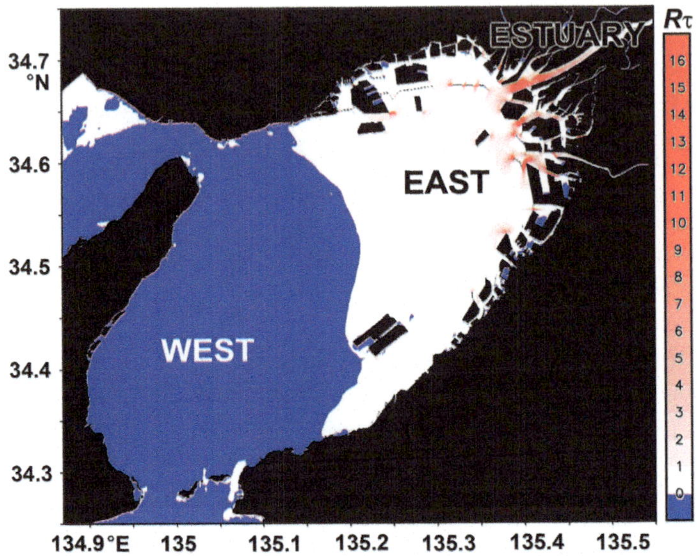

Fig. 5.1 Distribution of the non-dimensional bed shear stress, R_τ, in Osaka Bay estimated by Hayashi et al. (2015)

takes two months. Hayashi et al. (2015) discussed the possibility of the rolling up of the marine sediment of Osaka Bay based on the results of the numerical simulation (Nakada et al. 2015) of the tsunami caused by the Nankai Trough Earthquake. Figure 5.1 shows the distribution of the non-dimensional bed shear stress (R_τ) estimated by Hayashi et al. (2015).

It was clarified that although the rolling up will not occur ($R_\tau < 0$) in the western part of Osaka Bay (blue in Fig. 5.1), will happen ($R_\tau \geq 0$) in the eastern part (white and reddish in Fig. 5.1) where the water depth is less than 30 m and the moisture content (W) of the sediment is 57% and more as shown in Fig. 5.2.

In addition the suspension of sediment will be extremely strong around the Yodo River estuary where was named "hot spot" (red color in Fig. 5.1). Various substances and microorganisms are contained in marine sediment.

Natsuike et al. (2014) summarized the records of the maximum densities of *Alexandrium spp.* cysts and cells and of the maximum toxicity of bivalves observed at the coastal sea of Tohoku after the Tohoku Region Pacific Coast Earthquake on 11 March 2012. *Alexandrium spp.* is a harmful phytoplankton that causes shellfish poisoning. The reason of high values was considered that the accumulated cysts were released to the water column and germinated. *Alexandrium spp.* appears in Osaka Bay every year, formulates a red tide, and causes the shellfish poisoning. Nakayama (2011) reviewed a lot of academic papers, and concluded that the nutrient supply from the sea floor cannot be omitted when evaluating the low trophic level ecosystem and natural environment of the Seto Inland Sea. Much nutrient load is necessary for occurrence of red tide. And not only cysts but

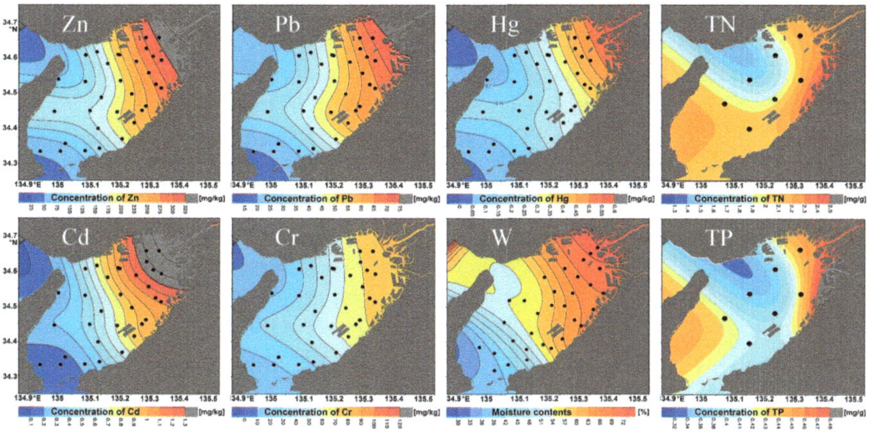

Fig. 5.2 Horizontal distributions of densities of zinc (Zn), cadmium (Cd), lead (Pb), chromium (Cr), mercury (Hg), moisture content (W), total nitrogen (TN) and total phosphorus (TP) in the marine sediment of Osaka Bay with the survey points

high-concentration N and P also is accumulated in the marine sediment of Osaka Bay (Ports and Airports Department, Chugoku Regional Development Bureau, Ministry of Land, Infrastructure, Transport and Tourism (MLIT), Government of Japan 2014). If cysts and nutrients are simultaneously released by tsunami from the marine sediment of Osaka Bay, large-scale red tide and shellfish poison may happen. Heavy metals are also contained in the marine sediment in the inner part of Osaka Bay close to the mega city. The densities of cadmium and mercury in the marine sediment of Osaka Bay are higher about 2 times of it in Sendai Bay, and are the high level same as Tokyo Bay (Shimizu et al. 2008). There are no environmental standards for heavy metals in the sediment. They in the seawater are defined by Environmental Basic Low.

Total load volumes of materials in Osaka Bay from the Nankai Trough Earthquake occurring for 10 h are estimated giving the realistic densities in the sediment to the estimated results of the rolling up Hayashi et al. (2015) based on the tsunami simulation (Nakada et al. 2015) which is most severe condition. The volumes are converted into the concentration in the seawater, and are standardized by the environmental standards. The possibility of change of the water quality by tsunami is shown and discussed.

5.2 Methods

5.2.1 Calculation

Flux of the marine sediment, F (kg m^{-2} s^{-1}), from the bottom to the water column is obtained by $= R_\tau$ (Mehta 1993). R_τ is the ratio of the bottom shear stress, which is produced by a flow, to the critical shear stress, which is determined by the characteristics of the marine sediment. R_τ in Osaka Bay by a tsunami have been calculated by Hayashi et al. (2015) as shown in Fig. 5.1, and by using the flow velocity of the tsunami, the elevation of the tsunami, the water depth and W of the marine sediment. W was surveyed by the Research Institute of Environment, Agriculture and Fisheries, Osaka Prefecture in 2013. The tsunami was simulated by Nakada et al. (2015) which added the tidal current and the Coriolis force to the tsunami simulation code based on non-linear wave theory developed at Tohoku University (Koshimura et al. 2009). The Japanese Government uses the code of Tohoku University for tsunami prediction (Central Disaster Prevention Council, Government of Japan 2003). It was judged that the reliability of the model is high based on the previous achievements of the model. The following conditions were assumed for the tsunami simulation, the fault model of case 3 of the Nankai Trough (Central Disaster Prevention Council, Government of Japan 2012) which gives most serious influence for Osaka Bay, the bathymetry dataset used in Kobayashi et al. (2005) was applied, the horizontal flow velocity is uniform in the direction of water depth, and the tsunami never bounces and does not run up to the land. The calculation result without the tide was used. The calculations of the tsunami and R_τ were carried out from the earthquakes occurring for 10 h every 0.5 s. The area of simulations covers the whole Osaka Bay (34.36 N ~ 34.94 N, 134.85E ~ 135.62E), and was divided into a 50 m grid horizontally with one vertical dimension. Details of R_τ and the tsunami simulation are expressed in Hayashi et al. (2015) and Nakada et al. (2015).

The sediment type of the eastern part of Osaka Bay is almost a clayey silt (Yokoyama and Sano 2015). Fujimoto (2011) decided M in the port of Osaka as 1.0 kg m^{-2} s^{-1} which is the average value based on the sensitivity analysis of the Murakami's model. Murakami et al. (1989) said that M is constant in each sediment type. However we should keep mind the range of M, because it is 2 ~ 27 times of Umita et al. (1988) which is an experimental study used a softy mud with 80% in W. We have applied to 2.3 to*n* which is the maximum number in the experiment of Umita et al. (1988). It is twice of the value indicated by Murakami et al. (1989), and applied to the port of Osaka by Kondou et al. (2011). We assume the worst situation, and was adopted a larger value.

The rolled up volume of sediment, V_s(kg s^{-1}), to the water column is obtained by multiplying F by the area of grid (50 m × 50 m). When the density of material in the sediment (mg kg^{-1}) is multiplied by V_s, the loaded volume of material V_m (mg s^{-1}), to the water column is obtained. V_s and V_m are calculated every 0.5 s because R_τ changes temporally. The summation of each volume for 10 h is the total loading volume of sediment (kg) and materials (mg) to each water column. When the total loading volume divided by the volume of each water column (l), the concentration of suspended solid (SS) and materials in the water (mg l^{-1}) after the tsunami are obtained. Stratification develops in the eastern part of Osaka Bay throughout the year because of the fresh water discharge form the Yodo River. However, since the tsunami will break the stratification (Nakada et al. 2016), it is supposed that the materials will be dispersed uniformly in the water column in immediately after the tsunami.

5.2.2 Data

The data sets of density of cadmium (Cd), leaden (Pb), chromium (Cr), mercury (Hg), zinc (Zn), total nitrogen (TN) and total phosphorus (TP) in the sediment with 50 m grids were constructed the survey data by using the Gaussian Interpolation. The horizontal distributions of densities of the heavy materials, TN and TP concentration and W with the survey points are shown in Fig. 5.2.

Two reports about heavy metals were used. One is Nagaoka et al. (2004) which was carried out at 22 points on April 2000, and was sampled the surface sediment with 5 cm thickness. Other is presented on the web site (Geological Survey of Japan, National Institute of Advanced Industrial Science and Technology; AIST, expressed by Imai 2010). The surface sediments with 3 cm thickness were samples at 11 points. The Sampling year is unknown. Japan Coast Guard (JCG) also survey the heavy metals every year (Japan Coast Guard 2016). However we did not use the data because the sampling points are limited in 5 and thickness of a sample is unknown. On the other hand, the temporal variations of Cd and Hg densities used JCG data are gave in Shimizu et al. (2008). The densities are steady after the 1990's. Therefore, it will be no problem that the year of survey may differ between Nagaoka et al. (2004) and AIST. Cd density of Nagaoka et al. (2004) is twice of the JCG data surveyed on January 2000 (Japan Coast Guard 2002). The large difference is not seen with other heavy metals. We should pay attention to the result of Cd. Any heavy metal has high density in the inner part of the bay.

TN and TP data surveyed on Oct. 2014 were obtained from the web site of the water quality survey of the Seto Inland Sea (Ports and Airports Department, Chugoku Regional Development Bureau, MLIT, Government of Japan 2014). Detail of the sampling methods of sediment has not been published. Sampling about 10 cm of a surface of sediment is recommended in the Sediment Monitoring Methods which the Ministry of Environment defined. Density in the eastern part of the bay are high along the coastal from Osaka to Sakai-Senboku, and are low toward the west (offshore). Since the place where densities are the highest changes from year to year, therefore the distribution pattern also changes. However, the average densities of the whole bay are almost constant in general in internal. Japanese government have not defined SS standard for the sea. Therefore, the standard for "protection of aquatic life" defined by Japan Fisheries Resource Conservation Association was used. The Ministry of Environment, Japan have defined the environmental standards about water pollution from the viewpoint of "health protection of people" and "living environment preservation". The standards of heavy metal, Cd, Pb, Cr and Hg have been defined in "health protection of people". The standards of Zn, TN and TP have been defined in "living environment preservation", and change with types of the sea as shown in Fig. 5.3.

The standard of the Organism-A type which occupy most of the eastern Osaka Bay as shown in Fig. 5.3a was applied to Zn. Osaka Bay is divided to the 3 types for TN and TP as shown in Fig. 5.3b, and the type III was applied.

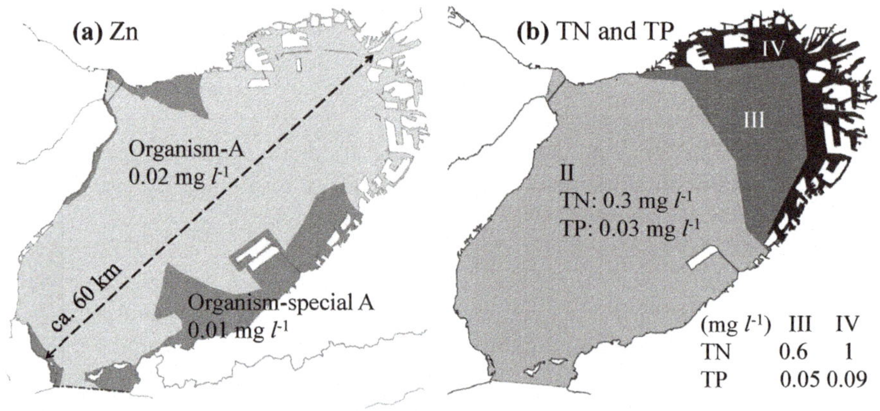

Fig. 5.3 Classification and values of the environmental standards of Zn (**a**) and TN and TP (**b**)

5.3 Results

Figure 5.4 shows the distribution of the standardized concentrations with the standards.

The western part of Osaka Bay had been leaved out because the rolling up of marine sediment will be not occurred in the region (Hayashi et al. 2015) which is

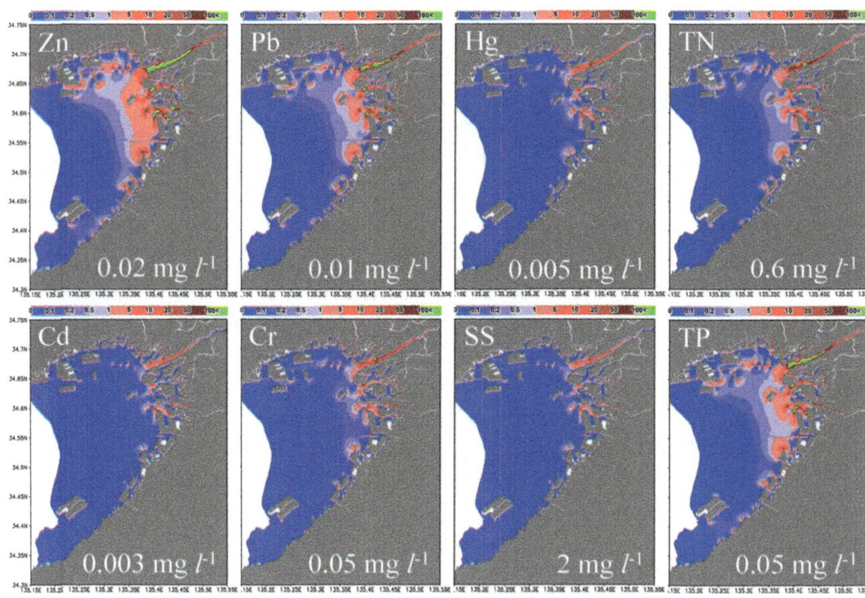

Fig. 5.4 Distributions of the standardized concentrations of Zn, Cd, Pb, Cr, Hg, SS, TN and TP with each environmental standard in seawater

indicated by white color. The concentrations exceed the criteria in the red and green-tinged regions.

The concentrations in the inner part are higher in any materials. It depends on the distribution of both R_τ and the concentration in the sediment. Especially the concentrations around the Yodo River estuary and the landfill of the port of Osaka Port exceed extremely the environmental standard more than 10 times. The area is the hot spot that the rolling up of marine sediment is strong and continually (Hayashi et al. 2015), and accumulates various materials in the sediment. Therefore the load volume of materials is large. Moreover because the water depth is shallow, the concentrations are very high.

The concentrations of SS, Cd, Cr and Hg are higher than the standers just around the hot spot. Since the concentration of Cd is low, it is not necessary to pay attention to the range of surveyed density of Cd. The concentrations of Zn, Pb, TN and TP are higher than the standers in the wide area, and are less than 5 times of the standers expect the hot spot. The hot spot and the type IV region for TN and TP overlap in general. When the standards of the type IV region are applied to TN and TP in the region, the reddish areas of TN and TP become narrow a little, and the concentrations of their become low. The area that the concentrations of are higher than the standers are just around the hot spot. Anyway, since the concentrations exceed the standards remarkably around the hot spot, the main source of material load from marine sediment to the water column by tsunami is the hotspot.

5.4 Discussion

The results mean that the importance of reducing the load in the hot spot. There will be no other options to reduce the concentration which is 50 times or more of the environmental standards except the remove of the sediment that contains a high density material. However it is not easy.

If we accept large load of materials in the hot spot, the next problem is an expansion of the materials from the hot spot. We estimate the spatial scale after 10 h of the distribution of grains, in other words the suspended sediment, which are rolled up in the hot spot. At first, the settling speed (cm s^{-1}) of the grain is calculated by the Stokes' equation, $S = [gd^2(\rho_s - \rho_w)]/18\eta$. g is the gravity acceleration(9.8×10^2 cms^{-2}). ρ_s is the density of the particulate, and is applied 2.67 g cm^{-3} which was used by Kondou et al. (2011). ρ_w is the density of the seawater, and is applied 1.023 g cm^{-3} which is calculated by the water temperature of 16 °C, salinity of 31 and depth of 10 m. d is the diameter of grain, and is applied 7.8×10^{-4} cm which is decided based on the median diameter investigated in Osaka Bay in 2013 (Yokoyama and Sano 2015). η is the dynamic viscosity of the seawater, and is applied 0.0119 $g(cms)^{-1}$ which is the case of the water temperature of 16 °C at the shallow water in the experiment of Suzuki and Nagashima (1980). The grains sink 166 cm in 10 h, and will continue to suspend even after the tsunami in most of Osaka Bay including the Yodo River. Since some material dissolve in the seawater, the residence time in that case becomes longer. Next, the horizontal transport by the tsunami considers. The speed of horizontal flow caused by the tsunami is more than 0.6 ms^{-1} in average of 10 h in the hot spot (Hayashi et al. 2015). Because there are backwash and anaseism, when it is assumed that the grain is transported during 5 h with 0.8 ms^{-1}, the drift distance is 14.4 km. Because the drift motion of particles is a nonlinear and the offshore flow velocity is weaker in fact, the drift distance will be shorter. In other words, the rolled up sediments in the hot spot will suspend in the eastern part of the bay during the tsunami even if the maximum condition.

SS concentration exceeds the environmental standards frequently in the inner part of the bay in the summer when a biological production is active (e.g. five times of the standards equivalent to the 10 mgl^{-1}, Joh 1985). Since an elevation of SS concentration in an ordinary environment has occurred depending on the season, an elevation of the concentration due to the tsunami might be acceptable range. Cd, Cr and Hg concentrations are reduced by spreading to the outside of the hot spot during the tsunami. It means the dispersion of environmental risk. On the other hand, Zn and Pb have the higher environment risk. It may be possible to reduce the concentrations of Zn and Pb in the seawater than the standards except the hot spot, if the density of Zn and Pb in the sediment in the eastern part lowered than the density of the western part. Although TN and TP concentrations exceed the standards almost in the type IV region, TN and TP may be transported to the type III region. TN and TP concentrations in the eastern part of Osaka Bay sometimes but still exceed the standards in the region of all the types (Osaka Bay renaissance Promotion Conference 2016). TN and TP concentrations will exceed the standards by the tsunami in all the eastern part of Osaka Bay where the concentrations are high potentially. And the possibility of falling into the hyper-eutrophic condition in the hot spot is suggested. It is necessary to estimate the influence on the primary production considering both the nutrient and SS concentrations because the photo environment in the seawater will change according to SS.

5.5 Conclusion

The environmental standards of water quality may be exceeded in the inner part of Osaka Bay by tsunami in the case of the worst conditions. And the suspend matters will drift during the tsunami.

Materials will be transported by the advection and diffusion during tsunami and be settled with particle, and also will be transported by the tidal and residual current after tsunami. Nakada et al. (2015) calculated the temporal and spatial variations of Zn distribution during one month after the occurrence of the earthquake in consideration of a flow and deposition. It was estimated that Zn was removed from the seawater of Osaka Bay within one month and more. Influence of the tsunami to primary production will be clarified quantitatively when same calculation is performed to TN, TP, SS and also cysts. The estimation by the three-dimensional calculation in consideration of stratification is desired in this case. Nakada et al. (2016) has tried the three-dimensional calculation. The application to an ecosystem model is also expected.

Acknowledgements This research was supported by Sumitomo Electric Industries Group CSR Foundation. The supercomputer of ACCMS, Kyoto University was used for the calculation. We appreciate the kind assistance of Mr. Y. Taniguchi in Kobe University, Dr. M. Sano in Osaka Prefecture and Professor T. Komai in Tohoku University.

References

Central Disaster Prevention Council, Government of Japan (2003, in Japanese) Disaster-measures to Tonankai-Nankai Earthquake. http://www.bousai.go.jp/jishin/tonankai_nankai/index.html

Central Disaster Prevention Council, Government of Japan (2012, in Japanese) The committee of mega earthquake model of Nankai Trough. http://www.bousai.go.jp/jishin/nankai/model/index.html

Fujimoto (2011, in Japanese) Influence of the port facilities by the marine sediment transport due to tsunami. In: Proceedings of the workshop of Kinki regional development bureau, no. 3

Geological Survey of Japan, National Institute of Advanced Industrial Science and Technology (2010) Geochemical map of sea and land of Japan. https://gbank.gsj.jp/geochemmap/index.htm

Goto K, Takahashi J, Oie T, Imamura F (2011) Remarkable bathymetric change in the nearshore zone by the 2004 Indian Ocean tsunami: Kirinda Harbor, Sri Lanka. Geomorphology 127:107–116

Hayashi M, Suzuki S, Nakada S, Koshimura S, Kobayashi E (2015) Estimation of the occurrence condition of sediment resuspension in Osaka Bay by tsunami. In: Proceedings of the 25th International Offshore and Polar Engineering Conference, vol 3, pp 783–787

Headquarters for Earthquake Research Promotion, Government of Japan (2013) Long-term evaluation of activity of Nankai trough earthquake. http://www.jishin.go.jp/evaluation/long_term_evaluation/subduction_fault/

Imai N (2010, in Japanese) Investigation of the distribution of elements of the whole of Japan and their applications - Geochemical map of land and sea in Japan. Synthesiology 3(4):281–291

Japan Coast Guard (2002, in Japanese) Report of marine pollution surveys, vol 28, 74 p

Japan Coast Guard (2016, in Japanese) Report and outline of marine pollution surveys. http://www1.kaiho.mlit.go.jp/KANKYO/OSEN/osen.html

Joh H (1985, in Japanese) Chemistry: the Seto Inland Sea. In: Coastal Oceanography of Japanese Inlands, Costal Oceanography research committee, the Oceanographical Society of Japan. Tokai University Press, Tokyo, pp 642–655

Kobayashi E, Koshimura S, Kubo M (2005, in Japanese) A basic study on ship drifting by Tsunami. Kansai Soc Nav Arch Jpn 243:49–56

Kondou T, Morimoto T, Fujimoto N, Tonomo K, Iemura K, Shikata T (2011, in Japanese) Influence of the port facilities by the sediment transport due to tsunami. J Jpn Soc Civ Eng, Ser B2 (Coastal Engineering) 67(2):I_261–I_265

Koshimura S, Oie T, Yanagisawaand H, Imamura F (2009) Developing fragility function for tsunami damage estimation using numerical model and post-tsunami data from Banda Aceh, Indonesia. Coast Eng J 51:243–273

Mehta AJ (1993) Nearshore and Estuarine Cohesive Sediment Transport, American Geophysical Union, Washington, DC, 581 pp

Murakami K, Suganuma F, Sasaki H (1989, in Japanese) Experimental investigation on erosion and deposition of fine cohesive sediments in an annular rotating channel. Report of the port and harbor research institute, vol 28, pp 43–76

Nagaoka C, Yamamoto Y, Eguchi S, Miyazaki N (2004, in Japanese) Relationship between distribution of heavy metals and sedimental condition in the sediment of Osaka Bay. Nippon Suisan Gakkaishi 70(2):159–167

Nakada S, Hayashi M, Koshimura S, Yoneda S, Kobayashi E (2015) Tsunami simulation generated by the greatest earthquake scenario along the Nankai Trough under consideration of tidal currents in a large bay. In: Proceedings of the 25th International Offshore and Polar Engineering Conference, vol 3, pp 811–816

Nakada S, Hayashi M, Koshimura S, Kobayashi E (2016, in press) Simulation of heavy metal transport induced by a giant tsunami based on Nankai-Trough Earthquake: application to Osaka Bay. In: Proceedings of the 26th International Offshore and Polar Engineering Conference

Nakayama A (2011) Review of characteristics on the current and nutrients distribution of Seto Inland Sea and development of the hydrodynamic model coupled with primary production model applied to this region. Bull Fish Res Agen 34:49–70

Natsuike M, Kanamori M, Baba K, Moribe K, Yamaguchi A, Imai I (2014): Changes in abundances of *Alexandrium tamarense* resting cysts after the tsunami caused by the Great East Japan Earthquake in Funka Bay, Hokkaido, Japan. Harmful Algae 39:271–279

Osaka Bay renaissance Promotion Conference (2016, in Japanese) Results of Osaka Bay general survey for water quality. http://www.kkr.mlit.go.jp/plan/suishin/index510.html

Ports and Airports Department, Chugoku Regional Development Bureau, Ministry of Land, Infrastructure, Transport and Tourism, Government of Japan (2014, in Japanese) Water quality multi-survey in the Seto Inland Sea. http://www.pa.cgr.mlit.go.jp/chiki/suishitu/

Shimizu J, Noguchi K, Miura Y, Tomohisa T (2008, in Japanese) Temporal trends of the concentrations of pollutants in the surface sediments from the major bays of Japan. Rep Hydrogr Ocean Res 44:57–66

Suzuki H, Nagashima A (1980, in Japanese): Viscosity of Seawater under High Pressure. Trans Jpn Soc Mech Eng Series B 408:1574–1582

Umita T, Kasuda T, Futawatari T, Awaya Y (1988, in Japanese) Study on erosional process of soft muds. J Hydraul Coast Environ Eng 398(II-9):33–42

Yokoyama H, Sano M (2015, in Japanese) Zoning of Osaka Bay based on principal component analysis of sediment parameters and comparison between the 2013 and previous surveys. Nippon Suisan Gakkaishi 81(1):68−80

Chapter 6
Litter in the Mediterranean Sea

François Galgani

Abstract In the Mediterranean Sea human activities generate considerable amounts of waste, and quantities are increasing, although they vary between countries. To support actions to be taken in order to minimize impacts on the marine environment, the European Commission included recently marine litter as one of the descriptor of the Marine Strategy Framework Directive (MSFD), establishing a framework within which Member States must take action to achieve or maintain Good Environmental Status (GES). Beside, the Barcelona convention with its Regional Action is coordinating the monitoring of marine litter and support the implementation of reduction measures. As necessary steps, a better management will then need to (i) increase our knowledge and further development of data analysis in all regions, enabling to map areas at risk and sources, (ii) develop supporting tools for larger scale assessments, including harmonized databases, (iii) a better definition of thresholds, baselines and targets, to better understand harm and finally (iv) a coordinated monitoring. Research, supported by many institutions has then become critical and priorities include (i) a better definition of standardized/harmonized protocols, (ii) research on nanoparticles at sea, (iii) a better understanding of circulation and transport (iv) an increased knowledge of the ecology of microbial life on plastic and consequences on degradation, species dispersion and release of chemicals, and (v) a better understanding of the interactions between plastic and marine organisms. Using some examples, the process of the implementation of monitoring is described and discussed in the context of reduction measures.

Keywords Mediterranean sea · Marine litter · Microplastics · Monitoring

F. Galgani (✉)
IFREMER, LER/PAC, ZI Furiani, 20600 Bastia, France
e-mail: Francois.galgani@ifremer.fr

T. Komatsu et al. (eds.), *Oceanography Challenges to Future Earth*,
https://doi.org/10.1007/978-3-030-00138-4_6

6.1 Introduction

Because of (i) large cities, rivers and shore uses, (ii) some of the largest amounts of Municipal Solid Waste that are generated annually per person (208–760 kg/Year), (iii) because of tourism, (iv) 30% of the world's maritime traffic and (v) a closed basin, the Mediterranean has been described as one of the most affected areas by marine litter in the world. Plastic, mainly bags and fishing equipment, is the largest part of debris and they may physically degraded into smaller fragments, the so-called microplastics. Most existing surveys of microplastics on the surface, worldwide, have found average densities lower than in the NW Mediterranean Sea (80–250000 items/km^2) and microplastics have been also found on beaches and sediments, including deep sea, reaching concentrations of 1000 pellets/m^2.

Research demonstrated (i) the importance of hydrodynamics, (ii) the impact of plastic at sea that include entanglement, physical damage and ingestion, the release of chemicals, the transport of species and the alteration of benthic community structures, and (iii) social and economic harm.

In fact, the issue of marine litter and related information on the amounts and types in the Mediterranean is rather complicated, as it is addressed principally by scientific institutions and sub-regional and local authorities in most countries on the one hand and by NGOs on the other.

6.2 Source and Distribution

Sources of marine litter are traditionally classified as either land-based or sea-based, depending on where the litter enters the water. Other factors, such as ocean current patterns, climate, tides, and proximity to urban centers, waste disposal sites, industrial and recreational areas, shipping lanes, and commercial fishing grounds, influence the type and amount of marine litter found in open ocean areas or collected along beaches and ocean, including underwater areas.

Studies on land based source pollution indicated inputs of the Po river only at a level of 50 billons particles every year (Vianello et al. 2015). Another study on this river (Tweehyusen 2015), demonstrated that 677 tons of microplastics were entering the Mediterranean Sea every year. Uncontrolled discharges also act as main sources of litter in the Mediterranean Sea as in some countries, a reduced percentage of coastal cities are controlling their waste discharges in adapted structures. Ocean-based sources for marine litter include merchant shipping, liners, fishing vessels, military fleets, and offshore installations. It is expected that garbage coming from ships to the Mediterranean may be in the range of 1 million tons.

Assessments of the composition of beach litter in different regions show that synthetic materials (bottles, bags, caps/lids, nets, unidentifiable plastic and polystyrene) make up the largest proportion of overall litter pollution, mostly from recreational/tourism activities. Smoking related wastes and litter from fishing in general is also a significant problem in the Mediterranean.

Circulation is the primary driver of marine litter transport. As a semi-enclosed sea surrounded by developed areas, the Mediterranean Sea has particularly high concentration of marine debris. There, studies have already documented the beaching of litter, its transport on the surface and its accumulation on the sea floor (review in UNEP/MAP 2015).

The variability of the surface circulation is very high as the basin has many constraints. Some scenarios could be hypothesized to evaluate a realistic distribution through simulations based on homogeneous and continuous deployment of litter. Only few large sub-basins appear as possible retention areas (the north-western and the Tyrrhenian sub-basins, the southern Adriatic, and the Gulf of Syrt) but no permanent gyres, lasting longer than a few months, are occurring because seasonal and inter-annual variability alter the water movements and the distribution of litter.

On the sea floor, compilation of data from 16 studies covering the entire basin of the Mediterranean Sea confirmed the importance of plastic, at 62.7% ± 5.47 of total debris (Pham et al. 2014). The study showed that in canyons, plastic was the dominant litter items (50%) followed by fishing gear (25%), On slopes, the dominant litter items recovered was fishing gear (59%), followed by plastic (31%). This was also confirmed by an analysis of data from regular monitoring of litter on the sea floor in the gulf of Lion (Fig. 6.1).

In deep analysis finally detected that the "Distance to the coast" variable accounted for less than 20% of the variance in the distribution of litter between

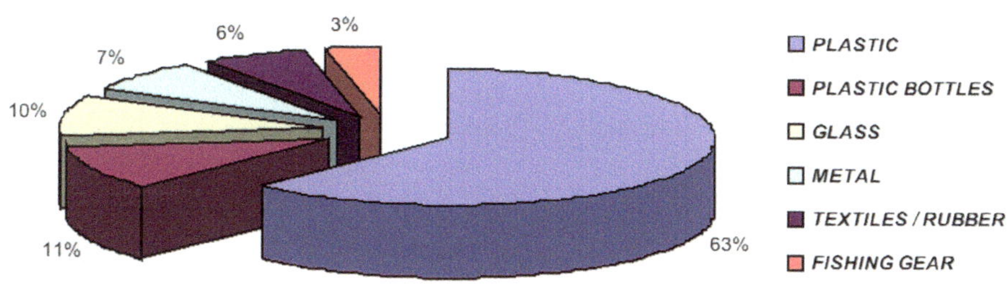

Fig. 6.1 Typology of debris collected between 30 and 800 m in the Gulf of Lion, France (MEDITS cruises, average from 70 stations/year and 15 years monitoring, 1994–2009, Galgani et al. 2011a, b)

canyons (see UNEP/MAP 2015). In some areas, fishing gears may account for the largest part of debris, depending on fishing activity. As an example dominant type of debris (89%) consisting of fishing gear were found on rocky banks in Sicily and Campania (Angiolillo et al. 2015)

Only a few studies have been published on the abundance of floating macro and mega debris in Mediterranean waters and the reported quantities measuring over 2 cm range from 0 to over 600 items per square kilometer (review in UNEP/MAP 2015).

On the sea floor, Counts from 7 surveys and 295 samples in the Mediterranean Sea and Black Sea (2,500,000 km^2, worldatlas.com) indicate an average density of 179 plastic items/km^2 for all compartments, including shelves, slopes, canyons, and deep sea plains (Pham et al. 2014).

In the Gulf of Lions, only small amounts of debris were collected on the continental shelf. Most of the debris was found in canyons descending from the continental slope and in the bathyal plain, with high amounts occurring to a depth of more than 500 m (Table 6.1, Fig. 6.2).

In the Mediterranean, static gear is an important part of ghost fishing and fishing can continue for years. Estimated ghost catches are generally believed to be well under 1% of landed catches and was estimated annually, for hake, between 0.27% and 0.54% of the total commercial landings.

In addition to large debris, there is growing concern with regards to Micro plastics, a very heterogeneous group, varying in size, shape, color, chemical composition, density and other characteristics. Most of them are resulting from the breakdown of larger plastic materials into increasingly small fragments rather from precursors (virgin resin pellets) for the production of polymer consumer products. Mean sea surface plastic was found in concentrations up to 100,000–200,000 particles/km^2 in the NW Mediterranean Sea (maximum 4,860,000 particles per km^2), giving an estimated weight over 1000 tons for the whole basin. Beach surveys revealed also an abundance of pellets on all Mediterranean beaches.

Table 6.1 Distribution of debris in the Gulf of Lion in relation to the depth (After Galgani et al. 2011a, b)

Depth (m)	Tows	Total area (km^2)	Total debris	Plastics	Debris (km^{-2})
<200	57	3.03	337	229 (68%)	111.2
200–1000	21	0.816	568	483 (85%)	696
>1000	10	0.17	631	537 (85%)	3712

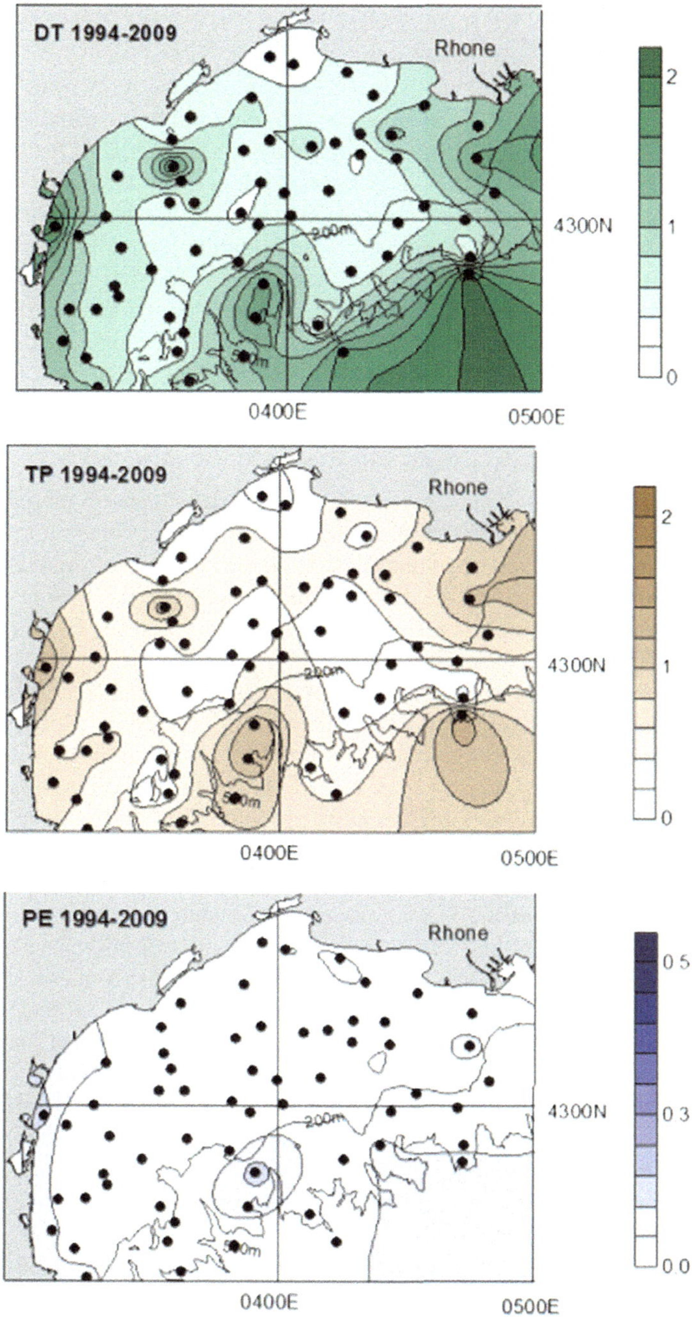

Fig. 6.2 Mean annual litter densities on the sea floor from the Gulf of Lion for a period of 15 years of sampling (1994–2009). Results are extrapolated densities expressed in items per hectare of the following categories: total Debris (DT), total plastics (TP) and fishing gears (PE). Data were from MEDITS cruise. (After Galgani et al. 2011a, b)

6.3 Impacts

As marine litter affects different ecological compartments, the study of its impact on marine biota of all trophic levels on the same temporal and spatial scale is of increasing importance. With regard to biodiversity, it is essential to focus research on sensitive species such as turtles, marine mammals, seabirds, and filter feeders, invertebrates or fish that may be ingest micro plastics. So far, 79 studies have investigated the interactions of marine biota with marine litter (mainly plastics) in the Mediterranean basin (Deudero and Alomar, in CIESM 2014). These studies cover a wide range of depths (0–850 m) and a large temporal scale (1986–2014), unveiling a vast array of species that are affected by litter, ranging from invertebrates (Polychaeta's, ascidians, bryozoans, sponges …), fish, and reptiles to cetaceans.

Effects from the studies are usually classified into entanglement, ingestion, and colonization and rafting.

About entanglement, there is a general lack of available data on marine wildlife in the Mediterranean but factors that may contribute to the entrapment of organisms in ghost gears have been described and include (i) the presence of organisms in the nets or in their proximity, (ii) the water turbidity, making the fishing gear less visible; (iii) the capability of organisms to detect the net filaments, (iv) the age, and for cetaceans (v) the ambient noise.

For ingestion, with more than 62 million of debris items estimated to be floating in the Mediterranean, these may affect a large number of species. Moreover, some species that are feeding on bottom may also ingest litter directly from the sea floor. Sub-lethal effects caused by marine litter ingestion may greatly affect populations in the long term because of possible reduced growth rates, longer developmental periods at sizes most vulnerable to predation, reduced reproductive output, and decreased survivorship. Highly affected fish species include *Boops boops*, myctophids, *Coryphaena hippurus, Seriola dumerilii,* (Deudero and Alomar, in CIESM 2014). Recently (Romeo et al. 2015), tunas and swordfish from the Mediterranean Sea were identified as targets species with occurrence of micro, meso, and larger plastics in more than 18% of the samples. Invertebrates such as *Mytilus galloprovincialis, Arenicola marina, and holothurids* are good indicator of harm due to their feeding habits as detritivorous or filter feeders.

The loggerhead sea turtle (*Caretta caretta*) may ingest plastic bags mistaken for jellyfishes (review in Galgani 2015) when they feed in neritic and offshore habitats. This is a very sensitive species to marine litter and one of the most studied.

In most cases, organisms are shown to utilize the debris items in oceans as habitats to hide in, adhere to, settle on, and move into new territories. The 250 billion microplastics floating in the Mediterranean Sea are then all potential carriers for alien, harmful species and so-called "invasive" species. However, because the Mediterranean Sea is a closed basin, species could potentially use litter to further expand their range (secondary invasion) after entering the basin by other means.

By sinking, debris may also have an impact on the deep sea environment, providing solid substrates and new habitats, impacting the distribution of benthic species even in remote areas (Pham et al. 2014).

Typically, large sized debris may affect humans from molecular (toxicity) to individual levels. Pieces of glass, discarded syringes, and medical waste all present possible harms to beach users. Moreover, plastic debris is now an abundant substrate for microbial colonization, physically and chemically distinct from natural substrates, and could support distinct microbial communities with a question of the transport of pathogens has now become crucial and may potentially support impact on human health. In recent years, secondary pollution from the leaching of pollutants from litter has been extensively studied. The transfer or enhanced bioaccumulation of persistent organic pollutants (POPs) may also occur as a consequence of the high sorption capacity of many plastics for lipophilic compounds. High concentrations of Di-(2-ethylhexyl) phthalate (DEHP) and nonylphenol have been also measured in small planktivorous fish, posing a long term risk to the environment. Beaches located downstream from industries and/or port facilities showed higher quantity of plastic debris and microplastics as well as higher concentrations of POPs (PAH, PCB and DDT).

Litter in the marine environment gives rise to a wide range of economic and social impacts, most often on small-scale, relying on anecdotal evidence, and negative environmental effects are often also interrelated and frequently dependent upon one another (Ten Brink et al. 2009). For the European commission, the total costs of marine litter is estimated at 263 million euros (Arcadis 2014), with a value for the closed Mediterranean Sea likely even more important due to the population in the region, maritime traffic, and tourism. In this basin, there is little or no reliable data on what the exact costs are. Furthermore, the loss of tourism and related revenues due to marine litter both on the beaches and in the sea, although recognized and considered, has not been quantified in detail. Economic impacts are most often described as including the loss of aesthetic value, impacts on fishing, the loss of non-use value, public health and safety impacts (extent and frequency of incidents), navigational hazards that are often unreported, and finally Ecosystem degradation, an extremely complex cost to evaluate.

6.4 Monitoring

Marine debris monitoring generally consists of various approaches, such as beach surveys, at-sea surveys, and estimates of the amounts entering the sea and impacts. Beach surveys are widely viewed as the simplest and the most cost effective, but they may not relate to true marine pollution and, because they may be affected by weather, the stranded debris may not necessarily provide a good indicator of changes in overall abundance.

There is actually no regular monitoring of micro particles in the Mediterranean Sea. Another approach to monitoring is to look at impacts directly. Entanglement data does suffer from not always being expressed as a proportion of the population, and the distinction between active gears and litter when sampling stranded organisms is too difficult to enable regular and consistent monitoring. Ingestion sampling provides consistent data but is restricted to deceased and stranded individuals as opposed to a sample from the population at large. Moreover, species that can be considered for monitoring purposes must meet a number of basic requirements, like (i) sample availability (adequate numbers of beached animals, by-catch victims or harvested species), (ii) regular plastic consumption (high frequency and amounts of plastic over time in stomachs), and (iii) feeding habits (stomach contents should only reflect the marine environment).

In the Mediterranean Sea, there is very little coverage of any marine compartment other than beach and stranded debris, the most mature indicator and the one for which most data is available because

Regular surveys on beaches have been made in many areas, often over a number of years, by various NGOs. Valuable information about the quantity and composition of marine litter found on beaches has been available in most of the countries, giving an overview of debris found in the Mediterranean countries. However, there is a lack of official statistics for most of the countries. The challenges in dealing with this problem are not due to lack of awareness or data from various regions but rather are due to the lack of standardization and compatibility between methods. If many programs that exist or have existed in most Mediterranean countries involve (d) NGOs, the institutional MEDITS survey program (International Bottom Trawl Survey in the Mediterranean, http://www.sibm.it/SITO%20MEDITS/principaleprogramme.htm) is sampling benthic and demersal species between 80 and 800 m, through systematic bottom trawl surveys and with a common standardized sampling methodology and protocols at a global Mediterranean scale. The recent collection of sea floor litter data on a regular basis will provide assessments at the basin scale, with a potential of 1280 sampling stations to be considered.

The use of sea turtles for monitoring ingested litter in the Mediterranean Sea was first suggested in 2010 by a MSFD task group after many years of research. Protocols were then implemented (Matiddi et al. 2011; Galgani et al. 2013a, b) providing support to monitoring. Only with sub regional experiments performed to date, reinforced coordination, capacity building, quality assurance, and harmonization are however still needed.

Due to the poor differences between the Mediterranean sub-regions in terms of litter densities, the unequal spread of available data-sets, and some countries belonging to two or more sub-regions (Italy, Greece), it was suggested that common baselines for the various litter indicators (beaches, sea surface, sea floor, microplastics, ingested litter) must be considered at the level of the entire basin (Mediterranean Sea) rather than at the sub-regional level. Definitive baselines may be adjusted after monitoring programs could provide additional data. It is also quite important to harmonize the monitoring programs with other Regional Seas Conventions (e.g. OSPAR) as much as possible.

Environmental targets are qualitative or quantitative statements that are important for management as they will enable regions to (i) link the aim of achieving objectives such as Good Environmental Status (GES) to the measures and effort needed, (ii) measure progress towards achieving the objective by means of associated indicator(s), and (iii) assess the success or failure of measures enacted to prevent marine litter from entering the seas and to support management and stakeholder awareness (Interwies et al. 2013). This author provided an overview of potential aspects to set targets on marine litter. They may consider (i) Location (Beaches, floating, estuaries, marine life, etc.), (ii) Composition or type (Plastic bags, cigarette bugs, microparticles, sanitary wastes, etc.), (iii) Sources and pathways (rivers, ship-based litter, landfills, etc.), (iv) Sectors (fisheries, recreation, industrial pellets, etc.), and (v) Measures (reduce urban waste production, improve waste collection of land-based sources/sectors, improve collection of ship-based waste in the port reception facilities, improve waste water treatment, reduce consumer littering, and improve inspections at sea, etc.). The Marine litter MEDPOL regional Plan also provides for strategic and operational objectives and lists a series of prevention and remediation measures that should be considered and implemented by the concerned actors. The establishment of both "state" and "pressure" complementary targets can then better reflect and support the effectiveness of specific operational objectives.

The different types of targets must be relevant to different types of information gaps (at-sea targets for improving the state of information about abundance, operational targets such as estuarine monitoring for improving information on pathway, source, and regional differences). However, due to a large set of factors affecting the quantities and distribution of marine litter in a certain area, it can be very challenging to detect clear reduction trends in the sea that can be associated to the implementation of measures in a particular area.

In regards to the coordinated monitoring strategy in the Mediterranean Sea and technical or scientific considerations, accessible targets were proposed (UNEP/MAP 2015) considering baselines that may be optimized after the 2015/2016 first results from monitoring. Targets may focus on the total amount of marine litter first, with some specific targets on individual items after impacts of reduction measures can be evaluated. For floating and sea floor litter, a significant decrease in amount requires overcoming the constraints of diffuses and uncontrolled sources (transboundary movements, influence of currents) and permanent accumulation processes on the sea floor. Targets on ingested litter in sea turtles will then focus on the number of affected animals and the amount of ingested debris by number or weight.

6.5 Management and Reduction Measures

Attempts to prevent marine litter require the inclusion of a vast amount of activities, sectors, and sources that cannot be addressed by a single measure.

Poor waste management, limited awareness of the public and inadequate interventions from industry and policy-makers are the main causes of the presence of litter at sea (Oosterhuis et al. 2014). As a result of the complexities caused by the diverse origin of marine litter, a wide range of instruments have been proposed to deal with it across multiple sectors

The Mediterranean MLRP and the Berlin Conference on Marine Litter, 2013, Berlin, Germany (http://www.marine-litter-conference-berlin.info/) provided the following guiding principles as well as an umbrella structure that serves as a guiding framework for any of the following marine litter measures:

- The principle of prevention establishes that any marine pollution measure should primarily aim at addressing the prevention at the source, as removal of already introduced waste is very costly and labour intensive, especially compared with prevention measures.
- The polluter-pays principle has a preventive function in that externalities from polluting activities should be borne by the polluter causing it, which puts more pressure on potential polluters to make better attempts to avoid polluting. However, the application of this principle is limited by the difficulty in determining the polluter and also the extent of (environmental) damage.
- The precautionary principle is based on the understanding that measures must not be postponed in the light of scientific uncertainties. This principle plays an important role in setting targets and addressing the issue of micro-particles, despite an incomplete scientific knowledge on the specific sources and consequences of marine litter.
- The ecosystem-based approach is an approach that ensures that the collective pressures of human activities are considered.
- The principle of public participation is an important aspect of creating awareness for the problem of marine litter.
- The principle of integration means that environmental considerations should be included in economic development. This principle constitutes a key element of the Protocol on Integrated Coastal Zone Management in the Mediterranean.

Then, measures and actions taken should respond to the major sources and input pathways, but they should also take into consideration feasibility and the specificity of this pollution in the Mediterranean Sea. The main groups of items found on beaches in the Mediterranean are sanitary items, cigarette butts, as well as packaging items and bottles, all related to coastal-based tourism and recreation. This indicates direct disposal, intentionally or negligently, on the beaches or inland (river banks, dumpsites, etc.) as the main input pathways.

The fishing and shipping industries are also considered major sources of marine litter. In the Mediterranean Sea, specific measures in specific areas must be most effective in tackling the problem, including "Fishing for Litter", one of the most important measures that would lead to the reduction and removal of marine litter from sea.

Further implementation is being considered within the Mediterranean Regional Action Plan, developing best practices adapted to the context of the basin (Gallego 2015).

Some of reduction measures are regulatory policy instruments that focus on adopting relevant legislation to help minimize marine litter, such as the EU Directive 2000/59/EC on port reception facilities for ship-generated waste and cargo residues. Other instruments, more economic in nature, influence the amount of marine litter through taxes, charges, or subsidies.

Effectiveness is a key determining factor for economic instruments. The cost of implementation is another important factor that influences which instrument to opt for, and it focuses on how to allocate scarce resources (e.g. public funds) to meet a certain environmental objective. This is the case in the cost of ghost gears.

Unfortunately for the Mediterranean Sea, there is no unique economic instrument, and the choice of an appropriate intervention is case specific, largely depending on the source and nature of pollution, the country's institutional characteristics and infrastructure, consumer preferences, perception and habitual behavior, and the economy's overall sectorial composition.

6.6 Setting Priorities

UNEP/MAP-MEDPOL (2015), MSFD (Galgani et al. 2011a, b), the European project STAGES (http://www.stagesproject.eu), and CIESM (2014) recently reviewed the gaps and research needs of knowledge, monitoring, and management of marine litter. This requires scientific cooperation among the parties involved prior to reduction measures due to complexity of issues.

Typically, more valuable and comparable data could be obtained by standardizing our approaches. In terms of distribution and quantities, identification of litter (size, type, etc.), evaluation of accumulation areas (bays, gyres, canyons, etc.), and identification of sources (rivers, diffuse inputs, etc.) need additional research prior to develop GIS and mapping systems to locate hotspots and to better understand transport dynamics. Further development and improvement of modelling tools must be considered for the evaluation and identification of both the sources and fate of litter.

A better understanding of rates of degradation and the development of appropriate methodologies to quantify degraded materials, including nanoparticles will be an important step.

Pilot-scale monitoring is now an important step towards monitoring litter harm in terms of determining baselines and/or adapting the strategy to local areas. A better understanding of entanglement (lethal or sub lethal) and of how litter is ingested by marine organisms are key questions. A more precise definition of target (GES) and the identification of Parameters/biological constraints and possible bias sources must be considered when defining the good environmental status. Work on other "sentinel" species (fishes and invertebrates) is also important, as it may provide additional protocols supporting the measurement of impacts. The (i) increase in the probability of translocation of species due to floating litter, (ii) the identification of species (including pathogens for both marine organisms and human) in the Mediterranean that settle on marine litter, (iii) the nature of constraints for the colonization of floating plastic, are also key questions to consider for a better understanding of harm.

Knowledge about the extent of ghost fishing is still very limited due to the costs and practical difficulties of underwater survey work and partial knowledge about fish stocks losses. There are actually no overall estimates of the extent of the problem for the Mediterranean as a whole.

Harmonization of sampling protocols for the water surface and for microplastics is highly recommended. Moreover, the comparability of available data remains highly restricted, especially with respect to different size class categories, sampling procedures, analytical methods, and reference values. From the economic/ management point of view (UNEP 2015) it is clear that (i) the lack of international legal instruments (except for IMO/MARPOL Annex V), (ii) the lack of coordination between actors, (iii) the poor management of coastal waste and (iv) the problems that are encountered in the application of economic instruments must be solved prior to implementation of measures. This must be based on the development of common methodologies to collect social and economic data, an assessment of socially acceptable levels of marine litter to the public and industry, the development of social and economic impact indicators and the education of the public.

In terms of measures, the development of tools to assess the effectiveness of monitoring, the implementation of measures intended to reduce the amount of marine litter and/or effectiveness programs, the development of port reception facilities (taking into consideration the Mediterranean maritime traffic), and the consideration/elimination of trans border marine litter, including the intervention in case of critical situation, are the main priorities in the management of marine waters that have to complement management measures to reduce inputs.

In conclusion, marine litter in the Mediterranean has become a critical issue. Management and reduction still need to be developed, implemented and coordinated. However, a number of points need to be addressed in order to better understand the issue. A number of key issues will have to be considered in order to provide a scientific and technical background for a consistent monitoring, a better management system, and science based reduction measures. For this, the improvement of basic knowledge, the support of monitoring and management has become critical.

References

Angiolillo M, di Lorenzo B, Farcomeni A, Bo M, Bavestrello G, Santangelo G, Cau A, Mastascusa V, Cau A, Sacco F, Canese S (2015) Distribution and assessment of marine debris in the deep Tyrrhenian Sea (NW Mediterranean Sea, Italy). Mar Pollut Bull 92:149–159

Arcadis (2014) Marine litter study to support the establishment of an initial headline reduction target- SFRA0025? European commission/DG ENV. Project number BE0113.000668, 127 p

CIESM (2014) Plastic Litter and the dispersion of alien species and contaminants in the Mediterranean sea. Ciesm Workshop No 46 (Coordination F Galgani), Tirana, 18–21 Juin 2014, 172 p

Galgani F, Henry M, Orsoni V, Nolwenn C, Bouchoucha M, Tomasino C (2011a) MACRO-DECHETS en Méditerranée française: Etat des connaissances, analyses des données de la surveillance et recommandations. Rapport IFREMER, RST.DOP/LER-PAC/2011, pp 42

Galgani F, Hanke G, Werner S, De Vrees L (2013a) Marine litter within the European Marine Strategy Framework Directive. ICES J Mar Sci 70(6):1055–1064. http://dx.doi.org/10.1093/icesjms/fst122

Galgani F, Piha H, Hanke G, Werner S, GES MSFD group (2011b) Marine litter: technical recommendations for the implementation of MSFD requirements. EUR 25009 EN. Luxembourg (Luxembourg): Publications Office of the European Union. JRC67300. (http://publications.jrc.ec.europa.eu/repository/handle/111111111/22826)

Galgani F, Hanke G, Werner S, Oosterbaan L, Nilsson P, Fleet D, Kinsey S, Thompson RC, van Franeker J, Vlachogianni T, Scoullos M, Mira Veiga J, Palatinus A, Matiddi M, Maes T, Korpinen S, Budziak A, Leslie H, Gago J, Liebezeit G (2013b) Monitoring guidance for marine litter in European seas. JRC Scientific and Policy Reports. Report EUR 26113 EN, 120 p. (https://circabc.europa.eu/w/browse/85264644-ef32-401b-b9f1-f640a1c459c2)

Galgani F, Hanke G, Maes T (2015) Global distribution, composition and abundance of marine litter. In: Bergmann M, Gutow L, Klages M (eds) Marine Anthropogenic Litter. Springer, Cham, pp 29–56 http://dx.doi.org/10.1007/978-3-319-16510-3_2

Interwies E, Görlitz S, Stöfen A, Cools J, Van Breusegem W, Werner S, de Vrees L (2013) Issue paper to the "International Conference on Prevention and Management of Marine Litter in European Seas". Final Version, 16 May 2013, 111 p. (http://www.marine-litter-conference-berlin.info/downloads.php)

Matiddi M, van Franeker JA, Sammarini V, Travaglini A, Alcaro L (2011) Monitoring litter by sea turtles: an experimental protocol in the mediterranean. In: Proceedings of the 4th mediterranean conference on sea turtles, 7–10 Nov, Naples, p 129

Oosterhuis F, Papyrakis E, Boteler B (2014) Economic instruments and marine litter control. Ocean Coast Manag 102:47–54

Pham C, Ramirez-Llodra E, Claudia HS, Amaro T, Bergmann M, Canals M, Company J, Davies J, Duineveld G, Galgani F, Howell K, Huvenne Veerle A, Isidro E, Jones D, Lastras G, Morato T, Gomes-Pereira J, Purser A, Stewart H, Tojeira I, Tubau X, Van Rooij D, Tyler P (2014) Marine litter distribution and density in european seas, from the shelves to deep basins. Plos One 9(4):e95839. Publisher's official version. http://dx.doi.org/10.1371/journal.pone.0095839

Romeo T, Battaglia P, Peda C, Consoli T, Andaloro F, Fossi C (2015) First evidence of presence of plastic debris in stomachs from large pelagic fish in the Mediterranean Sea. Mar Pollut Bull 95(1):358–361

Ten Brink P, Lutchman I, Bassi S, Speck S, Sheavly S, Register K, Woolaway C (2009) Guidelines on the use of market-based instruments to address the problem of marine litter. Institute for European Environmental policy (IEEP), Brussels, Belgium

Tweehuysen G (2015) Sampling River litter: preliminary results. Micro-2015. Seminar of the Defishgear project, Abstract book, Piran, 4–6 May 2015, p 29

UNEP/ MAP (2015) Updated report on marine litter assessment in the mediterranean sea. Meeting of the MAP focal points, Athens, Greece, 13–16 Oct 2015. UNEP(DEPI)/MED WG.421/Inf.18, 84 p

Vianello A, Acri F, Aubry F, Boldrin A, Camati E, Da Ros L, Merceta T, Moschind V (2015) Occurrence of floating microplastics in the North Adriatic Sea: preliminary results. Micro 2015. Seminar of the Defishgear project, Abstract book, Piran, 4–6 May 2015, p 43

Gallego Z (2015) Guide on best practices for fishing for litter in the Mediterranean. UNEP/MAP document, 17 p

Chapter 7
The European eel (*Anguilla anguilla*) in France: An Example of Close Cooperation Among Researchers and Fishers to Study and Manage an Endangered Species

Patrick Prouzet, Elsa Amilhat, Catherine Boisneau, Philippe Boisneau, Eric Feunteun and Nicolas Michelet

Abstract The European eel (*Anguilla Anguilla*) was a very valuable species for the economy of small scale fisheries in France both in marine and in inland waters. At the beginning of the 21st century the species was the third economical value of all the species caught in the Bay of Biscay despite a strong reduction in abundance since the middle of the 20th century and especially since the seventies. Right now, there is a sharp decline of the economic value of that fishery due to some constraining fishing regulations and due to the export ban of eel outside Europe. Due to a dramatic decrease of its abundance in all the European waters, the EU decided to implement an eel management plan in 2007, in a council regulation establishing measures for the recovery of the stock of the European eel. Each member states producing eels had to identify regulation measures and eel habitat and had to evaluate the effectiveness of these measures. They represented the basis of the national management plans defined by each member states that wish to exploit the eel resource. The objective of the eel national management plan is to reduce anthropogenic mortalities so as to permit with high probability the escapement to

P. Prouzet (✉)
SFJO France, 12 avenue, Antoine de Saint Exupéry, 64500 Saint-Jean de Luz, France
e-mail: patrick.prouzet@orange.fr

E. Amilhat
University of Perpignan, Perpignan, France

C. Boisneau
University of Tours, 60 rue du Plat D'Etain, 37020 Tours cedex 1, France

P. Boisneau
CONAPPED, Paris, France

E. Feunteun
MNHN/CRESCO, Dinard, France

N. Michelet
CNPMEM/CONAPPED, Paris, France

T. Komatsu et al. (eds.), *Oceanography Challenges to Future Earth*,
https://doi.org/10.1007/978-3-030-00138-4_7

the sea of at least 40% of the silver eel biomass relative to the best estimate of escapement that would have existed if no anthropogenic influences had impacted the stock. In that context, some large scientific programs such as INDICANG and EELIAD have been undertaken to improve our present knowledge on the migratory behaviour of eel at different phases of its biological cycle, to develop some methodologies to evaluate the abundance of the stock in a given river system and to assess the relative effects of both fisheries and others anthropogenic factors on the future of the stock. A large part of these studies were undertaken through a close cooperation between scientists and fishers in order to add their know-how and knowledge to develop some technical approaches on the field. Some examples are given to illustrate these field experiments.

Keywords European eel · Anguilla anguilla · Population trend · Habitat · Management

7.1 An Historical Overview

The European eel (*Anguilla anguilla*) has been historically a species commonly consumed in many French or European regions. This fish species is easily catch in continental and coastal waters where it occurs in a great diversity of ecosystem: freshwater, brackish and marine habitats in estuaries, lagoons, rivers and lakes.

It is easy to keep alive and to transport outside water on a long distance and during a long time as the sea lamprey (*Petromyzon marinus*). It is a reason why this species was eaten in a great number of areas and at different stage of its life cycle: glass-eel, yellow and silver eel (Fig. 7.1).

This species was taken by recreational and professional fishermen with a great diversity of fishing gears well adapted to the physical structure of the fishing areas such as: hooks and lines, eel pots, fyke nets, hand and push sieves (scoop nets), barrier and trap nets (Fig. 7.2).

The European Eel, in continental ecosystems, was at very high level of abundance and during the winter period at least until the seventies, the abundance of glass eels migrating though the estuaries was so high that, in some hydrological conditions, a fisher with a single hand sieve could catch during a night up to several hundred kg (Fig. 7.3). If this amount of catches per fisher was exceptional, it was not unusual to catch several tens of kg during the best period.

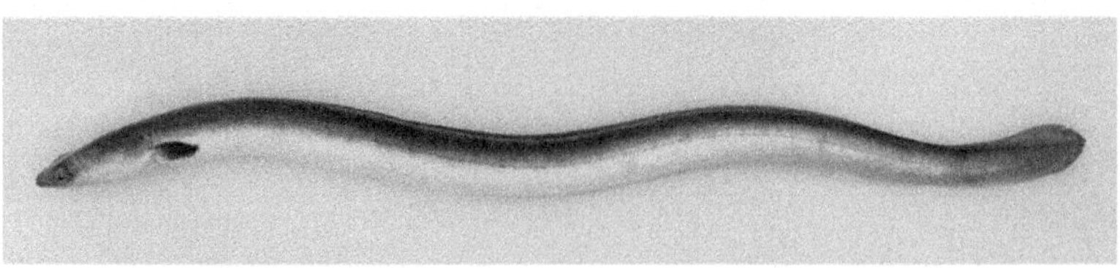

Fig. 7.1 European Eel (*Anguilla anguilla*)—silver eel stage—(copyright Porcher/Ifremer)

Fig. 7.2 On the left a plastic eel-pot used on the Adour estuary; on the right a barrier and fyke nets used on the Thau lagoon and called "capetchade". (Copyright Patrick Prouzet)

After the seventies, due to the development of the Asian markets (Japan then China) for glass eels and the rapid increase of the price per kg[1] (Nielsen and Prouzet 2008) (see Fig. 7.4), the eel fishery was very valuable for the marine and inland professional fisheries (glass-eel fisheries on the Atlantic coast and yellow eel fisheries on the Mediterranean coastal lagoons). For example, in 2000 (Léauté et al. 2002), the value of the glass eel fishery for the central and southern parts of the gulf of Biscay was estimated to 15.6 Meuros[2] for the French fleet, the third rank in value for this fishing area.

Considered as an harmful species, systematic destruction was carried out until 1986 by the French managers of inland waters. In two decades, the species changed status from a pest to endangered and the species has been added to the CITES Annex II (13 March 2009). Since 1984, scientists have raised the alarm concerning the eel resources. In 1996, professional fishers and scientists together rose a number of questions about the eel future in many river systems in France and asked for a better management of the species and its habitat's which were severely affected at least since the middle of the 20th century.

In 2007, the EU adopted an eel regulation[3] to restore at a European scale the eel population. The management plan may contain different measures, such as fishery regulations, restocking, improvement of rivers habitats, temporary switching off of hydroelectric power turbines. The final target of this restoration plan is to enable escapement to the sea of at least 40% of the biomass of pre-adult silver eel relative to the best estimate of natural escapement.[4]

---

[1]773 euros per kg for the mean annual price for the fishing season 2004/2005.

[2]2137 millions yens.

[3]EC regulation 1100/2007 of 18 September 2007.

[4]Defined according to historical data or by habitat-based assessment of potential eel production in the absence of anthropogenic effects or with reference to the ecology and hydrography of similar river systems.

Fig. 7.3 An exceptional catch of glass-eel just after the Second World War on the Adour river (collection A. Lataillade/Musée de Port de Lannes)

The article eight of this regulation states: "if a Member state operates a fishery on glass eels, it has to guarantee that 60% of all glass eels caught during the whole year are utilized as part of a restocking program in European inland waters having access to the sea, for the purpose of increasing the escapement levels of adult silver eels".

In 2007, the CITES[5] decided to restrict the eel export outside Europe to 40 tons, then 28 tons in 2008, 14 tons in 2009 and finally to ban the export outside Europe in 2010. As the Asian market was the most profitable for the French glass-eel and, without replacement market, the prices per kg collapsed (see Fig. 7.4). Fishing constraints, eel export ban outside Europe, depletion of the eel resource, non organised European market for eel restocking resulted in a strong decrease of the value of the glass-eel fishery (Fig. 7.5) on the Atlantic area.

The social consequence was a significant decrease of the number of fishers: 51.5% for the marine component and 66% for the inland component (Fig. 7.6) during the period 2009–2014.

The decrease of the number of fishermen is also recorded for the yellow and silver eel fisheries as shown in Table 7.1 below. It is very significant in the Mediterranean area and we observed since 2009 a decrease of 48.5% for the nominal catch effort on yellow eel and 27.9% for the silver eel. This decrease is mainly due to the slump of eel market and the PCB contamination on eel.

[5]Convention on International Trade in Endangered Species of Wild Fauna and Flora.

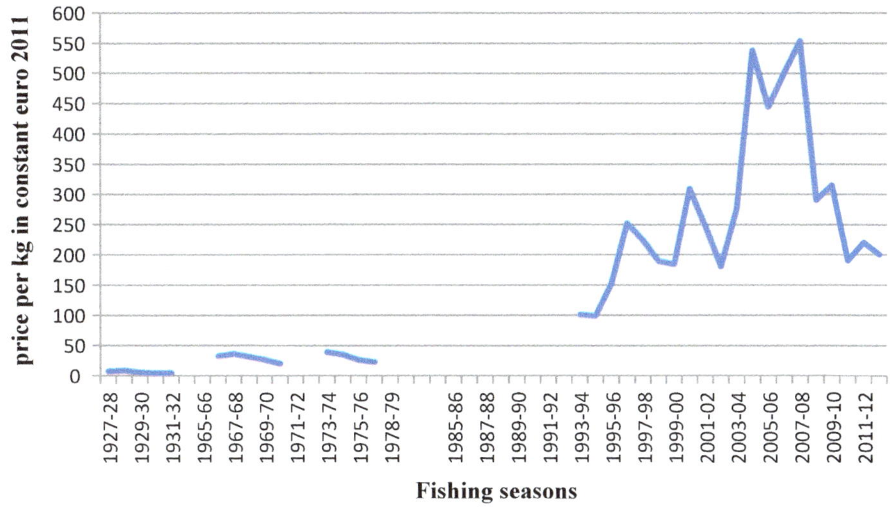

Fig. 7.4 Variation of the price per kg of glass-eel (landing prices) during the 20th century and the beginning of the 21st century

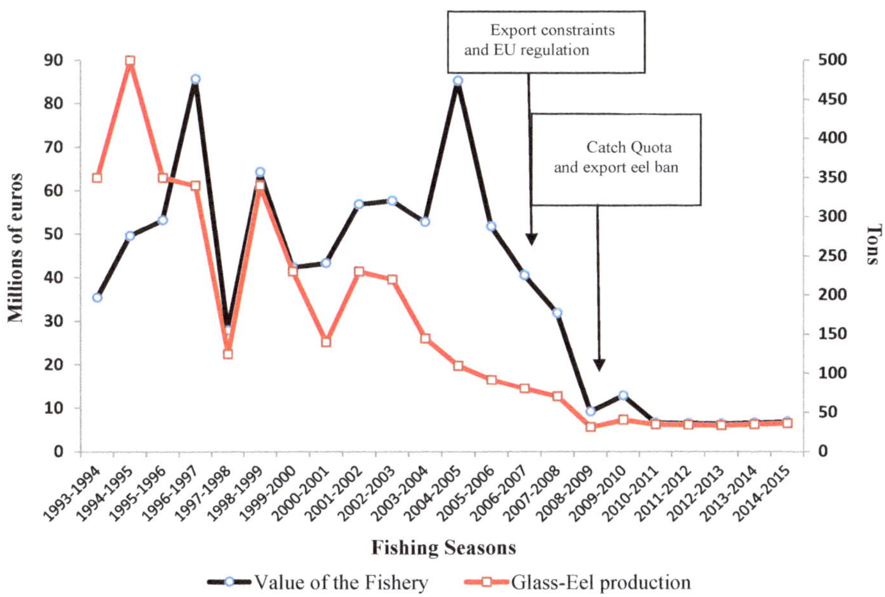

Fig. 7.5 Variations of the value and the production of the French glass-eel fishery since the beginning of the nineties (from IFREMER and CNPMEM data). Catch quota from the 2009/2010 fishing season

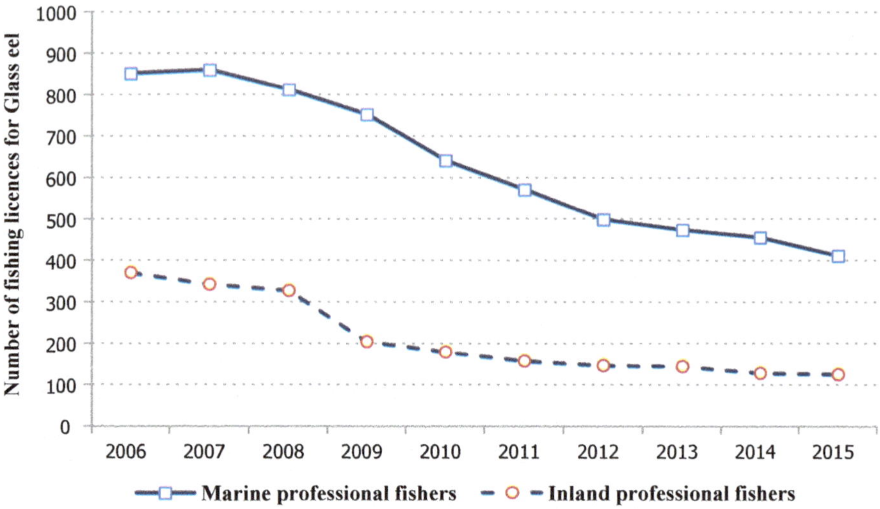

Fig. 7.6 Fluctuations over the last decade of the numbers of marine and inland professional fishers for Glass eel. (CNPMEM/CONAPPED data)

Table 7.1 Variation of the number of professional fishers for yellow and silver eel in France on the period 2009–2015 (CNPMEM/CONAPPED data)

		2009	2010	2011	2012	2013	2014	2015
Inland fishers	Yellow eel	169	171	170	169	172	146	143
	Silver eel	44	41	37	35	34	34	33
Marine fishers atlantic	Yellow eel	209	268	245	236	224	248	225
	Silver eel							
Marine fishers mediterranean	Yellow eel	390	288	258	264	230	218	201
	Silver eel	283	267	253	256	230	219	204

A notification[6] of the French Agency for Food, Environmental, Occupational Health and Safety (ANSES) advised "to consume eel exceptionally whatever the river system". This notification had a negative effect on the economy of the eel industry[7] and a further analysis is requested by the professional fishers in order to precise the eel origin because of the strong disparity in environmental PCB contamination of the river systems where the eel is caught.

Considering the complex life cycle of the species, the numerous questions concerning the migratory behaviour, the colonization process of different ecosystems, the main anthropogenic causes of the decline, the definition of abundance indicators at different stages of the life cycle, it was decided to add the knowledge

[6]Saisines 2014-SA-0122 et 2011-SA-039.

[7]810 fishing enterprises in 2015 (CNPMEM/CONAPPED data).

and know-how of professional fishers and scientists within two main programs of participatory sciences: INDICANG (www.ifremer.fr/indicang) and EELIAD (European eel in the Atlantic: Assessment of their Decline) (Righton 2012, 2015).

7.2 A Partnership Between Professional Fishers and Scientists to Improve the Knowledge on eel Biology

7.2.1 The eel Biological Cycle

The eel is an amphihaline catadromous species living alternatively in fresh and salt water with a spawning at sea and a long period of growth generally in freshwater.[8] The Fig. 7.7 gives a schematic diagram of the biological cycle of European eel. It is considered as a panmictic species with a common spawning area located in the vicinity of the Sargasso Sea (see below the migration path observed in the framework of the EELIAD project). From the reproduction area, *leptocephali* are transported eastwards by the Gulf Stream, then by the North Atlantic Drift towards the European continental shelf where the *leptocephali* are transformed into glass-eels after 1–3 years.[9]

The glass-eels migrate actively towards the coastal area and colonized bay, lagoon and estuaries. After a certain laps of time, the duration of which is linked to the temperature (more the temperature is high, more the time period is short), the non pigmented glass-eel, which is gregarious, develops its pigmentation and a territorial behavior. This is a very important phase to determine the size of the new recruitment which is dependent not only of the glass-eel abundance but also of the size of the suitable areas to produce elvers. Elver stage is the biological phase from which the diffusion into the different habitats shall be made. The yellow eel stage corresponds to the growth phase in different ecosystems and some recent studies made from the simultaneous variation of the ratio strontium: calcium and barium:calcium in otolith as markers of habitat, show the migration of yellow eel in environments of different salinities (Daverat et al. 2005; Fablet et al. 2007; Tabouret et al. 2010). After a period ranging between 3 and 24 years according to the sex and the growth rate,[10] the yellow eel metamorphose into silver eel. The silver stage corresponds to the migratory phase from rivers or estuary to sea and it is the

[8]A part of the population, unknown, spent their entire biological cycle in salt water and recent studies made with biological markers have shown the capacity of yellow eel to migrate in environment of various salinity during their growth period (Daverat et al. 2005; Tabouret et al. 2010).

[9]The duration of the migration of leptocephali is always a subject of debate. According to several authors the duration would be between 1 and 3 years (Lecomte-Finiger and Yahyaoui 1989; Tesch 1998; Bonhommeau et al. 2009).

[10]In the northern part of the colonization area, the growth rate is lower and the mean age of eel is higher than in the Southern part.

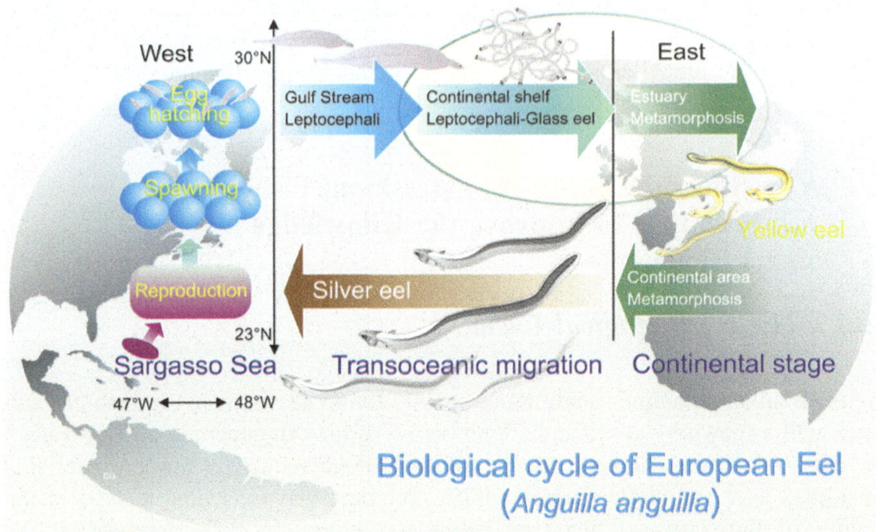

Fig. 7.7 Schematic diagram of the biological cycle of European eel representing its different phases: *Leptocephali*, glass-eel, yellow and silver eel. (From Indicang Project—www.ifremer.fr/indicang/)

beginning of the reproductive migration. This phase is characterized by a change in the behaviour, in the physiology and in some anatomico-morphological characteristics of the yellow eel. The most visible modification is the change in skin colouring as it becomes silver whilst the lateral line becomes pigmented. Downstream migration occurs in France, in the autumn with different triggers such as temperature, flow or atmospheric pressure.[11]

7.2.2 *The Need of a Cross-Disciplinary, Concerted and Multisectoral Approach to Address the Challenge of eel Recovery*

As mentioned previously, the eel biological cycle is complex and the dynamic of its panmictic population at the scale of its colonization area[12] result from the combination of the dynamic of a large number of eel subpopulations that depend on the characteristics of each ecosystem unit. We may consider that the appropriate scale for this ecosystem unit is the catchment area including the estuary and the immediate coastal zone, but for the management at the European scale it may be more efficient to consider three large areas where the dynamic of the eel groups differ during their

[11]See Tesch (1977), Adam et al. (2008), Feunteun (2012) for more detailed information.

[12]From Mauritania to Norway.

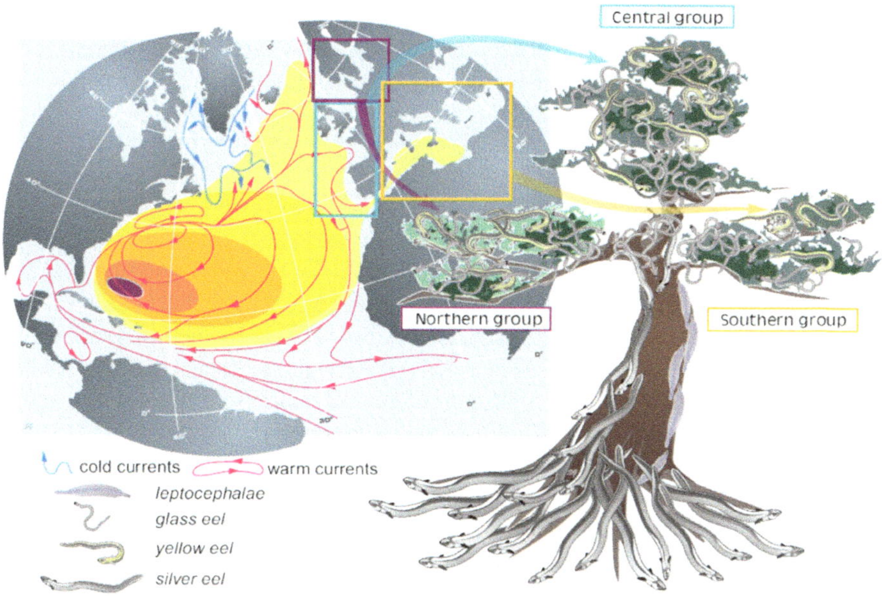

Fig. 7.8 The eel tree (from Indicang Project—www.ifremer.fr/indicang/). *Comment: This "eel tree" can only work if its roots, anchored in the Sargasso Sea, are rich in spawners, i.e. silver eels. It can only flourish if sap rises or falls along its trunk, this represents the oceanic circulation. This circulation cannot stop or even slow down, otherwise the "leptocephali" larvae (ascending sap) will not be carried eastwards at least with the same celerity, and nor will the silver eels (descending sap) be carried back to their spawning grounds. Hence the unanswered question: what will be the effect of climate change on oceanic circulation and hence on the functioning of this population? Finally, the tree can only prosper if glass eels, originating from larvae, colonise the different parts of its foliage (representing the river basins) and of course, if continuously thinned, the tree will eventually die. (From* Adam et al. 2008, *Indicang Guide, page 15)*

growth phase. The Fig. 7.8 named "the Eel tree" defined three different geographical groups of European eels. It shows the way in which the eel population functions.

This arborescent structure is the scientific and technical basis for the Indicang European project (Prouzet 2007) which took it into account by recommending local action to take care of the leaf (action at the level of the hydrographic unit) together with, inated actions undertaken on a significant number of leaves (Atlantic Arc) so that foliage restoration would be sufficient to have a significant impact on the future of the "eel tree". Another program coordinated by Righton (2008), had for objectives to assess the decline of the European eel stock and especially to collect information on the migration behaviour of silver eel in the ocean.[13] Both these projects used knowledge and know-how from professional fishers, particularly

[13]A new project coordinated by E. Amilhat and E. Falliex in 2013 have been implemented to collect information on the reproductive migration on eel from the Mediterranean Sea with the financial support of the MEDDE and the technical assistance of Mediterranean professional fishers.

those concerning the effects of hydro-climatic conditions on the migration behavior of glass-eel or silver eel and the way to catch the individuals with different techniques in order to collect samples for tagging or to sample efficiently the population migrating.

7.2.2.1 The Indicang Project: Information Collected on the Migration Behaviour of Glass-eel and on the Estuarine Recruitment

The aim of the Indicang project was to progress in large biological process solving that the "eel tree" suggests such as[14]: (i) the migration behavior of glass-eel in estuary and the development of a methodology to estimate their abundance when they migrate upstream; (ii) the indicator of habitat colonization of elvers and a methodology to assess the actual and potential productivity of the catchments (a leaf of the eel tree) concerning the eel production; (iii) a methodology to assess the abundance of silver eel in large river system such as the Loire axis for example (see Sect. 2.2.2); (iv) the knowledge enhancement on the migration pathway of silver eel in the Atlantic ocean; this subject will be further elaborated in the EELIAD program (see Sect. 2.2.3).

Simulation of the Migratory Behavior of Glass-eel in Estuary and Estimation of the Abundance of Glass-eel Recruitment in the Adour Estuary

From observations made during their fishing activities, professional fishers had collected a large amount of knowledge on the influence of water turbidity, rate of flow, tide coefficient, water temperature on the success of their fishing trips. They know the best period to catch the fish (November to January in the Adour River but later in catchments located further north such as the Loire River) and the influence of floods on the concentration of glass-eels close to the river banks. The information they provided was primarily qualitative for example about glass-eel migration period during daytime and behavior as the positioning of fish within the water column according turbidity and environmental luminosity. The observations made by the professional fishers during their fishing trips on glass-eel and their empirical knowledge were synthesized, then quantitatively or qualitatively specified on the Adour River by the scientists (Prouzet 2003; Bru et al. 2009). The influence of temperature variation between sea and river on the swimming activity of glass-eel in the water column: in winter, differential higher than 5 °C inhibits migration (Adam et al. 2008). The effect of the light intensity in the water column: during the period of full moon if the water turbidity is low, the major part of the glass-eel run migrate close to the bottom (de Casamajor et al. 1999) and, as a consequence, the gear effectiveness (pushed sieve net) is low as this gear filters the volume of water close to the surface (Bru et al. 2009).

[14]The influence of the climate change on the transoceanic migration of the *leptocephali* suggested by the «eel tree» was outside the scope of the Indicang program.

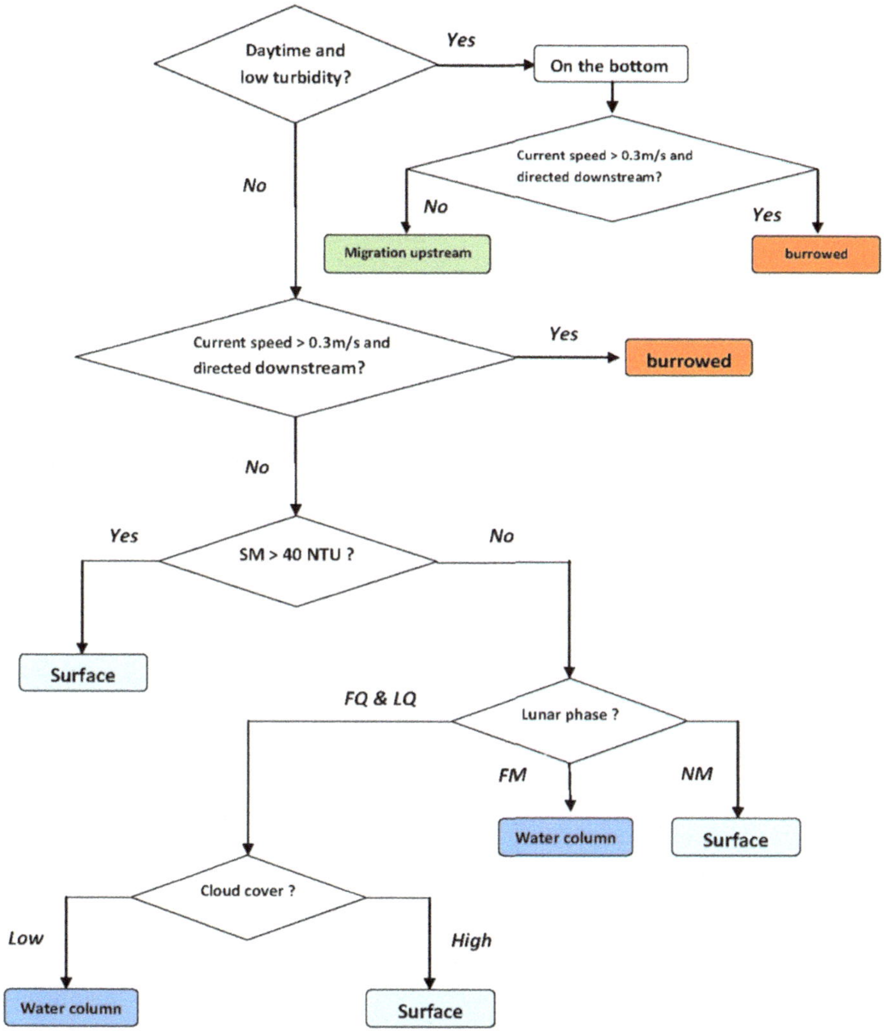

Fig. 7.9 The conceptual diagram of glass-eel migration in estuary based on the knowledge acquired by professional fishers and scientists. SM: suspended materials; NTU: Nephelometric unit; FQ: first quarter; LQ: last quarter; FM: full moon; NM: new moon

So from the fishers' observations and the different scientific studies, a conceptual model of the migration behaviour of glass-eel could be defined as mentioned in Fig. 7.9.

This Figure shows the influence of the hydro-climatic factors on the migration process of glass-eel during its estuarian crossing. Thus, the contribution of scientists was to quantify the most part of this qualitative information and to reply to the questions which are asked by the professional fishers. This allowed the definition of a full conceptual model of the behavior at the basis of a numerical model of glass-eel migration in estuary according the variability of hydroclimatic conditions (Prouzet et al. 2009; Odunlami and Vallet 2012). This numerical model is a useful tool to estimate the catchability of glass-eel and the fish vulnerability to fishing gear (pushed sieve net) used. That allows to specify, for a given day according the hydro-climatic conditions, the catchability of the glass-eel (***q***), product of 3 parameters such as: $\boldsymbol{q} = \boldsymbol{ac}.\ \boldsymbol{a}/\boldsymbol{A}.\ \boldsymbol{s}$ where ***ac*** is the accessibility: the probability that the fish will be present on the fishing grounds (best period between November and March for the Adour river); ***a/A*** the probability that the volume where the fish can be found will be swept by the fishing gear with ***a*** the area swept by a unit of fishing effort and ***A*** the total volume of the zone explored by the fishing fleet; ***s*** the effectiveness of the fishing gear.

For example, during the day with a low turbid water, as the pushed sieve filters only the volume of water between the surface and 1.5 m deep, the effectiveness of the fishing gear is nil as the fish migrates upstream close to the bottom.[15] During the night, the catchability depends on several physical factors allowing glass-eel to migrate upstream close to the surface of water as mentioned in Fig. 7.9. One situation is not illustrated on this conceptual diagram: the occurrence of floods that stop the ingress of the tide in the estuary. This hydro-climatic situation does not allow the use of the pushed sieve but during the period of floods some good catches with hand sieve are recorded on the river banks in area of calm water.

The use of this numerical model to simulate the movement of a group of glass-eel during its upstream migration allows to explain the displacement of the peak of catches from the lower part to the upper part of the estuary and the hydro-climatic conditions that have an effect on this displacement (Prouzet et al. 2009).

The Fig. 7.10 shows some outputs of the numerical model with an entry of a group of glass-eel in Adour estuary at 5:00 am (Fig. 7.10a) the 09/11/1999 at the end of the flood tide. The tide coefficient is 86 (medium level), the rate of flow is 140 m^3/s (low level) and the light intensity is low in the water column so the glass-eel run can migrate close to the surface.

The Fig. 7.10b shows a snapshot of the graphic output the same day at 15:00. The tide was flooding and the group of glass-eel is located close to the bottom and is migrating upstream during the daytime. The Fig. 7.10c and d show the group of fish migrating during the flood tide at night the 2 following days. The last figure corresponds to the progressive exit of the glass-eels from the upper part of the estuary influenced by the tide and located at 30 kms upstream of the river mouth. So we can estimate for a medium tide coefficient and a low or average rate of flow that the time spent by the glass-eel run to cross the Adour estuary (30 kms) is around 2 or 3 days.

Quantification of the Relationships Between Hydro-Climatic Parameters and Catch Intensity to Estimate the Rate of Exploitation of Glass-eel Runs

[15]In some river system such as the Gironde estuary, the high turbidity of water allows to catch the fish in surface even during the daytime.

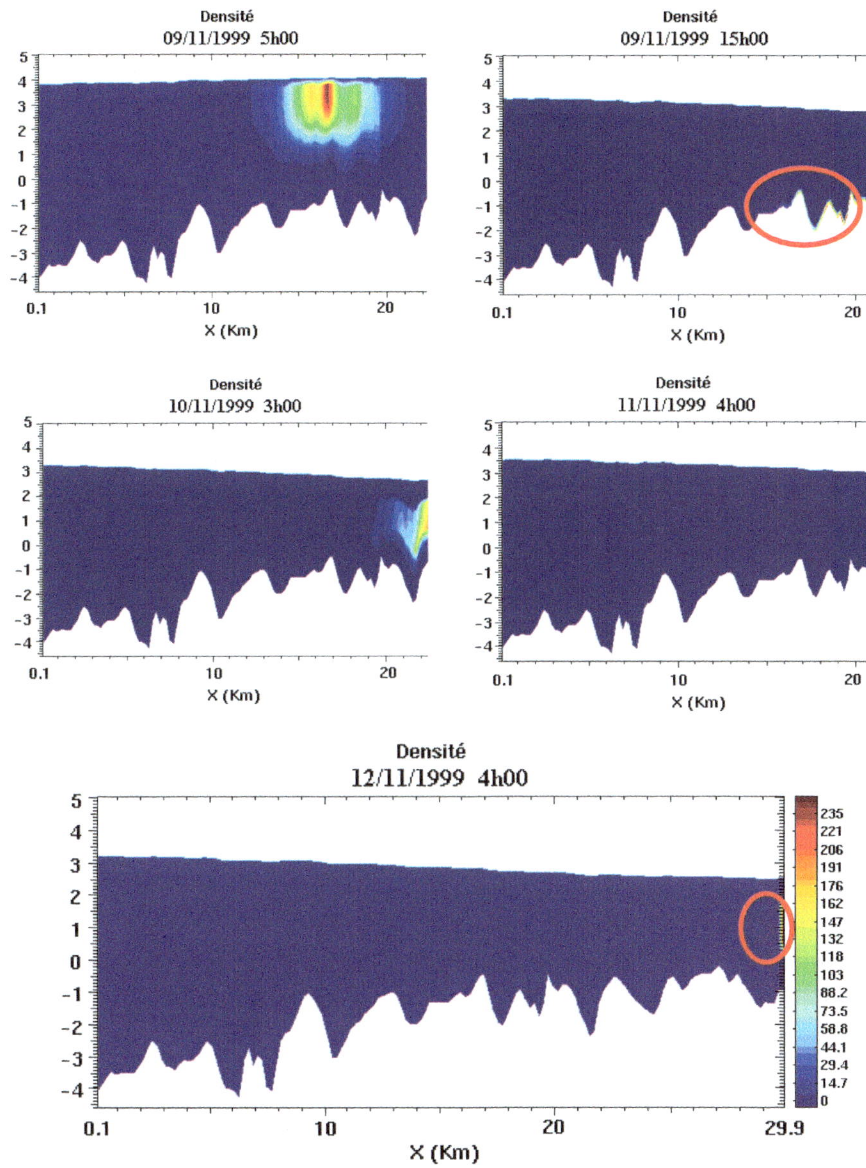

Fig. 7.10 Snapshots from a numerical model of the migration behaviour of glass-eel in the Adour estuary (Prouzet 2002, 2003)—see comment in the text: **a** top left; **b** top right; **c** middle left; **d** middle right and **e** bottom. Water depth (Y-axis)

To quantify these relationships between physical environment and catch level, scientific surveys have been undertaken on different French rivers[16] according a sampling design developed on the Adour River (Prouzet 2003). This sampling design allows to estimate the densities of glass-eel in the water column (and not only in surface) according three longitudinal transects: one near the right bank; the other one in the middle of the river and the last one near the left bank and thus at two depth *strata* (1 and 4 m) of the water column. The entire width of the river is sampled in thirty minutes allowing a maximum of 8 sampling cycles during the rising tide at a given sampling station. From 1998 to 2005, 54 scientific surveys have been made and 2,592 observations on glass-eel densities have been collected.

So, according the following formula defined by Bru et al. (2009), the biomass of glass eel transported during a rising tide through the fishing area is estimated such as:

$$B_{night} = \frac{1}{N} \times \sum_{1}^{N} \sum_{S=1}^{6} \left[\int_{tide\_begin}^{tide\_end} d_s(t) \times v_s(t) \right] \times S_n \qquad (7.1)$$

with:

B_{night} biomass of glass-eel migrating during a rising tide at night.
d density of glass-eel (g/100 m^3)
v speed of water (m/s)
S surface of the sector sampled (6 sectors for a river section: 2 on the left, 2 on the middle and 2 on the right side)
s qualitative index for the sector concerned; s = [1;6]
n qualitative index for the cutting of the river section; n = [1;N]

At the same time, professional fishers who had caught glass-eel on the site of sampling fill in a log-book to specify the catch they made. So the following table (see Table 7.2) could be defined.

From this data base it is possible to establish a relationship between exploitation rate versus catches and physical variables. For the Adour River, Bru et al. (2009) defined the following statistical model[17]:

$$Explrate \approx turbcod + catches^{0.678} + tidecod + debcod : lunarphase + catches : lunarphase \qquad (7.2)$$

The factor turbidity (after coding) explained the greatest part of the total variance which is logical as the Adour estuary is a clear estuary (low average turbidity), the

[16]See web site www.ifremer.fr/indicang/.

[17]This model explain 90% of the total variance. The exponent of catches is defined with a linear relationship between log(Explrate) and log(catches).

Table 7.2 Data base for the definition of the exploitation rate from catches and physical variables[a]

Date	B_{night} (kg)	Professional catches (kg)	Exploitation rate (in %)	Tide coeff	Turbidity (NTU)	Rate of flow (m^3/s)	Lunar phase	Tidecod	Turbcod	Debcod
12/01/99	138	6	4.35	40	32.2	403.5	NM	1	2	3
22/01/99	157	11.2	7.13	81	19.1	270.4	FQ	2	2	2
14/12/04	198	17.6	8.89	93	9.0	94.55	NM	3	1	1

Tidecod: 1 if Tide coeff < 60; 2 if 60 < Tide coeff > 86; 3 if Tide coeff > 86
Debcod: 1 if debcod < 250; 2 if 250 < debcod < 400; 3 if debcod > 400
Turbcod: 1 if turbcod <13; 2 if turbcod >13
Lunarphase: 4 codes: FM: full moon; FQ: first quarter; NM: new moon and LQ: last quarter
[a]See Annex 12 for the entire database on the web site www.ifremer.fr/indicang/

water column opacity has a great influence on the presence of glass-eels near the water surface where they are vulnerable to the fishing gear.

In the Loire River, the estuary is highly and permanently turbid so the turbidity is not a relevant factor as showed by the statistical analysis (Prouzet et al. 2007) and the best statistical model that explains the greatest percentage of the total variance (81%) is the following one:

$$Explrate \approx 3.36 * \log(catches) - 0.011 * deb + 0.058 * tidecoef \tag{7.3}$$

The tide coefficient is a highly significant factor. Its effect shows that stronger the tide, lower the rate of exploitation and inversely. In fact, the tide level has a strong effect on the time of residence of the run of glass-eel in the fishing area.[18]

The final phase of the estimation process uses the Eq. (7.3).[19] The Table 7.3 shows an example of database that allows to backtrack the fluctuation of the daily biomass from professional fishery statistics and physical data recorded. The estimated biomass (biomestim) is the ratio between *Explratestim* and catches.

The Fig. 7.11 shows the graphic output for the fishing season 2005/2006 on the Loire River. It shows the dispersal of observed biomasses, the fluctuations of estimated biomasses (smoothed curve), the variation of the rate of flow and the dispersal of the catches in the area sampled during the fishing season.

This reconstruction of the estimated daily biomasses was the result of a close cooperation between professional fishers who have provided the daily catches data and the scientists who have determined the sampling and mathematical design to provide the relationship between hydro-climatic conditions and variation of fish catchability. For that fishing season, the B_{night} was estimated to 25.7 tons corresponding to a total biomass of 54.1 tons; the total catches declared was 9.5 tons and thus, the exploitation rate by the professional fishery was estimated to 17.5% (Prouzet et al. 2007).

7.2.2.2 A Partnership Between Professional Fishers and Scientists to Estimate the Variation of the Silver eel Escapement on the Loire River

In a large river system, the estimate of the silver escapement with certain accuracy is almost impossible without the technical assistance and the traditional knowledge and know-how of the professional fishers.

18Irrespective of the values of the rate of flow and tide coefficient, the turbidity is always strong and the glass-eels migrate always near the surface of water on the Loire estuary. But, we know as on the Adour estuary that floods associated with low or medium tide coefficients prevent the glass-eels to migrate and concentrate the fishes on the river banks allowing to make very good catches.

19Or the (7.2) for the Adour River

Table 7.3 Extract from the data base relative to the fishery season 2005/2006 on the Loire River

Date	Catches	Tide coeff	Deb	Explratestim	Explratobserv	Stdeviation	Biomestim	B_{night}
01/12/05	0,6	83	242	0,044		0,009	13,6	
02/12/05	1,3	87	248	0,049		0,009	26,7	
03/12/05	1,7	88	264	0,051		0,010	33,5	
21/01/06	50,2	52	663	0,174		0,023	289,6	
22/01/06	49,8	44	660	*0,188*	*0,128*	0,028	*265,6*	*388,8*
23/01/06	45,6	41	646	*0,183*	*0,180*	0,029	*249,5*	*253,6*
24/01/06	28,7	37	663	*0,148*	*0,158*	0,026	*194,5*	*181,7*
25/01/06	22,5	40	656	0,129		0,022	174,9	
12/02/06	27,3	76	478	0,106		0,013	256,8	
13/02/06	59,1	81	473	*0,141*	*0,199*	0,019	*420,2*	*297*
14/02/06	85,2	84	452	*0,164*	*0,200*	0,030	*517,8*	*426,2*
15/02/06	118,0	84	452	*0,201*	*0,188*	0,051	*588,0*	*627,01*
16/02/06	121,8	82	509	0,213		0,056	572,1	
19/02/06	100,4	61	1140	0,291		0,040	345,2	
20/02/06	129,2	50	1400	*0,600*	*0,843*	0,169	*215,4*	*153,18*
21/02/06	83,3	39	1650	0,449		0,128	185,8	
19/03/06	12,9	75	1590	0,094		0,020	137,5	
20/03/06	44,6	65	1420	*0,164*	*0,116*	0,025	*271,8*	*384,58*
21/03/06	41,6	53	1320	0,174		0,028	238,8	
26/03/06	31,7	68	1590	0,138		0,027	229,3	
27/03/06	72,5	87	1770	*0,187*	*0,210*	0,036	*388,6*	*345,53*
28/03/06	106,7	103	1850	*0,204*	*0,191*	0,050	*523,0*	*557,7*
29/03/06	68,0	113	1860	*0,143*	*0,143*	0,031	*475,3*	*477,02*
30/03/06	62,2	115	1890	0,136		0,029	458,6	

With *catches, biomestim and Bnight* in kg; *deb* in m^3/s—Bnight is the biomass estimated from the scientific survey with (7.1); Explratestim is estimated from (7.3) and Explrateobserv is the ratio catches/B_{night}

The sampling design relies on the activity of successive anchored stow-net fisheries located in the medium and lower courses of the Loire River such as shown in Fig. 7.12 (see Fig. 7.15 for the type of fishing gear used).

These fisheries are used to catch the fish and to tag the fish with different injection of red ink and acrylic paint (see Acou et al. 2015 for details on the tagging operations). Only, the silver eels characterized by the ocular hypertrophy, a sharp contrast between the black above and the white below and a lateral line well defined have been selected for tagging. During the season 2012/2013, 675 eels have been tagged and released. On each fishery, a specific coding has been used. The recapture is made downstream in the successive fisheries by the professional fishers themselves who recorded the information needed: (date and hours of catches, code of tagging, physical parameters). The estimate of the migrating population is made by the Pooled-Petersen method.[20] The Fig. 7.13 gives the variation of the escapement

[20]See Acou et al. (2015) for details and Schwarz and Taylor (1998) for mathematical basis.

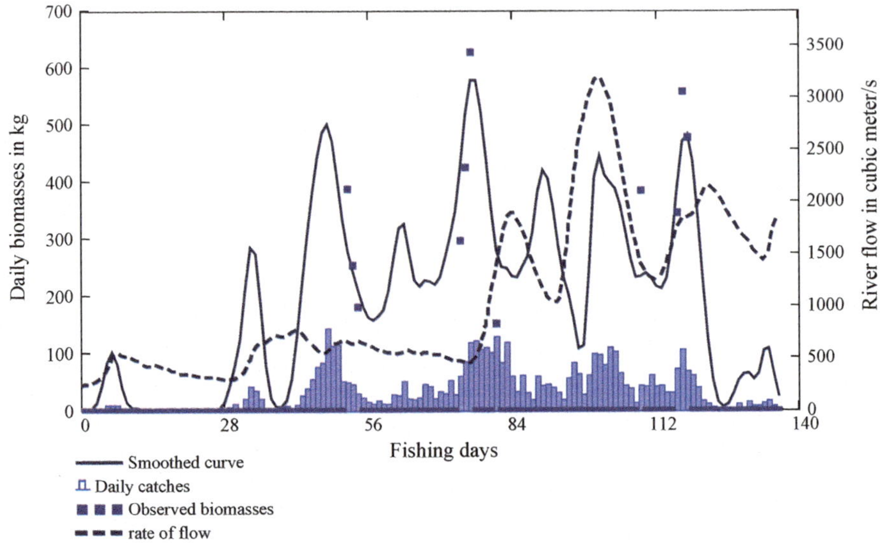

Fig. 7.11 Fluctuations of estimated biomasses (smoothed curve) and catches of glass-eel on the Loire River for the fishing season 2005/2006 (from Indicang data)

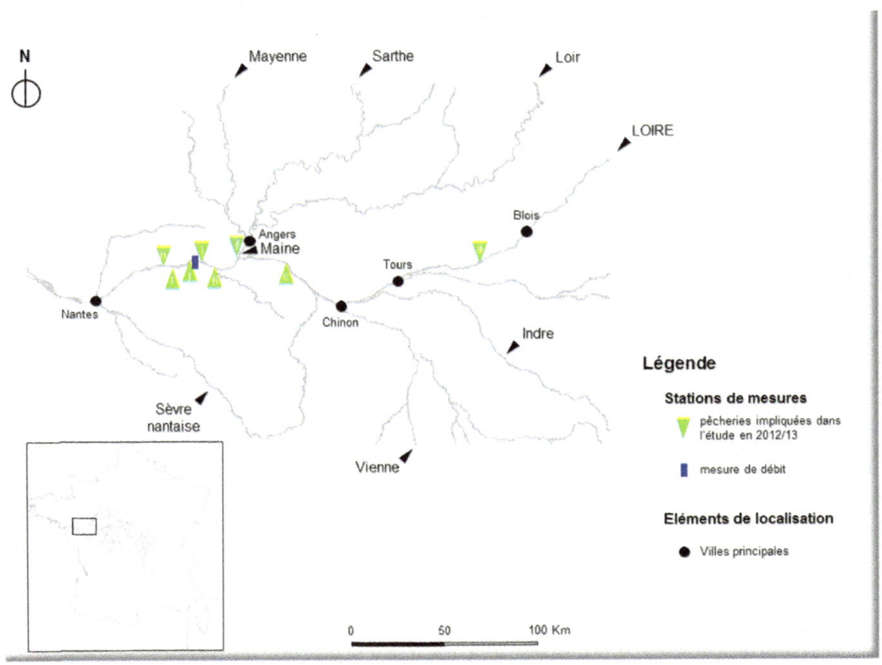

Fig. 7.12 Locations of the 8 anchored stownet fisheries used in the Loire river to estimate the silver eel escapements (after Acou et al. 2015)

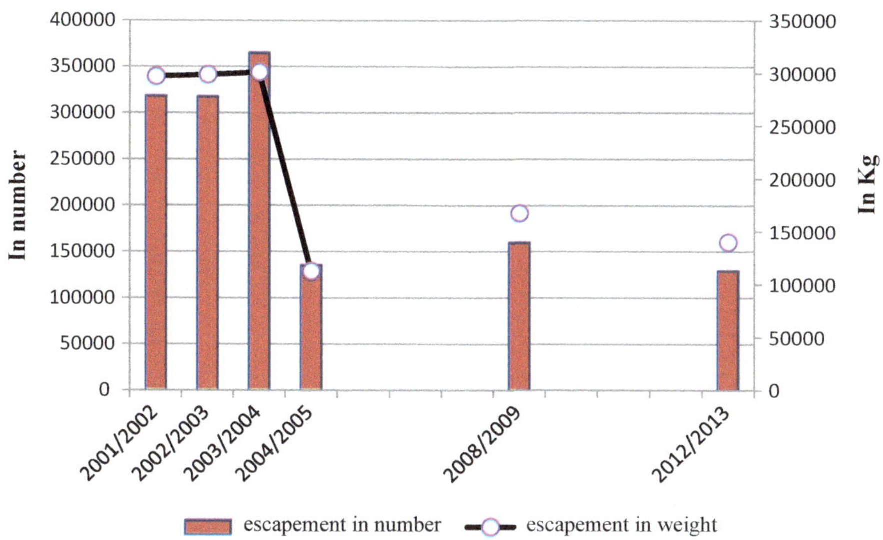

Fig. 7.13 Fluctuation of the silver eel escapements observed on the Loire River (After Acou et al. 2015)

for the Loire (this estimate doesn't take into account the estuarine part as we can see in the Fig. 7.12).

A longer series of escapement indices has been obtained by Boisneau and Boisneau (2014, 2015) since 1987. The yearly index is the mean of the decimal logarithms of catches per unit of effort for all the fisheries (Adam et al. 2008).

The Fig. 7.14 shows a significant breakdown in the series of silver eel abundances for the season 2003. The evolution of this escapement index is coherent with that recorded in Fig. 7.13 concerning the absolute escapement for the river Loire (estuary not included). The reasons for such a new level after 2003 are not known yet, but probably linked to the decrease in glass-eel recruitments that we have observed during the beginning of the nineties (see Fig. 7.5).

7.2.2.3 The Eeliad Project: A Partnership Between Professional Fishers and Scientists to Study the Migration of Silver eel from the Loire River and the Mediterranean Lagoons

This European project was coordinated by Dave Righton (Cefas) in association with scientists from different European countries and professional fishers to catch the fish for tagging experiments (Fig. 7.15). The objective of the EELIAD project was to improve our knowledge on the biology and marine ecology of European eel.

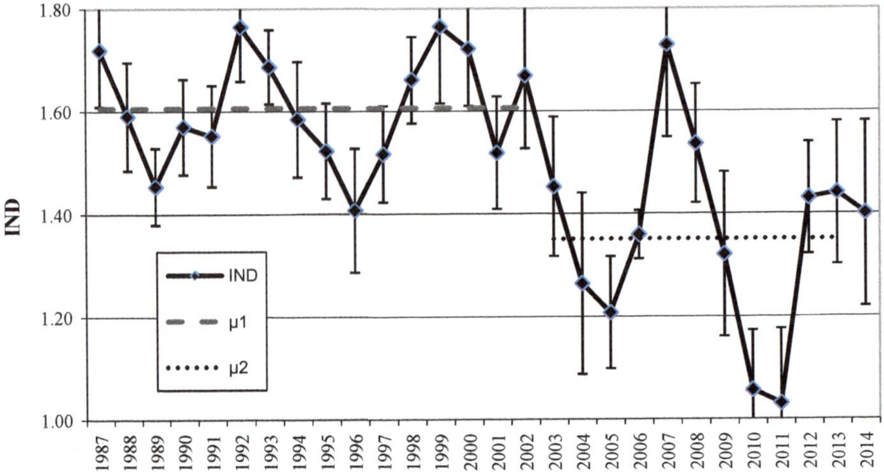

Fig. 7.14 Fluctuation of the silver Eel escapement index on the Loire river from 1987 to 2014 with $\mu 1 = 1.61$ and $\mu 2 = 1.35$ (From Boisneau and Boisneau 2015)

A part of the project was dedicated to the study of Eel migrating behaviour at sea. For this purpose, 362 electronic tags were attached or implanted to eels to record data on the migration at sea (Righton et al. 2016). Two types of tags were used: 287 internal flotsam tag called "*i-DST*" consisted of an electronic unit and three floats mounting on a wire and 75 external pop-up tags "Microwave telemetry PSAT" released from the west coast of Ireland and France and Sweden (Fig. 7.16). This PSAT houses sensor measuring temperature and depth.

Another experiment included in the project EELIAD used PSAT tags to track silver eels from a site of release located in the French Mediterranean coast. That study was also undertaken in cooperation with professional fishers to catch the fish of an appropriate size (Amilhat et al. 2014).

Figure 7.17 shows a synthesis of the locations of silver eels when the PSATs have transmitted their information. Silver eels coming from different parts of the colonization area seem to migrate towards the Azores but no eel have been followed farther than the Azores.

All these tagging experiments show that eel moved into the mesopelagic zone with diel vertical migrations between 200 and 1000 m in the Atlantic Ocean as in the Mediterranean Sea (Aarestrup et al. 2009; Amilhat et al. 2014). The migration speed varied from 5 to 25 km per day.[21]

[21]Much lower than the 35 km per day required to reach the Sargasso Sea for spawning in April.

Fig. 7.15 Fishing gear used to catch the silver eel in the Loire River. It is an anchored stownet called "Guideau" (E. Feunteun)

Fig. 7.16 Silver eel with an attached PSAT (From EELAD project—Life in the Big Blue Box—D. Righton 2008)

7.3 Towards the Certification Process of the eel Fishery Industry

In the framework of the Indicang project (see above www.ifremer.fr/indicang/), different technical documents have been produced:

- A methodological guide entitled "The European Eel: Indicators on abundance and colonization".

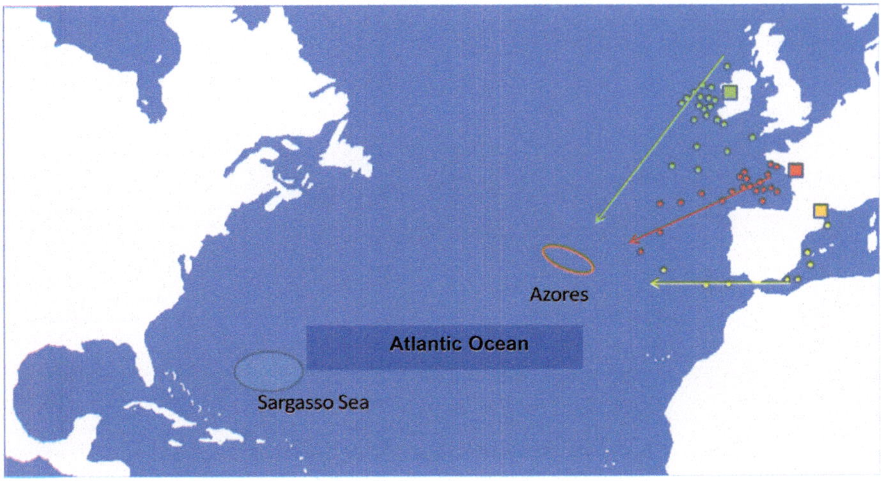

Fig. 7.17 Approximate locations of PSATs attached on eel released from 3 different geographical areas. For precise information see Aarestrup et al. (2009), Feunteun (2012) and Amilhat et al. (2014 and 2016), Righton et al. (2016)

- A guide to the completion of field study forms on "obstacles to migration" and "accidental mortality" (www.ifremer.fr/indicang/documentation/pdf/liste-annexes-guide-method.pdf)
- A sanitary guide (www.ifremer.fr/indicang/documentation/pdf/guide-sanitaire.pdf)
- A guide to the completion of field study forms and recommendation for the restocking and the transfer of individuals[22]).

All these documents have been approved by the Indicang partners from 4 European countries (Portugal, Spain, France and UK) involved in the project and constitute a common technical basis sharing scientific and traditional knowledge and know-how.

In addition to these documents professional fishers in cooperation with some scientists involved in the Indicang project decided to produce a document on "the code of practice for the capture and restocking of glass-eels and elvers".[23] This code details different topics:

- Determining the catchment area's specific environment and fishing context in order to assess the level of fishing effort exerted on the run of glass-eel.

[22]www.ifremer.fr/indicang/documentation/pdf/liste-annexes-guide-method.pdf.

[23]Contribution of French Professional Fishermen to Eel Restocking: Good practice guide for the glass-eel fishing and implementation of a community-wide restocking program. Document CNPMEM/ARA France/CONAPPED with the support of the WWF—France, 19 pages.

- Determining the quality of glass-eel caught for restocking.
- Minimizing the impact of glass-eel fishery on other species.
- Optimizing the equipment on board for transportation of glass-eel alive.
- Guarantee of optimal transport conditions.
- Regulate the use of products for restocking in Europe and especially include the eel restocking in Europe in the framework of the ecosystemic approach for fishery and aquaculture[24] with an assessment of its effects on the wild population since the juveniles used are taken exclusively from the natural environment and transferred to areas sometimes very far from the sites of captures.
- Monitoring the effectiveness of the restocking system and establish management rules. In particular, restocking must not be a substitute for a more ambitious and more integrated program of restoration of eel and its habitat. Restocking actions with the sole aim to support fisheries should be prohibited.

As such, the requests of the French professional fishers for a consistent and comprehensive program for eel restoration are:

- Short-term restoration objectives for the free migration of eels in order to have a colonization of the potential eel habitat both in the lower part (hydrological annexes) and in the upper part of the catchments.
- Implementation of a system to ensure the traceability of fishing products and regulation of the commercial sector.
- Establishment of an eel restocking program in France to increase the number of eel in "density deficient areas".

This code of good practices is taken into account now in the Sustainable Eel Standard defined by the Sustainable Eel Group (www.sustainableeelgroup.com). This Standard has been designated to (SEG 2013):

- Enable operators to demonstrate high standards and their commitment to sustainability;
- Encourage high and responsible standards through the supply chain, from fishery to market;
- Encourage sustainable practices and sustainable markets;
- Discourage unsustainable practices and unsustainable markets;
- Provide confidence for customers who wish to buy responsibly.

The Sustainable Eel Standard has been developed in line with the principles of the Brundtland Convention's definition of sustainability.

[24]See the FAO code of conduct for responsible fishing (1995).

References

Aarestrup K, Okland F., Hansen MM, et al (2009) Oceanic spawning migration of the European eel (*Anguilla anguilla*). Science 325. Brevia

Acou A, Boisneau C, Bodin M, Bultel E, Boury P, Feunteun E (2015) Evaluation du flux d'anguilles argentées produit en Loire fluviale (amont d'Ancenis) pour 6 saisons d'avalaison (de 2001/02 à 2004/05, 2008/09 et 2012/13) et étude du comportement d'avalaison jusqu'à l'estuaire par télémétrie acoustique (saison2011/2012). Rapport final. MNHN, Univ. Tours, AAIPPBLB, Service des Staions Marines, 54 p

Adam G, Feunteun E, Prouzet P, Rigaud C (2008) L'anguille européenne – Indicateurs d'abondance et de colonisation. Editeur QUAE, série Savoir Faire, 393 p

Amilhat E, Simon G, Faliex E (2014) Rapport technique du suivi scientifique des relâchers d'anguilles argentées menés en Méditerranée en 2013. Rapport Cefrem-CNRS- Univ. Perpignan, 46 p

Amilhat E, Aarestrup K, Faliex E, Simon G, Westerberg H, Righton D (2016) First evidence of European eels exiting Mediterranean Sea during their spawning migration. Sci Rep 6:21817. https://doi.org/10.1038/srep21817, 9 p

Boisneau C, Ph. Boisneau (2014) Long-term evolution of silver European eel indicator of abundance in the Loire watershed (1987–2013), a unique data set for Europe. In: Oral communication of the 144th American Fisheries Society (AFS) Annual Meeting "International Eel Symposium: Are eelsclimbing back up the slippery slope?", Aug 17–21, Quebec City, Canada

Boisneau C, Ph. Boisneau (2015) Indice annuel d'abondance des anguilles d'avalaison du bassin de la Loire à partir des captures des pêcheurs professionnels au guideau. Rapport Univ. Tours, Plan Loire Grandeur Nature III, 15 p

Bonhommeau S, Gascuel D, Blankeet B, et al (2009) Estimates of the mortality and the duration of the trans-Atlantic migration of European eel (*Anguilla anguilla*) leptocephali using a particle tracking model. J Fish Biol 74(9):1891–1914

Bru N, Prouzet P, Lejeune M (2009) Daily and seasonal estimates of the recruitment and biomassof glass-eel runs (*Anguilla anguilla*) and exploitation in the Adour open estuary (Southwestern France). Aquat Living Resour 22(4):509–524

Daverat F, Tomas J, Lahaye M, Palmer M, Elie P (2005) Tracking continental habitats shifts of eels using otolith Sr/Ca ratios: validation and application to the coastal, estuarine and riverine eels of the Gironde-Garonne-Dordogne watershed. Mar Freshw Res 56:619–627

De Casamajor M-N, Bru N, Prouzet P (1999) Influence de la luminosité nocturne et de la turbidité sur le comportement vertical de la civelle d'anguille (*Anguilla anguilla*) dans l'estuaire de l'Adour. Bull Fr Pisciculture 355:327–347

Fablet R, Daverat F, De Pontual H (2007) Unsupervised Bayesian reconstruction of individual life histories from otolith signatures: case study of Sr:Ca transects of eel (*Anguilla anguilla*) otoliths. Can J Fish Aquat Sci 64(1):152–165

Feunteun E (2012) Le rêve de l'anguille, une sentinelle en danger – Petite encyclopédie sur un poisson extraordinaire. Editeur Buchet – Chastel, série Ecologie, 266 p

Lecomte-Finiger R, Yahyaoui A (1989) La microstructure de l'otolithe au service de la connaissance du développement larvaire de l'anguille européenne *Anguilla anguilla*. CR Acad Sci Paris 308:1–7

Léauté J-P, Caill-Milly N, Leblond E, Prouzet P, Mayer N, Garabey J (2002) Caractéristiques des petites pêches côtières et estuariennes de la côte atlantique du sud de l'Europe. Contrat PECOSUDE, rapport final EC/DG FISH (DG XIV): 99/024, 178 p + annexes

Nielsen T, Prouzet P (2008) Capture-based aquaculture of the wild European eel (*Anguilla anguilla*). In: Lovatelli A, Holthus, PF (eds), Capture-based aquaculture, Global overview. FAO Fisheries Technical Paper, No 508. FAO, Rome, pp. 141–168

Odunlami M, Vallet G (2012) Modelling and mathematical analysis of the glass-eel migration in the Adour river estuary. Math Model Nat Phenom 7(3):168–185

Prouzet P (coord.) (2002) Historique des captures de civelles, intensité actuelle de leur exploitation, variation de leur capturabilité par la pêche professionnelle maritime et indices de colonisation sur le bassin versant de l'Adour. Contrat EC/DG Fish (DG XIV) 99/023, 150 p + annexes

Prouzet P (coord.) (2003) Etude sur la civelle (*Anguilla anguilla*) dans l'estuaire de l'Adour – Pêche, biologie, comportement, modélisations hydrodynamique et comportementales, estimation des flux de civelles en estuaire. Contrat région Aquitaine, Agence de l'Eau Adour-Garonne. R. INT.DRV/RH/LHA, 295 p

Prouzet P (coord.) (2007) Website: https://www.ifremer.fr/indicang/

Prouzet P, Bouvet J-C, Bru N, Boussouar A (2007) Quantification de la biomasse saisonnière de civelles (*Anguilla anguilla*) dans l'estuaire de la Loire et estimation du taux d'exploitation saisonnier de la pêche professionnelle pour les saisons de pêche 2003 à 2005, Rapport Indicang, UE – Ifremer – UPPA – ADERA, 45 p

Prouzet P, Odunlami M, Duquesne E, Boussouar A (2009) Analysis and visualization of the glass-eel behavior (*Anguilla anguilla*) in the Adour estuary and estimate of its upstream migration speed. Aquat Living Resour 22(4):525–534

Righton D (2008) Life in the big blue box: studying the marine life of European eels. Poster-ICES CM 2008/P:18

Righton D (coord.) (2012) EELIAD: European eels in the Atlantic: Assessment of their Decline. Project final report – Grant agreement number 212133 – UE FP7 program, 40 p

Righton D (2015) EELIAD report summary, 2 p. http://cordis.europa.eu/result/rcn/55872_en.html

Righton D, Westerberg H, Feunteun E, Økland F, Gargan P, Amilhat E, Metcalfe J, Lobon-Cervia J, Sjöberg N, Simon J, Acou A, Vedor M, Walker A, Trancart T, Brämick U, Aarestrup K (2016) Empirical observations of the spawning migration of European eels: the long and dangerous road to the Sargasso Sea. Sci Adv 2:e1501694

SEG (2013) The Sustainable Eel Standard, Sustainable Eel Group report, 29 p

Scharwz CJ, Taylor CG (1998) The use of the stratified Petersen estimator in fisheries management with an illustration of estimating the number of pink salmon (*Oncorhynchus gorbusha*) that return to spawn in the Fraser river. Technical paper, 38 p

Tabouret H, Bareille G, Claverie F, Pécheyran C, Prouzet P, Donard OFX (2010) Simultaneous use of strontium: calcium and barium:calcium ratios in otoliths as markers of habitat: Application to European Eel (*Anguilla Anguilla*) in the Adour basin, South West France. Mar Environ Res 70(1):35–45

Tesch FW (1998) Age and growth rates of North Atlantic eel larvae (*Anguilla* spp.) based on published length data. Helgol Meeresunt 52:75–83

Tesch FW (1977) The eel. Biology and management of anguillid eels. Chapman & Hall, London, p 434p

Chapter 8
Trophic Cascade in Seaweed Beds in Sanriku Coast Hit by the Huge Tsunami on 11 March 2011: Sea Urchin Fishery as a Satoumi Activity Serving for Increase in Marine Productivity and Biodiversity

Teruhisa Komatsu, Shuji Sasa, Hiroki Murata, Shuhei Sawayama, Masahiro Hamana, Minami Asada, Ryo Tsujimoto, Genki Terauchi and Tetsuo Yanagi

Abstract The Great East Japan Earthquake produced a huge tsunami, which struck and changed the coastal ecosystems along the Sanriku Coast in front of the epicenter. It is important to monitor changes in the ecosystems caused by this huge tsunami and their succession due to natural processes and human impacts, because tsunami events occur at intervals of less than 100 years in Japan. We have monitored the seaweed beds in Shizugawa Bay on the Sanriku Coast since October 2011. Our observations showed that the tsunami didn't seriously damaged the seaweed beds on the rocky substrate. The sea urchins on the rocky coast were also not affected by the tsunami. These sea urchins were able to spawn in the summer of 2011, because the sea urchin fishery suspended work in 2011 and 2012 due to a lack of fishing boats and a possibility that sea urchins fed corpses of victims of tsunami on the sea bottom. In 2014, the sea urchins born in summer of 2011 started to feed on the kelp forests, particularly from the deeper *Eisenia bicyclis* beds. In 2015, they consumed most of the *E. bicyclis* beds, which have disappeared from the head of the bay. The kelp forests in the Pacific involve a trophic cascade consisting of sea otters, sea urchins, and kelp. The sea urchin fishery corresponds to sea otters,

T. Komatsu (✉) · S. Sasa · H. Murata · S. Sawayama · M. Hamana · M. Asada
Atmosphere and Ocean Research Institute, The University of Tokyo, 5-1-5 Kashiwanoha, Chiba 277-8564, Japan
e-mail: komatsu@aori.u-tokyo.ac.jp

R. Tsujimoto · G. Terauchi
Northwest Pacific Region Environmental Cooperation Center, 5-5 Tower111 6th Floor, Ushijimashin-Machi, Toyama City, Toyama 930-0856, Japan

T. Yanagi
International EMECS Center, DRI East 5F 5-2 Wakinohama-Kaigandori 1-Chome, Chuo-Ku, Kobe 6510073, Japan

T. Komatsu et al. (eds.), *Oceanography Challenges to Future Earth*,
https://doi.org/10.1007/978-3-030-00138-4_8

which maintain healthy giant kelp beds along the west coast of the USA. Therefore, the sea urchin fishery is a *satoumi* activity defined as the human management of coastal seas that results in high biological productivity and biodiversity. It is necessary to maintain kelp beds through the sea urchin fishery for realizing sustainable fisheries and a healthy coastal ecosystem.

Keywords Satoumi · Sea urchin fishery · Trophic cascade · Tsunami · Kelp forest

8.1 Introduction

Coastal ecosystems provide many ecological services to society. In coastal ecosystems, seagrass and seaweed beds are important ecosystems for coastal fisheries (e.g., Komatsu et al. 1982) and affect water temperature (Komatsu et al. 1985, 1994), pH (Komatsu and Kawai 1986), dissolved oxygen (Komatsu 1989), water flow (Komatsu and Murakami 1994; Komatsu 1996; Komatsu et al. 2004), and ecosystem services for human societies (Costanza et al. 1997). Seaweed and seagrass beds are broadly distributed along the Sanriku Coast facing the Pacific Ocean in north Honshu Island.

The Great East Japan Earthquake, which had its epicenter off the Sanriku Coast in the Northwestern Pacific Ocean, occurred on March 11, 2011 and had a magnitude of 9.0. The huge tsunami caused by this earthquake brought catastrophic damages along the Sanriku Coast. The wave and run-up heights were approximately 10–20 m along the coast (The 2011 Tohoku Earthquake Tsunami Joint Survey Group 2012). This huge tsunami affected not only the land, but also coastal ecosystems under the sea (Komatsu et al. 2015). Broad seagrass beds and seaweed beds are distributed along the Sanriku Coast, which has many rias-type bays as a semi-enclosed bay. Shizugawa Bay located on the southern Sanriku Coast is also a typical rias-type bay. There were many seaweed and seagrass beds in Shizugawa Bay before the tsunami. Seaweed beds are very important habitat for abalones and sea urchins, commercial target species in this area, particularly in Shizugawa Bay. Sakamoto et al. (2012) reported that the seaweed beds had recovered immediately after the tsunami in 2011–2012, while the seagrass beds were severely damaged (Sasa et al. 2012). It is very important to monitor the succession of the seaweed beds in Shizugawa Bay to examine the restoration of the coastal ecosystem after the perturbation caused by the tsunami.

Remote sensing is a practical tool for monitoring the distributions of seagrass and seaweed beds (Komatsu et al. 2012). Remote sensing using satellite images has the advantage that archived images give past information. We can trace spatial changes in the distribution of seagrass and seaweed beds. Google Earth freely provides archived satellite images through the Internet. It is convenient to use

Google Earth images for monitoring seaweed beds. This study examined the succession of the seaweed beds in Shizugawa Bay before and after the huge tsunami on March 11, 2011 by analyzing satellite images.

8.2 Materials and Methods

8.2.1 Shizugawa Bay

Shizugawa Bay is located on the southern Sanriku Coast (Fig. 8.1). Shizugawa Bay is classified as an enclosed bay by the Ministry of the Environment of Japan (2010). It has a wide bay mouth (6.6 km), a relatively short longitudinal length (7.7 km), a maximum bottom depth of 54 m, and a bottom depth of 54 m at the bay mouth (Fig. 8.2). The 2011 Tohoku Earthquake Tsunami Joint Survey Group (2012) estimated that the mean run-up height of the tsunami in the bay on March 11, 2011 was 14.4 m, based on field observations.

8.2.2 Field Surveys

Field surveys were conducted at 6-month intervals from October 2011 to October 2013, and at approximately 3-month intervals since 2014. The field surveys consisted of visual and drop camera observations from a small fishing boat, visual observations from the shore, and interviews with fishermen.

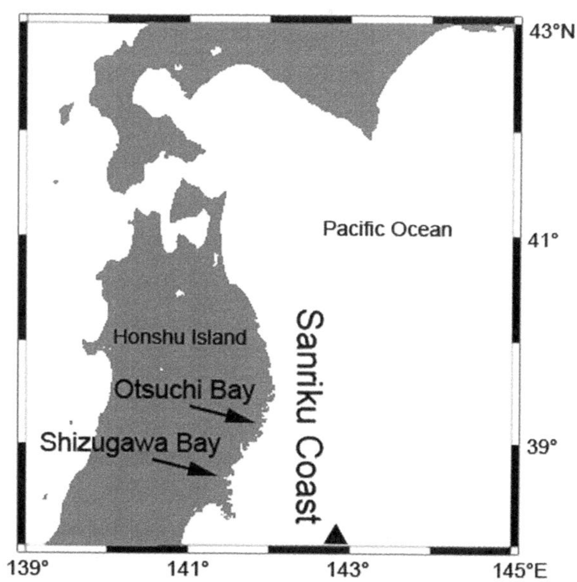

Fig. 8.1 Map showing Shizugawa Bay on the Sanriku Coast and the epicenter of the Great East Japan earthquake (black triangle) on March 11, 2011

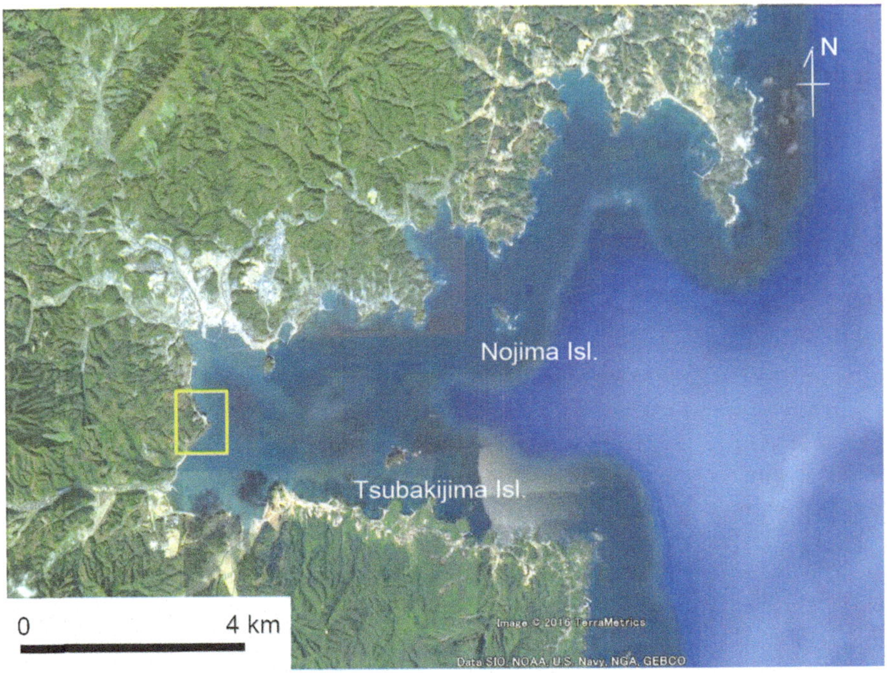

Fig. 8.2 Shizugawa Bay and the study area (yellow rectangle) shown on a satellite image provided by Terra Metrics for Google Earth

8.2.3 Satellite Images and Analysis

We used satellite images of the Digital Globe provided by Google Earth from 2010 to 2016 via the Internet. We downloaded satellite images of the head of Shizugawa Bay as JPEG images and processed them with Adobe Photoshop CS2 to extract seaweed beds under the sea. We used a function in Photoshop CS2 that can select neighboring pixels that have values of RGB within a range of values arbitrary decided on a chosen pixel corresponding to seaweeds. The range of values was also tuned to detect seaweed beds. We selected an area located in the west of the bay, because *Eisenia bicyclis* (Kjellman) Setchell that abalones prefer, grows on the rocky coast in this area (Fig. 8.2).

8.3 Results

Seaweeds were distributed along the rocky coast in Shizugawa Bay. The seaweed beds were classified into three types: *Saccharina japonica* (Areschoug) C. E. Lane, C. Mayes, Druehl & G. W. Saunders; *Sargassum* species; and *Eisenia bicyclis*

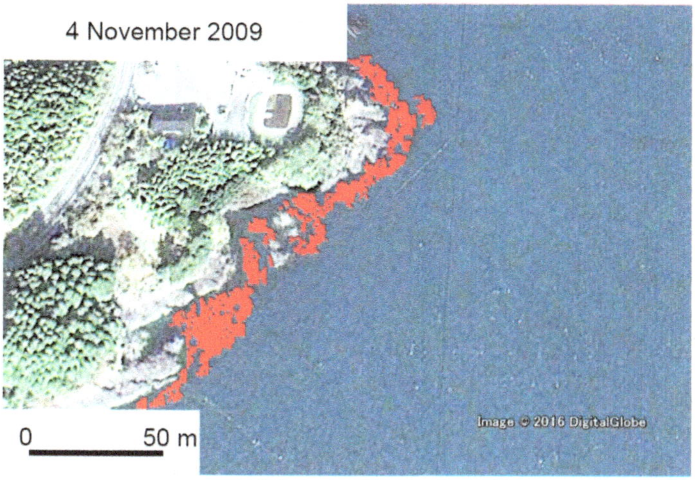

Fig. 8.3 Seaweed beds (red) extracted from and overlaid on a satellite image obtained on November 4, 2009 provided by NASA and Digital Globe for Google Earth

(Kjellman) Setchell beds. Southern limit of *S. japonica* which is a boreal species and northern limit of *E. bicyclis* which is a temperate species are located in the southern Sanriku Coast. In 2009, *E. bicyclis* was distributed on the rocky bottom at bottom depths from 1 to 7 m around the head of Shizugawa Bay (Fig. 8.3).

It survived the tsunami on March 11, 2011. However, the *E. bicyclis* beds nearly disappeared in the bay head area within 1 year in 2015 (Figs. 8.4 and 8.5). Many sea urchins were observed in place of the seaweed. *E. bicyclis* also disappeared from around Tsubakijima and Nojima Islands (Figs. 8.2 and 8.6).

8.4 Discussion

Estes and Duggins (1995) surveyed coastal habitats in many areas of the North Pacific Ocean and showed that kelp forests are usually extensively deforested where sea otters are absent, whereas such a condition is rare where sea otters occur. The deforestation of kelp forests has a profound, lasting impact on species-depauperate systems (Steneck et al. 2002). This phenomenon is explained by a single system: sea otters, sea urchins, and kelp. The view that has emerged is that of a trophic cascade, with sea otter predation regulating herbivore populations and thereby protecting kelp forests from destructive grazing (Estes and Palmisano 1974). The function of the sea otter involves top-down control of the ecosystem. Top-down ecosystem control is very important for maintaining marine biodiversity.

Along the Sanriku Coast, fishermen use a water glass, which consists of a box with a glass bottom, for viewing underwater to find commercial-size sea urchins;

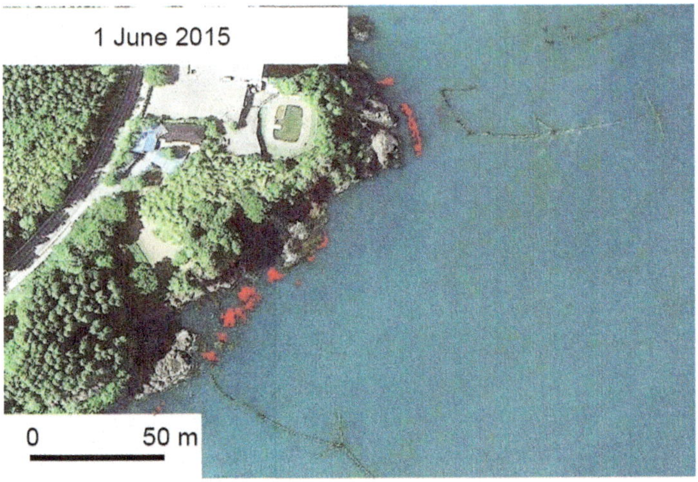

Fig. 8.4 Seaweed beds (red) extracted from and overlaid on a satellite image obtained on June 2, 2015 provided by NASA and Digital Globe for Google Earth

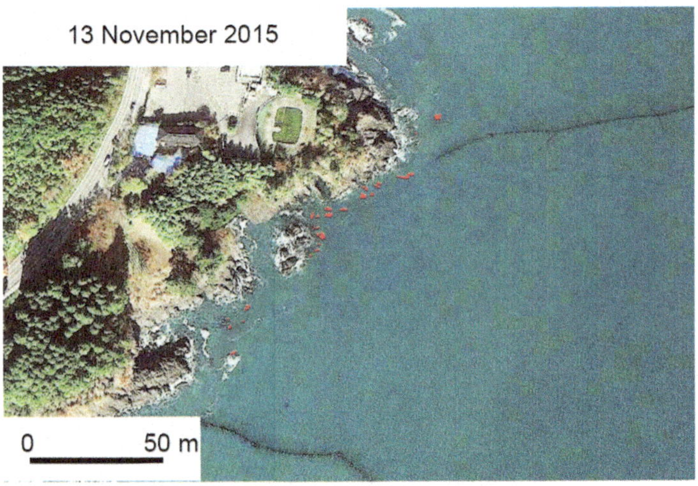

Fig. 8.5 Seaweed beds (red) extracted from and overlaid on a satellite image obtained on November 13, 2015 provided by NASA and Digital Globe for Google Earth

they catch them from a small boat with a small net attached to the end of long pole (Miyagi Prefectural Fisheries Experimental Station 1998). They select large sea urchins that have large gonads in summer, when the sea urchins are mature. Fishermen catching sea urchins fill an important predatory role that controls the coastal ecosystem. The total catch of sea urchins in Miyagi Prefecture, where Shizugawa Bay is situated, was about 700 t y^{-1} before the tsunami and decreased to 100 t in 2011 and about 400 t in 2014 after the tsunami (Fig. 8.7).

Fig. 8.6 Pictures taken in January 2015 showing sea urchins and the devastated seaweed forest around Tsubakijima Island. The barren rocky bed caused by sea urchins (**a**) and *Sargassum* species surrounded by sea urchins (**b**)

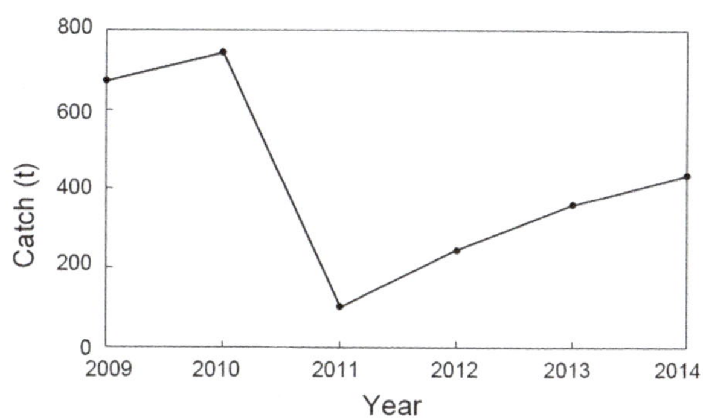

Fig. 8.7 Catch of sea urchins in Miyagi Prefectures, including Shizugawa Bay *Source* Tohoku Regional Agricultural Administration Office (2011, 2012, 2013, 2014, 2015, 2016)

Komatsu et al. (2017) observed that sea urchins survived the tsunami along the rocky coast in Otsuchi Bay (Fig. 8.1), which is a rias-type bay near and similar to Shizugawa Bay. It is thought that the sea urchins that survived the tsunami spawned in the summer of 2011, when there was no strong fishing pressure due to the lack of small boats for the sea urchin fishery, luxuriant growth of seaweeds after the tsunami in 2011 (Komatsu et al. 2017) and decrease in the number of fishermen after the tsunami. Since sea urchins are omnivorous, some of them could feed corpses of victims of tsunami on the sea bottom. Therefore, fishermen suspended sea urchin fishery (Masataka Kurosawa, personal communication 2011). The suspension of sea urchin fishery lasted around three years depending on individual's thought. The resulting sea urchins have been recruited to the adult populations that were born in the summer of 2011 since 2014. In 2010, before the tsunami, there were 60 boats smaller than 3 gross tons in Shizugawa Bay and in 2013 after the tsunami there were seven (Tohoku Regional Agricultural Administration Office 2012, 2015). Adult sea urchins that were spawned in 2011 under the low fishing pressure were massively recruited and started to feed actively on the kelp forests along the Sanriku Coast in 2014 (Aoki et al. 2013) and changed the kelp forests into barren rock beds. The absence of a sea urchin fishery triggered a trophic cascade in the seaweed beds along the Sanriku Coast. When seaweed is not available, sea urchins do not develop gonads, and fishermen do not catch sea urchins with small gonads since they have no commercial value. Along the Sanriku Coast, most of the commercial sea urchins are northern sea urchins, *Strongylocentrotus nudus*. This species can live 14–15 years (Agatsuma 1991). Therefore, to enable the kelp forest to recover, it is necessary to harvest sea urchins that have even no commercial value. The Minami Sanriku Town and Fishermen's Cooperative have started to collect sea urchins from the barren rocky coast in Shizugawa Bay. Divers have routinely removed sea urchins southwest of Nojima Island (see Fig. 8.2).

Yanagi (2005) proposed the term *satoumi* for the human management of coastal seas to increase productivity while maintaining high biodiversity. Sea urchin fishery is considered as a *satoumi* activity serving to increase marine productivity and biodiversity (Komatsu and Yanagi 2015). In Japan, this fishery plays the important role that sea otters play along the west coast of the USA to maintain kelp forests that foster marine productivity and biodiversity. We need to remove sea urchins that have no commercial value to enable the recovery of kelp forests. After the recovery of kelp forests, sea urchin fishery can maintain kelp forests sustainably and lead a coastal ecosystem consisting of kelp forests to increase biodiversity and productivity.

Acknowledgements This study was partly supported by the Mitsui and Co., Ltd. Environment Fund (FY2011 Restoration Grants; R11-F1-060), by Tohoku Marine Science of Ministry of Culture, Sports and Education and also by S(9) and S(13) of Strategic Research Funds of the Ministry of Environment of Japan. The authors express sincere thanks to the late Mr. Norio Sasaki, former President of Shizugawa Bay Management Council of Miyagi Fisheries Cooperative, and Mr. Toshimitsu Sato and Mr. Fujio Abe, Head of Shizugawa and Tokura Branches of Miyagi Fisheries Cooperative for their help to conduct field survey.

References

Agatsuma Y (1991) *Strongylocentrotus nudus*. In: Nagasawa K, Torisawa MA (eds) Pictorial book of fisheries biology on fish in northern Japan, Marine Center of Northern Japan, Sapporo, pp 330–333 (in Japanese)

Aoki M, Endo H, Horikoshi A, Agatsuma Y (2013) Dynamics of the rocky subtidal ecosystem after the Tohoku Pacific earthquake. J Integr Field Sci 10:32

Costanza R, d'Arge R, De Groot R, Faber S, Grasso M, Hannon B, Limburg K, Naeem S, O'Neill RV, Paruelo J, Raskin RG, Sutton P, Belt van den M (1997) The value of the world's ecosystem services and natural capital. Nature 387:253–260

Estes JA, Duggins DO (1995) Sea otters and kelp forests in Alaska: generality and variation in a community ecological paradigm. Ecol Monogr 65:75–100

Estes JA, Palmisano JF (1974) Sea otters: their role in structuring nearshore communities. Science 185(4156):1058–1060

Komatsu T (1985) Temporal fluctuations of water temperature in a *Sargassum* forest. J. Oceanogr Soc Jpn 41:235–243

Komatsu T (1989) Day-night reversion in the horizontal distributions of dissolved oxygen content and pH in a *Sargassum* forest. J Oceanogr Soc Jpn 45:106–115

Komatsu T (1996) Influence of a *Zostera* bed on the spatial distribution of water flow over a broad geographical area. In: Kuo J, Phillips RC, Walker DI, Kirkman H (eds) Proceedings of an International Workshop on Seagrass Biology, The University of Western Australia, Nedlands, Western Australia, pp 111–116

Komatsu T, Kawai H (1986) Diurnal changes of pH distributions and the cascading of shore water in a *Sargassum* forest. J Oceanogr Soc Jpn 42:447–458

Komatsu T, Murakami S (1994) Influence of a *Sargassum* forest on the spatial distribution of water flow. Fish Oceanogr 3:256–266

Komatsu T, Yanagi T (2015) An integrated approach for sustainable use of coastal waters, lessons from human-nature interactions during the Edo period of eighteenth-century Japan. In: Ceccaldi HJ, Hénocque Y, Koike Y, Komatsu T, Stora G, Tusseau-Vuillemin M-H (eds) Marine productivity: perturbations and resilience of socio-ecosystems. Springer International Publishing Switzerland, Cham, pp 283–290

Komatsu T, Ariyama H, Nakahara H, Sakamoto W (1982) Spatial and temporal distributions of water temperature in a *Sargassum* forest. J Oceanogr Soc Jpn 38:63–72

Komatsu T, Umezawa U, Nakaoka M, Spanwanid C, Kanamoto Z (2004) Water flow and sediment in *Enhalus acoroides* and other seagrass beds in the Andaman Sea, off Khao Bae Na, Thailand. Coast Mar Sci 29:63–38

Komatsu T, Ohtaki T, Sakamoto S, Sawayama S, Hamana Y, Shibata M, Shibata K, Sasa S (2015) Tsunami on seagrass and seaweed beds in Otsuchi Bay, Sanriku Coast, Japan. In: Ceccaldi HJ, Hénocque Y, Koike Y, Komatsu T, Stora G, Tusseau-Vuillemin M-H (eds) Marine productivity: perturbations and resilience of socio-ecosystems. Springer International Publishing Switzerland, Cham, pp 43–53

Komatsu T, Murakami S, Kawai H (1994) Some features of jump of water temperature in a *Sargassum* forest. J Oceanogr 52:109–124

Komatsu T, Sagawa T, Sawayama S, Tanoue H, Mohri A, Sakanishi Y (2012) Mapping is a key for sustainable development of coastal waters: examples of seagrass beds and aquaculture facilities in Japan with use of ALOS images. In: Ghenai C (ed) Sustainable development. Tech Publishing Co., Rijeka, Croatia, pp 145–160. ISBN 979-953-307-649-1

Komatsu T, Ohtaki T, Sasa S, Sawayama S, Sakamoto XS, Gonzalvo Maro S, Asada M, Hamana M, Murata H, Tanaka K, Tsujimoto R, Terauchi G (2017) Restoration process of blue infrastructures supporting coastal fisheries along the Sanriku Coast after the huge tsunami. Bull Coast Oceangr 54:117–127 (in Japanese with English abstract and figure captions)

Ministry of Environment of Japan (2010) Guide book of environments in enclosed sea areas (88 areas). Ministry of Environment of Japan, Tokyo, pp 407. (in Japanese)

Miyagi Prefectural Fisheries Experimental Station (1998). Traditional fishing gears and methods II. Central area. Miyagi Prefectural Fisheries Experimental Station, Ishiomaki, 83 pp. (in Japanese)

Sakamoto S X, Sasa S, Sawayama S, Tsujimoto R, Terauchi G, Yagi H, Komatsu T (2012) Impact of huge tsunami in March 2011 on seaweed bed distributions in Shizugawa Bay, Sanriku Coast, revealed by remote sensing. In: Proceedings of SPIE, 8525, 85251B-7. https://doi.org/10.1117/12.999308

Sasa S, Sawayama S, Sakamoto SX, Tsujimoto R, Terauchi G, Yagi H, Komatsu T (2012) Did huge tsunami on 11 March 2011 impact seagrass bed distributions in Shizugawa Bay, Sanriku Coast, Japan? In: Proceedings of SPIE, 8525, 85250X-6. https://doi.org/10.1117/12.999307

Steneck RS, Graham MH, Bourque BJ, Corbett D, Erlandson JM, Estes JA, Tegner MJ (2002) Kelp forest ecosystems: biodiversity, stability, resilience and future. Environ Conserv 29 (04):436–459

Tohoku Regional Agricultural Administration Office (2011). 57th annual statistics report of agriculture, forestry and fisheries of Miyagi Prefecture from 2009 to 2010, Ministry of Agriculture, Forestry and Fisheries of Japan, Tokyo, 207 pp. (in Japanese)

Tohoku Regional Agricultural Administration Office (2012). 58th annual statistics report of agriculture, forestry and fisheries of Miyagi Prefecture from 2010 to 2011, Ministry of Agriculture, Forestry and Fisheries of Japan, Tokyo, 212 pp. (in Japanese)

Tohoku Regional Agricultural Administration Office (2013). 59th annual statistics report of agriculture, forestry and fisheries of Miyagi Prefecture from 2011 to 2012, Ministry of Agriculture, Forestry and Fisheries of Japan, Tokyo, 242 pp. (in Japanese)

Tohoku Regional Agricultural Administration Office (2014). 60th annual statistics report of agriculture, forestry and fisheries of Miyagi Prefecture from 2012 to 2013, Ministry of Agriculture, Forestry and Fisheries of Japan, Tokyo, 240 pp. (in Japanese)

Tohoku Regional Agricultural Administration Office (2015). 61st annual statistics report of agriculture, forestry and fisheries of Miyagi Prefecture from 2013 to 2014, Ministry of Agriculture, Forestry and Fisheries of Japan, Tokyo, 236 pp. (in Japanese)

Tohoku Regional Agricultural Administration Office (2016). 62nd annual statistics report of agriculture, forestry and fisheries of Miyagi Prefecture from 2013 to 2014, Ministry of Agriculture, Forestry and Fisheries of Japan, Tokyo, 434 pp. (in Japanese)

The 2011 Tohoku earthquake tsunami joint survey group (2012). http://www.coastal.jp/tsunami2011/. Accessed 10 Oct 2017

Yanagi T (2005) "Sato-Umi"; New concept for the coastal sea management. reports of research institute for applied mechanics, Kyushu University, 129, 109–111

Chapter 9
The English Channel: Becoming like the Seas Around Japan

Jean-Claude Dauvin, Jean-Philippe Pezy and Alexandrine Baffreau

Abstract The European seas are known to be the recipient of several hundreds of Non-Native Species (NNS). Two main origins have been identified: from shipping, including discharge of ballast waters and biofouling, and through voluntary introduction in aquaculture. There are more than one hundred NNS recorded in the English Channel, which remains low in comparison with the number of marine invertebrate species known in the EC (>3000). The main sites of introduction are the harbours, especially Le Havre. Among these NNS, 54 species come from the seas off Japan, and are now present in 48 established populations. In this study, we provide comments on the taxonomic groups, introduction pathways, distributions and population dynamics of listed NNS in the EC. Three specific examples of invasive species are described in detail, the oyster *Magallena gigas* and the two crabs *Hemigrapsus sanguineus* and *H. takanoi*. These species have changed the dynamics and functioning of the foreshore and coastal ecosystems, where they make up abundant populations. In fact, the oyster is now a key species for the French economy, while the rapid expansion of both crabs could be a problem for the development of oyster and mussel aquaculture since they are predators of young bivalves.

Keywords Non-native species · English channel · Japanese origin · Hemigrapsus spp · Magallena gigas

9.1 Introduction

The English Channel (EC) is a temperate sea located at the entrance of the North Sea on the wide North-Eastern European continental shelf, bordered by the coasts of England (UK) to the north and France to the south. Due to its strategic location,

J.-C. Dauvin (✉) · J.-P. Pezy · A. Baffreau
Laboratoire Morphodynamique Continentale et Côtière, Normandie Univ—UNICAEN, UMR CNRS 6143, 24, Rue Des Tilleuls, 14000 Caen, France
e-mail: jean-claude.dauvin@unicaen.fr

T. Komatsu et al. (eds.), *Oceanography Challenges to Future Earth*,
https://doi.org/10.1007/978-3-030-00138-4_9

there is an intensive maritime traffic of ships coming from all over the world, mainly from the United States and Asia (Buléon and Shurmer-Smith 2007; Buléon 2013). Moreover, there are several cross-Channel ferry routes linking the UK to the European continental countries. Moreover, the EC is one of the seas most affected by cumulative anthropogenic impacts arising from a large number of human activities (Halpern et al. 2008; Dauvin 2012).

The three main maritime harbours along the French coast of the EC are Le Havre at the mouth of the Seine, Rouen located 120 km upstream from the Seine estuary and Dunkirk on the Opal Coast near the Dover Strait. These harbours account for the main shipping movements at the scale of the EC (800 movements per day including 250 with hazardous goods) (Buléon and Shurmer-Smith 2007; Buléon 2013). Other smaller harbours along the French and English coasts also contribute, but to a lesser degree, to the international impact of shipping movements, as well as proportionally to the introduction of ballast waters containing NNS (mainly micro-algae and invertebrate larvae) in the EC (David and Gollasch 2008; Nunes et al. 2014; Stiger-Pouvreau and Thouzeau 2015). Biological invasion remains a fundamental question in marine ecology (Carlton 1996; Molnar et al. 2008), and NNS in some areas of the World Ocean such as the Mediterranean Sea represent a high percentage of the marine diversity (see review of Coll et al. 2010), and have a significant impact on marine ecosystem "services" (Sousa et al. 2009; Katsanevalis et al. 2013, 2014).

A non-native species (NNS) is defined as a species appearing in a region outside its original geographical distribution, and always linked with a human activity. When a NNS proliferates and disturbs the functioning of an ecosystem, the term 'invasive' is also used. Such species are characterized by a wide distribution and an opportunistic behaviour, along with a high reproduction and growth rate. Williamson and Fitter (1996) point out that out of 1000 species moving into a new ecosystem, 100 survive transport and become non-native species. Among these latter, ten will colonize the environment and also become naturalized. Finally, one out of these ten species will proliferate and become invasive.

Several studies have listed the NNS observed along the north-eastern Atlantic coasts including the EC, while also discussing their introduction pathways and impact on the native species and marine ecosystems especially in the shallow waters and extensive intertidal zone of the EC (Goulletquer et al. 2002; Daisie 2009; Blanchard et al. 2010; Dewarumez et al. 2011; Minchin et al. 2013; Stiger-Pouvreau and Thouzeau 2015). Breton (2014) working at the Le Havre Natural History Museum, have reported many observations in the Le Havre Basins, which represent the mains sites of introduction of NNS into the EC, and have presented a review of NNS recorded in the harbours of Le Havre and Antifer in the eastern part of the Bay of Seine.

In this paper, we report a list of NNS originating from the north-western Pacific, including the waters around Japan (Table 9.1). The list was compiled using the data from previous major studies (Goulletquer et al. 2002; Daisie 2009; Blanchard et al. 2010; Minchin et al. 2013; Breton 2014). Apart from giving a list of NNS originating from waters around the Japanese archipelago, the aim of the present study is focused on the impact on three intertidal invasive species along the coastlines of the EC, i.e. the oyster *Magallana gigas* (Thunberg 1793) and the crabs *Hemigrapsus sanguineus* (De Hann 1835) and *H. takanoi* Asakura and Watanabe (2005) (for numerous observations, see Dauvin 2009a, b; Dauvin and Delhay 2010; Dauvin and Duffosé 2011; Dauvin et al. 2013; Gothland et al. 2013, 2014; Pezy and Dauvin 2015). In a last part of this paper, we explore some possibilitiesfor developing future projects in the shallow waters and intertidal zone of the EC, which involve (1) monitoring the expansion of NNS and (2) evaluating the impact of these NNS on the functioning of ecosystems.

9.2 Results and Discussion

9.2.1 Comments on NNS Originating from the North-Western Pacific and Recorded in the EC

A total of 54 NNS coming from Japanese waters is recorded along the UK and French coasts of the EC (Table 9.1). Among these species, we count 22 algae (7 micro-algae and 15 macro-algae) and 42 invertebrates. Among the invertebrate fauna, we recognize one Protozoan, two Hydrozoans, two Anthozoans, one Nematode, two Platyhelminthes, two Polychaetes, eight Crustaceans, one Insect, five Bryozoans, five Molluscs and four Ascidians. Most of the species (32 species, i.e. 60%) are sessile and fixed to a substratum, while the remainder are free living. Among the 54 species, 48 (89%) are established; they can provoke phytoplankton blooms, enter in competition with native species and have an impact on habitat functioning and the regional economy, such as in the case of the macro-algae *Sargassum muticum* and the bivalves *Magallena gigas* and *Ruditapes philippinarum*. The impact of NNS is known to be important on the marine environment, mainly in the intertidal zone and in shallow water depths most often <10 m (Brusati and Grosholz 2006; Daisie 2009; Sousa et al. 2009; Grabowski et al. 2014).

The main pathways are via ballast waters (32 species) and from aquaculture (21 species), mainly through cultivation of the oyster *Magallena gigas*. Thus, due to the deliberate introduction of *M. gigas*, many undesirable NNS have arrived accidentally in the EC, mainly comprising sessile species such as the four Ascidians and *Styela clava,* which is a serious nuisance causing biofouling of cables and boat hulls. Recently, Vincent (2015) warned about the potential hazard of the importation of wild Asian worms for fish bait, which can be infected by microfauna and parasites, causing an infestation of autochthonous species if some individuals are released into the natural environment.

Today, among the 54 species, 6 (11%) have colonized the entire EC, 36 (67%) have been recorded along the north-east coast of France between the Cotentin and the Dover Straits, 25 (46%) along the north-western coast of France between the

Table 9.1 List of alien species recorded in the brackish and marine waters of the English Channel (EC) with a north-western Pacific origin (Japanese and surrounding waters). NE: north-east sector of the EC, along the English coast from the Isle of Wight towards the Dover Strait; NW: north-west sector of the EC, along the English coast from the Isle of Wight towards the Isles of Scilly; SW: south-west sector of the EC, west of the Cotentin and along the French coast towards Ushant; SE: South to South-east of the Cotentin and along the French coast towards the Dover Strait. Vector/Pathway: AQ Aquaculture, V: Vessels; BB: Bait Buckets; D = Natural Drift; ET = Eel Transport (from Minchin et al. 2013). For the western basin of the English Channel, Brittany coast, see 1. Blanchard et al. (2010) and references therein; for the eastern basin of the English Channel see 2. Dewarumez et al. (2011) and references therein, and for the English coast see 3. Minchin et al. (2013) and references therein

Group	Species	Region	Vector/ Pathway	Current status	Impact	Ref.
Microalgae	*Alexandrium affine* (Inoue et Fukuyo 1985) Balech 1985	SW	Unknown	Not established	Unknown	1
	Alexandrium leei Balech 1985	SW	V	Established	Unknown	1
	Coscinodiscus wallessi Gran and Angst 1931	SE, NW	V, AQ	Established	Bloom	2
	Fibrocapsa japonica Toriumi and Takano 1973	SW, SE	Unknown	Established	Unknown	1
	Karenia brevisulcata Chang (1999), Hansen and Moestrup 2000	SW	Unknown	Not established	Unknown	1
	Odontella sinensis (Greville) Grunow 1884	SE	V	Established	Unknown	2
	Thalassiosira punctifera (Castracane) Hasl 1983	SE	AQ	Unknown	Bloom	2
Macroalgae	*Antithamnion densum* (Suhr) Howe 1914	SE, SW	Unknown	Unknown	Unknown	1; 2
	Antithamnionella spirographidis (Schiffner) Wollaston 1968	SE, SW, NW	V	Established	Unknown	2; 3
	Bonnemaisonia hamifera Hariot 1891	NW	V	Established	Competition	1; 3
	Caulacanthus ustulatus (Mertens ex Turner) Kützing 1843	NW	Unknown	Established	Unknown	3
	Codium fragile ssp *fragile* (Suringar) Harriot 1889	SE	Unknown	Established	Competition	1; 2; 3

(continued)

Table 9.1 (continued)

Group	Species	Region	Vector/ Pathway	Current status	Impact	Ref.
	Colpomenia peregrina Sauvageau 1927	SE, SW, NW, NE	D, AQ		Competition	1; 2; 3
	Griffithsia corallinoides (Linnaeus) Trevisan 1845	SE	Unknown	Not established	Unknown	2
	Heterosiphonia japonica Yendo 1920	NW, SW	Unknown	Established	Competition	1; 3
	Grateloupia turuturu Yamada 1941	NW, SW	AQ	Established	Unknown	3
	Lomentaria hakodatensis Yendo 1920	SW	AQ	Established	Unknown	1
	Polysiphonia harveyi (J.W. Bailey) Kim et al. 2001	SE, NW	Unknown	Established	Unknown	2; 3
	Polysiphonia senticulosa Harvey 1862	SE	AQ, V	Not established	Unknown	2
	Pylaiella littoralis (Linnaeus) Kjellman 1872	SE	AQ	Established	Competition	2
	Sargassum muticum (Yendo) Fensholt 1955	SE, SW, NW, NE	AQ, D	Established	Habitat modification, affecting navigation, competition	1; 2; 3
	Undaria pinnatifida (Harvey) Suringar 1873	SE, SW, NW, NE	V	Established	Fouling, competition	1; 2; 3
Protozoa	*Haplosporidium nelsoni* Haskin, Stauber and Mackin 1966	SW	AQ	Established	Unknown	1
Hydrozoa	*Gonionemus vertens* Agassiz 1862	SE	AQ, V	Established	Unknown	1; 2; 3

(continued)

Table 9.1 (continued)

Group	Species	Region	Vector/ Pathway	Current status	Impact	Ref.
Anthozoa	*Diadumene cincta* Stephenson 1925	SE	V	Established	Unknown	1: 2
	Diadumene lineata (Verrill 1869)	SE, NW	AQ, V	Established	Unknown	1: 2; 3
Nematoda	*Anguillicoloides crassus* (Kuwahara, Niimi and Itagaki, 1974)	SE, NW	ET, V	Established	Endoparasite of *Anguilla anguilla*	2; 3
Plathelmintha	*Pseudodactylogyrus anguillae* (Yin and Sproston 1948)	SW	ET	Established	Endoparasite of *Anguilla anguilla*	1
	Pseudostylochus ostreophagus (Hyman 1955)	SW	AQ	Established	Unknown	1
Polychaeta	*Clymenella torquata* (Leidy 1855)	SE	AQ	Established	Unknown	2; 3
	Hydroides ezoensis Okuda 1934	SE, SW	V	Established	Fouling	2
Cirripedia	*Amphibalanus reticulatus* (Utinomi 1967)	SE	V	Established	Unknown	2
Copepoda	*Acartia (Acartiura) omorii* Bradford 1976	SE	Unknown	Established	Unknown	2
Amphipoda	*Caprella mutica* Schurin 1935	SE, NW, NE	V	Established	Fouling	2
	Grandidierella japonica Stephensen 1938	NW	V	Established	Unknown	3
Decapoda	*Eriocheir sinensis* Milne Edwards 1853	SE, NW, NE	V	Established	Bank erosion	2
	Hemigrapsus sanguineus (De Hann 1835)	SE	V	Established	Competition	2
	Hemigrapsus takanoi Asakura and Watanabe 2005	SE	V	Established	Competition	2
	Palaemon macrodactylus Rathbun 1902	SE	V	Established	Unknown	2
Insecta	*Telmatogeton japonicus* Tokunaga 1933	SE	V	Established	Unknown	2

(continued)

Table 9.1 (continued)

Group	Species	Region	Vector/ Pathway	Current status	Impact	Ref.
Bryozoa	*Bugula stolonifera* Ryland 1960	SW, NW	V	Established	Unknown	1
	Pacificincola perforata (Okada and Mawatari 1937)	SE	Unknown	Not established	Unknown	2
	Schizoporella unicornis (Johnston in Wood 1844)	SE	V, AQ	Established	Unknown	2
	Tricellaria inopinata d'Hondt and Occhipinti Ambrogi 1985	SW, NW	V	Established	Fouling	1; 3
	Watersipora subtorquata (d'Orbigny 1852)	SE, NW, NE	V	Established	Competition	1; 2; 3
Bivalvia	*Corbicula fluminea* (O. F. Müller 1774)	SE	BB, V, AQ	Established	Competition	2
	Magallana gigas (Thunberg 1793)	SE, NW, NE, SW	AQ	Established	Fouling, Trophic competition, Aquaculture	1; 2; 3
	Mizuhopecten yessoensis (Jay 1856)	SW	AQ	Established	Unknown	1
	Ruditapes philippinarum (Adams and Reeve, 1850)	SE, SW, NW, NE	AQ	Established	In cultivation, Competition	1; 2; 3
Gastropoda	*Ocenebra inornata* (Récluz 1851)	SW	AQ	Established	Competition	1
Ascidea	*Botrylloides violaceus* Oka 1927	SE, NW	V	Established	Fouling	1; 2; 3
	Didemnum vexillum Kott 2002	NW	V, AQ	Established	Fouling	3
	Perophora japonica Oka 1927	SW, NW	V	Established	Unknown	1: 3
	Styela clava Herdman 1881	SE, SW, NW, NE	V, AQ	Established	Fouling	1; 2; 3

Cotentin and Ushant, 23 (42%) have been found along the southern coast of England from the Isle of Wight to the Isles of Scilly and only 9 (17%) between the Isle of Wight and the Dover Strait. A total of 19 (35%) species are recorded only in the eastern basin of the EC, while 17 (31%) species are known only in the western basin of the EC. It is clear that the Le Havre and Antifer harbours play an essential role in the introduction of NNS and their dispersal within the eastern basin of the EC, especially along the French coast.

9.2.2 *The Oyster* Magallena gigas *Reefs*

The oyster *Magallena gigas*, a native species of south-east Asia, was introduced into France in the late 1960s, after the disappearance of the Portuguese oyster *C. angulata* (Lamarck 1819) due to two viral diseases in the oyster farms of Arcachon and Marennes-Oléron on the Atlantic seaboard (Stiger-Pouvreau and Thouzeau 2015). The first oyster recruitments in the intertidal zone were observed in 1975 at Marennes-Oleron; from the mid-1990s, settlement was observed in North Brittany and extended from the Normano-Breton Gulf to the Spanish frontier (Stiger-Pouvreau and Thouzeau 2015). Since the end of the 1970s, oyster farms have been established in Normandy, and then along the Opal coast (Dover Strait).

As a result, Normandy is now the most important centre for oyster farming in France with a production of >20,000 t/y. Due to the increase in sea-water temperature since the 1980s (Genner et al. 2004[1]), the spring and summer temperatures are sufficient to ensure reproduction and wild colonisation of the intertidal zone by oysters in the North of Brittany, as well as in Normandy (Granville, Blainville-sur-Mer, Saint-Vaast-La-Houge, Le Havre harbour) and along the Opal coast (Boulogne sur-Mer and Dunkirk harbours, and the Aa estuary) (personal observations). The species *M. gigas* has also colonized most of the English coast, with a more intense colonization towards the west from the Isle of Wight to the Plymouth area.[2] At present, wild oysters are only absent along the French coast of the eastern Basin of the EC from Le Havre to Boulogne-sur-mer (Fig. 9.1).

Nevertheless, apart from Brittany, oyster reefs are observed only on rocky shoresoutside the harbour at Granville and in the Aa estuary, not just on boulders but also on soft-bottom sediment (personal observations).

The species also forms wild populations outside the EC in the North Sea along the Belgian, Dutch and German coasts and, more recently; in Scandinavia estuaries in relation to the exceptionally warm summers and mild winters over the last decade (see Troost 2010). In these northern areas, *M. gigas* forms extensive and dense reef structures which induce changes in plankton composition, habitat heterogeneity and biodiversity, as well as in carrying capacity of the ecosystem, foods webs and parasite life cycles (Troost 2010). Similarly, in Brittany, oyster reefs in the intertidal zone of the Bay of Brest are associated with major changes in the structure of the biodiversity and abundance of the intertidal macrofauna, with the trophic dominance of the macrofauna in muddy environments changing from suspension feeders to

---

[1]http://www.societe.org.gg/planetguernsey/download/5_Impacts.pdf.

[2]http://www.marlin.ac.uk/species/detail/1676, www.nonnativespecies.org.

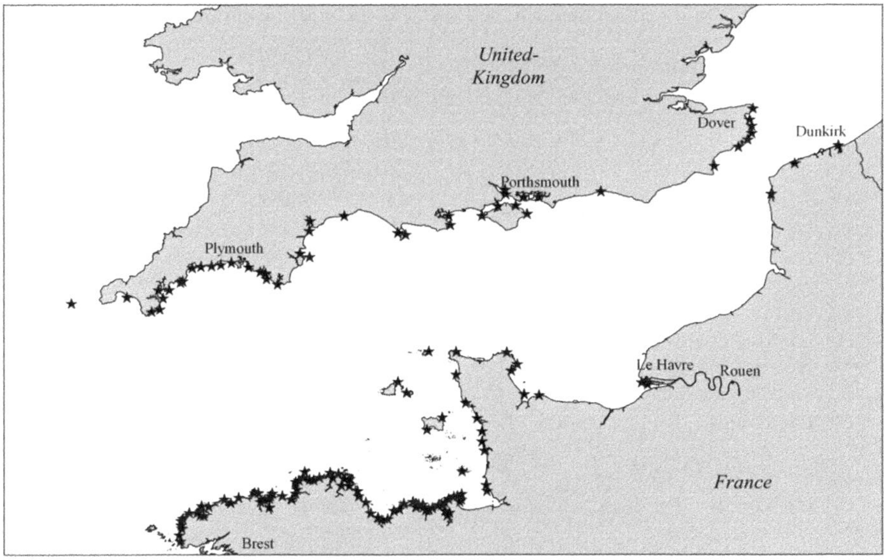

Fig. 9.1 Present-day colonization of the English Channel by *Magallena gigas* oyster. From Stiger-Pouvreau and Thouzeau (2015), http://www.marlin.ac.uk/species/detail/1676, www.iobis.org, www.nonnativespecies.org, and personal observations

carnivores (Lejart and Hily 2011). On rocky shores, *M. gigas* reefs are suspected to cause a homogenisation of the intertidal communities, with an impoverishment of habitat quality corresponding to enrichment in the organic content of the sediments (Lejart and Hily 2011). Owing to habitat changes in the Wadden Sea induced by the formation of *M. gigas* reefs, the Eurasian oystercatcher *Haematopus ostralegus* (Del Hoyo and Collar 2014) and Eurasian curlew *Numenius arquata* (Linnaeus 1758) have learnt to exploit the food supply offered by this new intertidal habitat (Markert et al. 2013). Long-living oyster reefs (>30 years) in the Oostershelde estuary in the Netherlands show a positive reef accretion rate of up to 17 mm per year and thus contribute to coastal protection in the face of sea-level rise (Walles et al. 2015). In this way, the action of *Magallana gigas* as an engineer species leads to many changes of the colonized foreshore and produces positive and negative effects on the EC and European coastal ecosystem (Table 9.2).

9.2.3 *Hemigrapsus Crabs*

Two species of Asian crabs of the genus Hemigrapus colonized the French coasts of the EC in the 1990s (Dauvin 2009a, b; Dauvin et al. 2009). The brush-clawed shore crab H. takanoi Asakura and Watanabe 2005 was first reported in Le Havre harbour in 1997 (Breton et al. 2002), while the Japanese shore crab H. sanguineus (De Haan 1835) was first observed on the Atlantic coast of Europe at Le Havre in 1999 (Breton

Table 9.2 Negative and positive impacts of *Magallena gigas* reefs in the north-eastern European coastal ecosystem

Positive impacts	Negative impacts
Increase inhabitat heterogeneity on soft-bottom habitats: increase in diversity and abundance of the macrofauna; increase of ecosystem carrying capacity, food webs and food supply for birds	Changes of the ecosystem Homogenization of the rocky shore Increase of organic matter content and decrease in quality of the habitat
Oyster exploitation on wild reefs	Competition on oyster culture tables with filter feeders consuming phytoplankton Competition with native bivalves for space and food, including larvivorous organisms
Eco-engineering species: accumulation of dead oyster shells and protection of tidal from erosion; accretion and coastal protection faced with sea-level rise	Effects of trophic changes: suspension feeders versus carnivores Effects on other trophic levels
Genetic resources for the species with wild oyster spat resistant to *Vibrio*-regulated summer mortalities. Reducing the parasite load in final hosts	Increases maintenance activities (damage of nautical gear) Reduction in fishing from the shore and sea bathing
Recreational oyster harvesting	

et al. 2002). H. sanguineus occurred in large numbers on the rocky shore of the open sea, whereas H. takanoi was abundant in sheltered harbours (Dauvin et al. 2009).

In 2008, we initiated an annual survey of the Asian crab colonization in two areas, at about 30 sites around the Cotentin Peninsula and along the Opal Coast. In 2011, the monitoring was extended to the Calvados coastline and, since 2013; the survey also covers the coasts of the Seine-Maritime and Somme departments, to obtain annual information (spring sampling) from the western part of the Cotentin to the Belgian frontier. In 2015, no Asian crabs were recorded along the English coast of the EC, but some specimens had been reported from Jersey and Guernsey in the Channel Islands; similarly, no Asian crabs had been observed along the Brittany coast from Ushant to the Mont-Saint-Michel Bay. At present, both crabs are extending their populations northwards into Danish waters (Dauvin et al. 2009; Gothland et al. 2013, 2014).

Figure 9.2 illustrates the presence (even if only a few specimens have been recorded) and the maximum densities recorded along the French coast of the EC for both species of *Hemigrapus*.

It appears that three main zones have been colonized by *H. sanguineus*: the North Cotentin, the western part of the Calvados as well as the Opal coast (mainly in the harbours of Boulogne-sur-mer, Calais and Dunkirk) and the Wimereux site. *H. takanoi* is present at fewer sites and shows lower abundance (maximum of 60. ind. m^{-2}) than *H. sanguineus*, and is recorded mainly on the eastern part of the Calvados coast, the basins of Le Havre harbour and in three harbours (Boulogne-sur-Mer, Calais and Dunkirk) of the North of France.

In the North of the Cotentin, densities reach 100 ind. m^{-2} at La Hougue, but remain lower than 70 ind. m^{-2} at the two other stations (Fig. 9.3). After a phase of

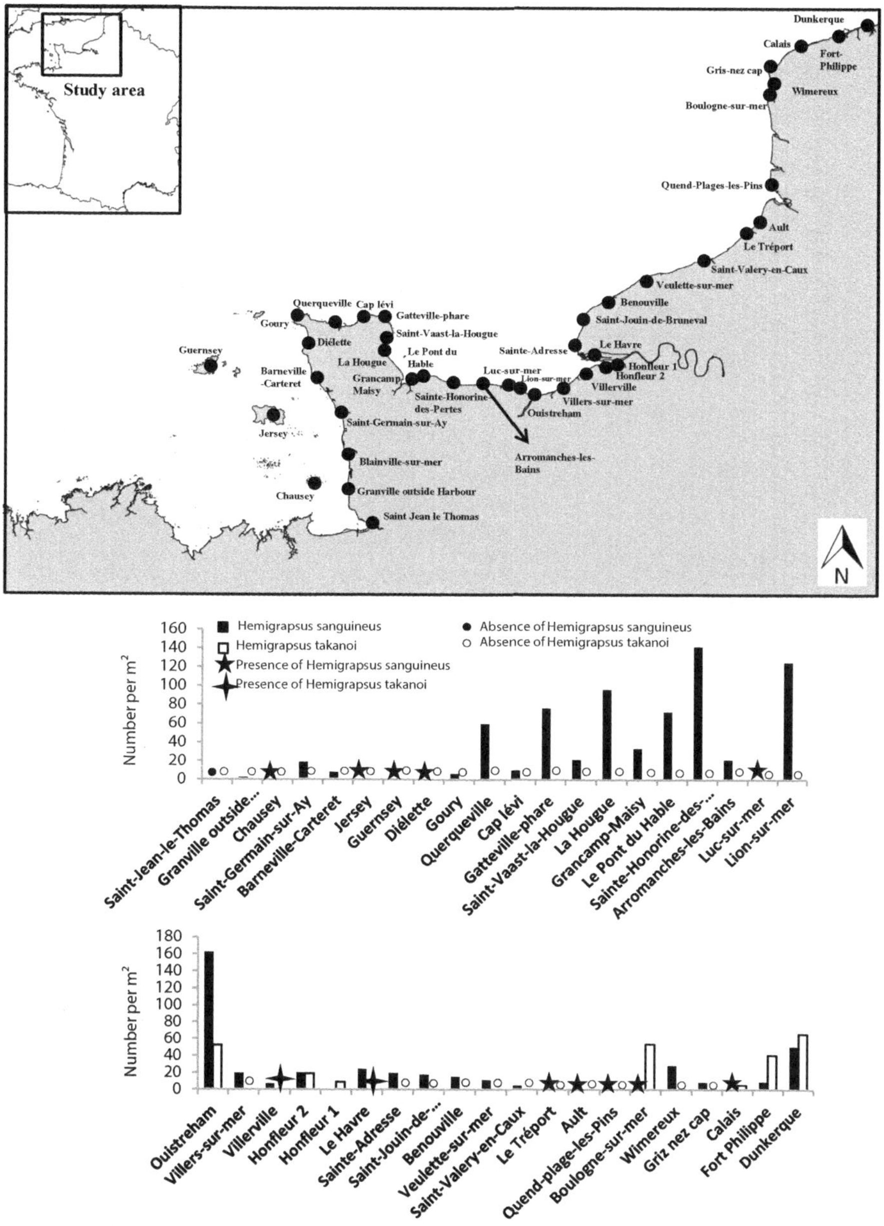

Fig. 9.2 Colonization of the English Channel coast by the Asian crabs *H. sanguineus* and *H. takanoi* (bold points on map indicate sampling stations). Four-pointed and five-pointed stars on diagrams indicate presence of a few specimens of the species, and bars represent maximum densities observed during the 2008–2015 survey

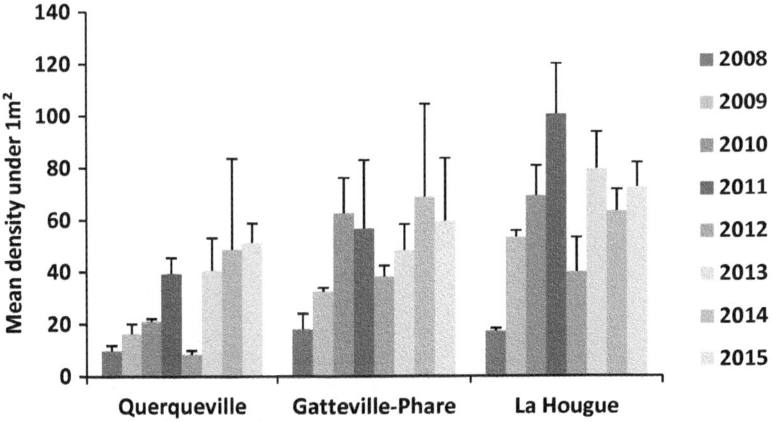

Fig. 9.3 Evolution of the density (number of individuals per m^2) of the crab *Hemigrapsus sanguineus* at three sites on the North Cotentin coast, Normandy, France, from 2008 to 2015

increase in density from 2008 to 2011, the populations at the three stations show a marked decline in 2012 probably due to the cold winter conditions and the sensitivity of the species to temperature <5 °C (Rocroy 2013). The populations then increase to show a stabilization of the densities during the three last years of the survey (Fig. 9.3).

Along the coast of Calvados, in both the Ouistreham and Honfleur harbours, the populations of H. *takanoi* show inconsistent trends with a maximum at the beginning of the survey in 2011 at Honfleur and a maximum in 2015 at Ouistreham (Fig. 9.4). There are no cases where the densities exceed 30 ind. m^{-2}. In spite of the presence of ovigerous females during the period from May to November

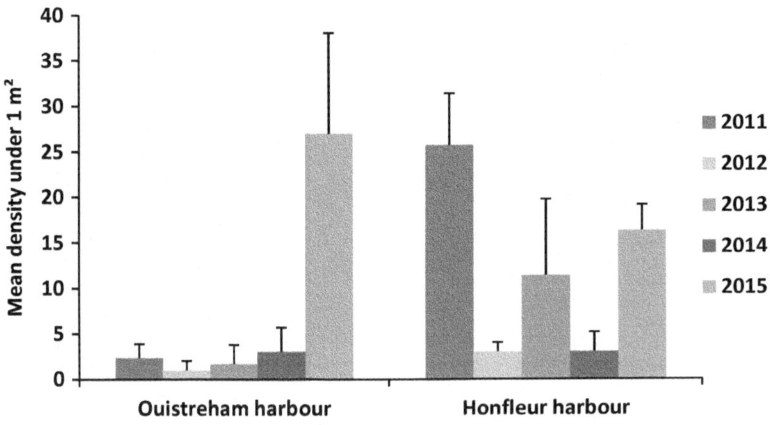

Fig. 9.4 Evolution of the density (number of individuals per m^2) of the crab *Hemigrapsus takanoi* at three sites on the North Cotentin coast, Normandy, France, from 2011 to 2015

(Gothland et al. 2013), *H. sanguineus* individuals with a carapace-width less than 6 mm are very rare in under-boulder samples collected from April to August, and are not sufficiently abundant to explain the renewal of adult populations found under the boulders. Thus, summer observations on mussel beds along the Calvados coast show that this habitat is an important settlement habitat for *H. sanguineus*, and densities higher than 500 ind. m^{-2} have been recorded (Pezy and Dauvin 2015). After their settlement on mussel beds, the larger specimens migrate under foreshore boulders where the adults can then be found.

9.3 Concluding Remarks

The number of NNS present in the EC remains low in comparison with the Mediterranean Sea (Coll et al. 2010) or the North Sea (Gollash et al. 2009). Among the one hundred NNS currently recorded, more than 50% (54) have an origin in the north-western Pacific, including Japanese waters, which is noteworthy in terms of introduction. The biodiversity of the flora and fauna in the EC increases with the arrival of NNS, but there is no correlative disappearance of autochthonous species. Some of the NNS species with an Asiatic origin remain inconspicuous or rare, while others show an invasive tendency, e.g. the macroalgae *Sargassum muticum* and *Undaria pinnatifica*, but only in marinas for this latter species, the amphipod *Caprella mutica* in estuarine ecosystems, *Hemigrapsus* spp., the bivalves *Magallana gigas* and *Ruditapes philippinarum* and the ascidian *Styela clava*.

In foreshore and coastal open ecosystems, it is not possible to eradicate the NNS; some local measures can reduce the negative effect of invasive species such as the destruction of wild *M. gigas* reefs. Nevertheless, it is essential to assess the long-term impact of such introduced species and especially those which proliferate. For the crabs, which show a rapid extension to the coasts of northern Europe as far as the south of Denmark (Dauvin et al. 2009; Gothland et al. 2013), their impact on oyster farming zones and on natural and cultivated mussel beds appears important. Future studies could be oriented towards the knowledge of biological traits of NNS living in the EC, such as their resistance to cold and high temperatures, larval dispersal in a high-energy megatidal sea, the trophic competition between intertidal crabs and the role of *Hemigrapsus* in trophic chains (i.e. are these Asian crabs preyed on by fishes and birds or other invertebrates, such as other crabs found on the foreshore of the EC?). In the future, we need to take note of the presence of *Hemigrapsus* crabs on oyster culture tables, and also in the subtidal zone. Indeed, some authors have recorded these species in shallow waters such as Dunkirk harbour and in enclosed bodies of sea water in the Netherlands.

For the long-term survey (recognition of the presence and abundance of NNS) along the intertidal zone of the EC, the involvement of volunteers from naturalists' associations, as carried out on the east coast of the United States, should be an interesting alternative to research monitoring (see Delaney et al. 2008). This associative and participative approach to scientific research also aims to raise the awareness of citizens about the identification of NNS and their impact on the marine ecosystem. This idea is being promoted by the UK non-native species secretariat[3] as well as Biolit in France (Les observateurs du littoral http://www.biolit.fr/). However, training and scientific supervision would be essential for this type of activity, which would also require validation of the observations and the database.

Acknowledgements The authors thank M. Carpenter for the English language revision, and the "Agence de l'Eau Seine Normandie" for its financial support of the 2016–2017 project "Réalisation d'un guide des espèces marines animales et végétales non indigènes dans les eaux normandes de la baie du Mont-Saint Michel à l'estuaire de l'Authie". They are also grateful to S. Jobert, M. Rocroy, B. de Boisgelin and L. Lefort for their help during *Hemigrapsus* sampling and C. Dancie and T. Ruellet for their participation in the inter-regional Normandy Picardy project COHENOPI. This study received also financial support from the French national programme EC2CO entitled 'Colonisation des côtes par deux espèces du genre *Hemigrapsus*: origine, état des lieux et compétition interspécifique en Manche/Mer du Nord'. Jean-Claude Dauvin acknowledges the Japanese-French Oceanographic Society in Japan and its President, Professor Teruhisa Komatsu (SFJO-J), for his invitation to take part in the 16th Japanese-French Oceanography Symposium, held in Tohoku/Tokyo, Japan, from 17–21 November 2015, on the topic 'The sea under human and natural impacts: Challenge of oceanography to the future earth'.

References

Blanchard M, Goulletquer P, Hamon D, LeMao P, Nezan E, Gentil F, Simon N, Viard F, Ar Gall E, Grall J, Hily C, Le Duff M, Stiger-Pouvreau V, Acou A, Derrien-Courtel S, Feunteun E, D'ondt JL, Canard A, Ysnel F, Perrin B, Cabioc'h J, Gruet Y, Le Roux A (2010) Liste des espèces marines introduites dans les eaux bretonnes et des espèces introduites envahissantes des eaux périphériques. Synthèse élaborée par l'bservatoire du Patrimoine Naturel de Bretagne, mars 2010: www.bretagne-environnement.org/especes-invasives

Breton G, Faasse M, Noël P, Vincent T (2002) A new alien crab in Europe: *Hemigrapsus sanguineus* (Decapoda: Brachyura: Grapsidae). J Crust Biol 22:184–189

Breton G (2014) Espèces introduites ou invasives des ports du Havre, d'Antifer et de Rouen (Normandie, France). Hydroécol Appliq 18:23–65

Buleon P, Shurmer-Smith L (2007) Espace Manche, un monde en Europe. Channel Spaces, a world within Europe. Editions Armand Colin, Paris, France, p 198

Buleon P (2013) Prospective espace Manche. Transports—energies. CAMIS Project Report, Rouen, France, p 20

Brusati E, Grosholz ED (2006) Native and introduced ecosystem engineers produce contrasting effects on estuarine infaunal communities. Biol Inv 8:683–695

Coll M, Piroddi C, Steenbeck J, Kashner K, Lasram FBR, Aguzzi J, Ballesteros E, Bianchi CN, Corbera J, Dailianis T, Danovarp R, Estrada M, Froglia C, Galil BS, Gasol JM, Gertwagen R, Gil J, Guihaumon F, Kesner-Reyes K, Kitsos MS, Koukouras A, Lampadario N, Laxamana E, Lopez De La Cuadra CM, Lotze HK, Martin D, Mouillot D, Oro D, Raicevich S, Rius-Barile J, Saiz-Salinas JI, San Vicente C, Somot S, Templado J, Turon X, Vafidis D, Villanueva R, Voultsiadou E (2010) The biodiversity of the Mediterranean Sea: estimates, patterns, and threats. PLoS ONE 5:1–36

[3]http://www.nonnativespecies.org/home/index.cfm.

Carlton JT (1996) Pattern, process, and prediction in marine invasion ecology. Biol Conserv 78:97–106
Daisie (Delivering Alien Invasive Species Inventory for Europe) (2009): Handbook on Alien Species in Europe. ISBN 978-1-4020-8279-5. www.springer.com/life+sci/ecology/book/978-1-4020-8279-5. Accessed 17 Nov 2008
Dauvin JC (2009a) Records of the Asian shore crab *Hemigrapsus sanguineus* (Crustacea: Brachyura: Grapsoidea) in the Cotentin, Normandy, English Channel. Aquat Inv 4:467–472
Dauvin JC (2009b) Asian Shore Crabs *Hemigrapsus* spp. (Crustacea: Brachyura: Grapsoidea) continue their invasion around the Cotentin Peninsula, Normandy, France: Status of the *Hemigrapsus* population in 2009. Aquat Inv 4:605–611
Dauvin JC, Tous Rius A, Ruellet T (2009) Recent expansion of two invasive crabs species *Hemigrapsus sanguineus* (De Haan 1835) and *H. takanoi* (Asakura and Watanabe 2005) along the Opal Coast. France Aquat Inv 4:451–465
Dauvin JC, Delhay JB (2010) First record of *Hemigrapsus takanoi* Asakura and Watanabe 2005 (Crustacea: Decapoda: Grapsidae) on the western coast of northern Cotentin, Normandy, western English Channel. Mar Biodiv Rec 3: e101. (Published online)
Dauvin JC, Dufosse F (2011) *Hemigrapsus sanguineus* (De Haan 1835) (Crustacea: Brachyura: Grapsoidea) a new invasive species in European waters: the case of the French English Channel coast (2008–2010). Aquat Inv 6:429–438
Dauvin JC (2012) Are the eastern and western basins of the English Channel two separate ecosystems? Mar Poll Bull 64:463–471
Dauvin JC, Dancie C, Jego Y, Lecornu B, Rocroy M, Ruellet T (2013) Etat de la colonisation des décapodes invasifs du genre *Hemigrapsus* sur le littoral normand-picard (COHENOPI). Rapport final juin 2013 au Réseau d'Observation du Littoral Normand Picard, UMR Morphologie Continentale et Côtière, Caen, France, 18 pp
David M, Gollasch S (2008) EU shipping in the dawn of managing the ballast water issue. Mar Poll Bull 56:1966–1972
Delaney D, Sperling C, Adams C, Leung B (2008) Marine invasive species: validation ofcitizen science and implications for national monitoring networks. Biol Inv 10:117–128
Dewarumez JM, Gevaert F, Masse C, Foveau A, Desroy N, Grulois D (2011) Les espèces marines animales et végétales introduites dans le bassin Artois-Picardie. UMR CNRS 8187 LOG et Agence de l'Eau Artois-Picardie, Wimereux, France, 140 pp
Gothland M, Dauvin JC, Denis L, Jobert S, Ovaert J, Pezy JP, Spilmont N (2013) Additional records and current distribution (2012) of *Hemigrapsus sanguineus* (De Haan 1835) along the French Coast of the English Channel. Manag Biol Inv 4:305–315
Gollasch S, Haydar D, Minchin D, Wolff WJ, Reise K (2009) Introduced aquatic species of the North Sea coasts and adjacent brackish waters. Ecol. Stud. 204:507–528
Gothland M, Dauvin JC, Denis L, Duffossé F, Jobert S, Ovaert J, Pezy JP, Tous Rius A, Spilmont N (2014) Habitat preference and population characteristics explain the distribution and colonisation ability of the invasive shore crab *Hemigrapsus takanoi*. Estuar Coast Shelf Sci 142:41–49
Goulletquer P, Bachelet G, Sauriau PG, Noël P (2002) Open Atlantic coast of Europe-a century of introduced species into French waters. In: Leppäkoski E, Gollasch S, Olenin S (eds) Invasive aquatic species of Europe: distribution, impacts and management. Kluwer Academic Publishers, Dordrecht, pp 276–290
Halpern BS, Walbridge S, Selkoe KA, Kappel CV, Micheli F, D'Agrosa C, Bruno JF, Casey KS, Ebert C, Fox HE, Fujita R, Heinemann D, Lenihan HS, Madin EMP, Perry MT, Selig ER, Spalding M, Steneckand R, Watson R (2008) A global map of human impact on marine ecosystems. Science 319:948–952
Katsanevakis S, Gatto F, Zenetos A, Cardoso AC (2013) How many marine aliens in Europe? Manag Biol Inv 1:37–42
Katsanevakis S, Wallentinus I, Zenetos A, Leppäkoski E, Çinar ME, Oztürk B, Grabowski JH, Golani D, Cardoso AC (2014) Impacts of invasive alien marine species on ecosystem services and biodiversity: a Pan-European review. Aqua Inv 9:391–424

Lejart M, Hily C (2011) Differential response of benthic macrofauna to the formation of novel oyster reefs (*Crassostrea gigas*, Thunberg) on soft and rocky substrate in the intertidal of the Bay of Brest, France. J Sea Res 65:84–93

Markert A, Esser W, Franck D, Wehrmann A, Exo KM (2013) Habitat change by the formation of alien *Crassostrea*-reefs in the Wadden Sea and its role as feeding sites for waterbirds. Estuar Coast Shelf Sci 131:41–51

Minchin D, Cook EJ, Clark PF (2013) Alien species in British brackish and marine waters. Aqua Inv 8:3–19

Molnar JL, Gamboa RL, Revenga C, Spalding MD (2008) Assessing the global threat of invasive species to marine biodiversity. Front Ecol Environ 6:485–492

Nunes AL, Katsanevakis S, Zenetos A, Cardoso AC (2014) Gateways to alien invasions in the European seas. Aqua Inv 9:133–144

Pezy JP, Dauvin JC (2015) Are mussel beds a favourable habitat for settlement of *Hemigrapsus sanguineus*? Aqua Inv 10:51–56

Rocroy M (2013) Etat de la colonisation des crustacés décapodes du genre *Hemigrapsus* sur le littoral bas-normand et implication du facteur abiotique dans l'évolution de leur invasion. Rapport de stage de Master 2, Université Pierre et Marie Curie, Paris VI, France, 32 pp

Sousa R, Gutierrez JL, Aldridge DC (2009) Non-Indigenous Invasive bivalves as ecosystem engineers. Biol. Inv. 11:2367–2385

Stiger-Pouvreau V, Thouzeau G (2015) Marine species introduced on the French Channel-Atlantic coasts: a review of main biological invasions and impacts. Open J Ecol 5:227–257. https://doi.org/10.4236/oje.2015.55019

Troost K (2010) Causes and effects of a highly successful marine invasion: case-study of the introduced Pacific oyster *Crassostrea gigas* in continental NW European estuaries. J Sea Res 64:145–165

Vincent T (2015) L'ethnologie du ver de vase sur le littoral de la Manche orientale, en particulier au Havre (Seine Maritime): une histoire sociale mouvementée et une perspective écologique inquiétante. Bull Soc Sci Nat Ouest France, n.s 37: 219–231

Walles B, Mann R, Ysebaert T, Troost K, Herman PMJ (2015) Demography of the ecosystem engineer *Crassostrea gigas* related to vertical reef accretion and reef persistence. Estuar Coast Shelf Sci 154:224–233

Williamson M, Fitter A (1996) The varying success of invaders. Ecology 77:1661–1666

Part III
Physical Oceanography

Chapter 10
Recent Research Results and Future Project in the Antarctic Ocean by Umitaka-Maru Research Group for Physical Oceanography

Yujiro Kitade, Keishi Shimada, Shigeru Aoki and Kay I. Ohshima

Abstract Training research vessel Umitaka-Maru belonging to the Tokyo University of Marine Science and Technology has implemented the research cruises in the Southern Ocean, counting the 19th times in the 2015 cruise. Particularly in recent years, Umitaka-Maru has greatly contributed to various findings in oceanography in the Antarctic Ocean, such as clarification of the bottom water formation and monitoring in Lutzow Holm Bay, cooperating with the Japanese Antarctic Research Expedition. This article reviews such scientific progresses implemented by Umitaka-Maru, and further describes the recent trend of Antarctic Bottom Water from the newly obtained data. Most important findings by Umitaka-Maru are the discovery of the Antarctic bottom water (AABW) formation off the Cape Darnley with a high sea ice production polynya and the subsequent discovery of AABW formation off the Vincennes Bay with a medium sized polynya. The base oceanographic observation for physical and chemical properties has been carried out continuously along 110°E line in Australian Antarctic Basin, providing the monitoring of water mass transformation. Freshening of AABW in Australia–Antarctica Basin has been caught in a series of the observation in recent years, it is suggested that the AABW originating from the Vincennes Bay Polynya gives significant effects on the AABW in the basin. To grasp an effect of freshening of AABW on the global ocean circulation, intense hydrographic observations and deployment of a huge mooring system are now in consideration.

Keywords Antarctic Bottom Water · AABW freshening · Vincennes Bay · Australian Antarctic Basin · Cape Darnley

---

Y. Kitade (✉) · K. Shimada
Tokyo University of Marine Science and Technology, 4-5-7 Konan Minato-ku, Tokyo 108-8477, Japan
e-mail: ykitade@kaiyodai.ac.jp

S. Aoki · K. I. Ohshima
Institute of Low Temperature Science, Hokkaido University, Kita-19, Nishi-8, Kita-ku, Sapporo 060-0819, Japan

T. Komatsu et al. (eds.), *Oceanography Challenges to Future Earth*,
https://doi.org/10.1007/978-3-030-00138-4_10

10.1 Introduction

Training Research Vessel Umitaka-Maru IV belonging to the Tokyo University of Marine Science and Technology was born in 2000. Umitaka-Maru voyage to the Southern Ocean in January 2016 was the 19th time in successive Umitaka-Maru cruise to there. At the first voyage to Antarctic Ocean, Umitaka-maru contributed to transporting supplies as the associated ship for Antarctic ice breaker Soya in 1956. Then, it has been promoting the research project, such as a voyage of biomass (1983) as a research vessel. As an opportunity of the International Polar Year in 2007–2008, Umitaka-Maru IV voyage to Antarctic Ocean has been increased to support many research projects, such as the project for Japanese Antarctica Research Expedition, Cape Darnley project, and now counted as the 9th voyage.

Antarctic Bottom Water (AABW) is the densest water in the global ocean, and its production at the Antarctic margin is very important for the global overturning circulation. Historically there have been three well-known source regions in the Weddell (Gill 1973), Ross Seas (Jacobs et al. 1970) and off Adélie Land (Rintoul 1998) (Fig. 10.1).

In general, AABW is produced over the continental slope and rise by the mixing of cold, Dense Shelf Water (DSW) descending into warm Circumpolar Deep Water (CDW) (e.g., Foster and Carmack 1976). DSW is formed on the continental shelf by regionally varying combinations of brine rejection from sea-ice growth and ocean-ice shelf interactions. The AABW originating from Weddell Sea that is called Weddell Sea Bottom Water (WSBW) flows into Weddell-Enderby Abyssal Plain which faces to Atlantic Ocean and the west of Indian Ocean, whereas the AABW from Ross Sea that is called Ross Sea Bottom Water (RSBW) flows into

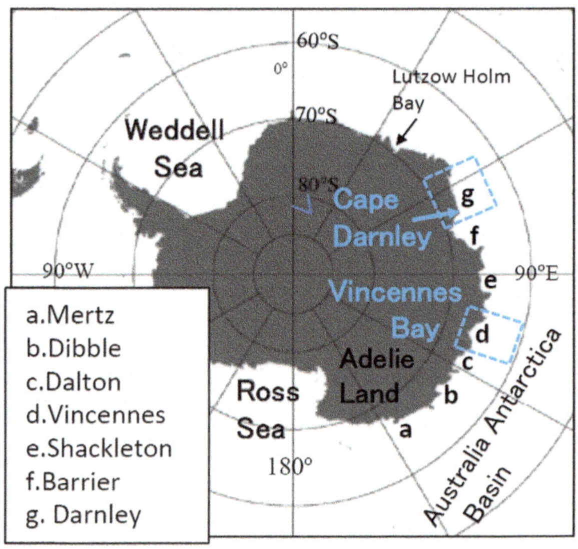

Fig. 10.1 Locations of major polynya in the East Antarctica and location of the study area

Amundsen Abyssal Plain in the south of Pacific Ocean and Australia–Antarctica Basin in the southeast of Indian Ocean (Orsi et al. 1999). On the other hand, the bottom water that is supplied from Mertz Polynya is called Adélie Land Bottom Water (ADLBW) mixed with RSBW which comes from the east and spread to Australia Antarctic Basin (Rintoul 1998).

Flow rate of bottom water formed in the Antarctic margin has been estimated by various researchers (Jacobs 2004), however, reported total volume are largely inconsistent among them. This would mean that the total flow is not known exactly. Uncertainty of the total flow rate is not only a problem for physical oceanography and being bottleneck for accurate the global model and hence, the climate change prediction. Thus, they are currently underway observations from various aspects.

Tamura et al. (2008) estimated sea-ice production in the majority of coastal polynyas around Antarctica and showed that the Ross Sea Polynya (390 $\pm$ 59 km^3/yr), the Cape Darnley Polynya (181 $\pm$ 19 km^3/yr) west of the Amery Ice Shelf, and the Mertz Polynya (120 $\pm$ 11 km^3/yr) off Adélie Land were the three most productive polynyas. While the Ross and Mertz polynyas had already been connected to AABW formation, the Cape Darnley result was new. Thus, Cape Darnley project, including long term mooring observation, was began to clarify the connection between sea-ice production and AABW formation and to identify the another potential formation area of AABW. The success of the observations there leads to the subsequent Vincennes project.

In this article, we introduce new discoveries and global warming and related problems studied by physical oceanography group in the later years of Cape Darnley Project in a point of view of physical oceanography, and finally introduce plan and research issues of the future.

10.2 Discovery of New Formation Regions of Antarctic Bottom Water

Taken out of the sea ice information by making full use of satellite data, Tamura et al. (2008) showed the volume of sea ice production around Antarctica. They revealed that Cape Darnley Polynya (CDP) is the second largest sea-ice production area, which locates 1200 km east of Syowa Station, Antarctic. Ohshima et al. (2013) expected that CDP was potentially the fourth generation area of the AABW, intensive observation was carried out in order to confirm it (Fig. 10.2). Mooring systems were deployed by the Hakuho-Maru in February 2008, and they were recovered by Umitaka-Maru in January 2009. Along with recovering of mooring systems, detailed hydrographic observations were carried out of the CDP by TR/V Umitaka-Maru.

In the vertical section of potential temperature, salinity and dissolved oxygen (DO) obtained off Cape Darnley by Umitaka-Maru (Fig. 10.3), newly found AABW distributes on the continental slope bottom. The AABW is characterized by

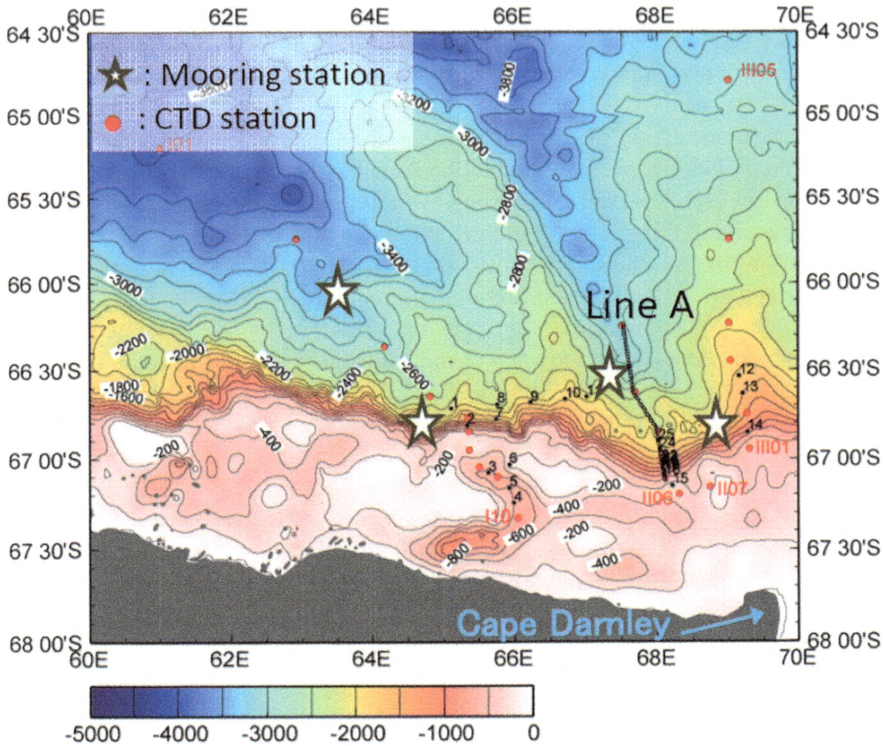

Fig. 10.2 Observation site off Cape Darnley. Stars indicate mooring location and red dots indicate hydrographic observation point

cold, fresh and high DO water in comparison with Modified Circumpolar Deep Water (MCDW) which forms salinity maximum and DO minimum layer around 600 m depth.

From the mooring record, rapid decrease in temperature (0.5 °C) and salinity (0.25), indicating newly formed dens bottom water, has been captured in the beginning of the middle of May and early in June (Fig. 10.4). Ohshima et al. (2013) called the AABW generated in this area as Cape Darnley Bottom Water. They have analysed together with the CTD data by Umitaka-Maru and the bio-logging data by elephant seal, and estimated formation volume in this area to be 0.65–1.5 Sv. This is equivalent from 13 to 30% of the Antarctic bottom water production of the Atlantic sector (Orsi et al. 1999; Meredith et al. 2001; Garabato et al. 2002).

Previously, it had been thought that a wide continental shelf and large depression was necessary for the sufficient storage and salinification of DSW capable of producing AABW. However, Ohshima et al. (2013) showed that despite its relatively narrow shelf region, the enhanced sea ice production in the CDP was forming one of the most saline varieties of DSW around Antarctica. This has opened the door to the possibility of other polynya-based AABW sources, in particular in East

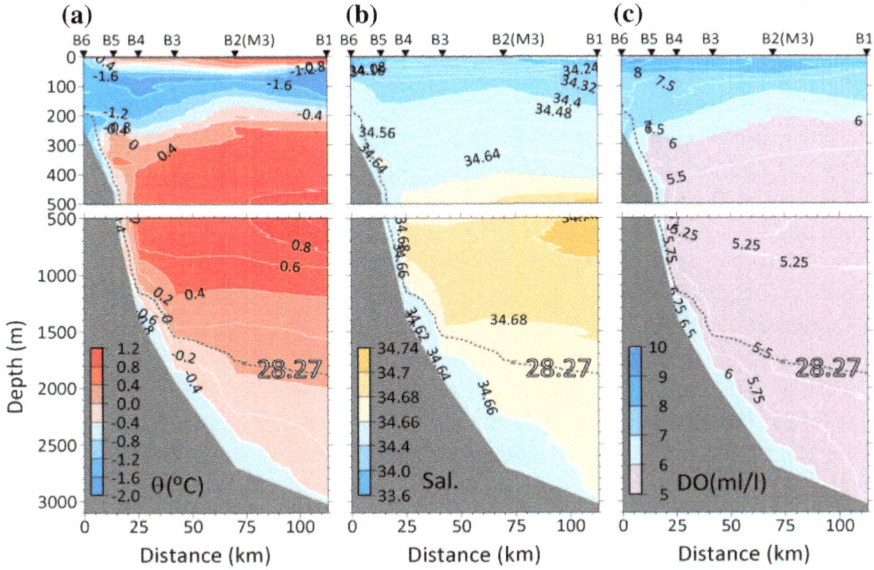

Fig. 10.3 Vertical section of **a** potential temperature, **b** salinity and **c** dissolved oxygen along Line A in Fig. 10.2 (after Supplemental information Figure S4 of Ohshima et al. (2013)). Dashed line in each panel indicates neutral density of 28.27 kg m^{-3}

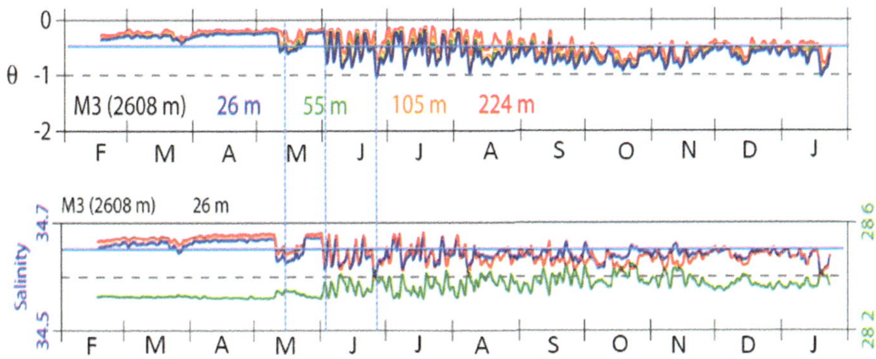

Fig. 10.4 Time series of potential temperature, salinity and neutral density rearranged from Ohshima et al. (2013)

Antarctica. While many of these are significantly smaller than the CDP in terms of annual sea ice production (Tamura et al. 2008), many have similar geophysical features.

Meredith (2013) has reported in News&View of Nature geoscience about impact of discovery of CDBW. The Weddell Sea site was the first discovered in 1940s, and was believed to be the sole source until other sites were discovered in the 1960s of

Ross Sea site and 1970s of Adélie Land site. Dense water formation and export in the region of Cape Darnley relatively reduced the contribution of WSBW on total amount of the AABW in Weddell-Enderby Abyssal Plain.

Although there are five coastal polynyas around Antarctica facing to Australia–Antarctica Basin (Fig. 10.1), only the AABW source associated with Mertz Polynya as mentioned in Sect. 10.1. It is natural to consider that there is another AABW source like a case of CDP. Thus, we have chosen Vincennes Polynya as the typical middle sized polynya to benchmark its possibility of AABW production.

The Vincennes Bay Polynya (VBP) forms every year in the coastal embayment southwest of Cape Poinsett (Fig. 10.5) with sea-ice production of 73.3 ± 9.9 km^3/yr (Tamura et al. 2008), which ranks it as a 'medium' class polynya relative to all polynyas around Antarctica. A small depression exists over the continental shelf at the west of VBP and sill depths from the depression to the continental slope are 500–600 m. The storage capacity of the depression is much smaller than the sizable Adélie Depression and with its modest amount of sea-ice production it has traditionally been ruled out as a source of DSW for AABW production.

We carried out longterm moorings at A and B, and hydrographic observations by TR/V Umitaka-Maru off VBP from 2011 to 2012 and found strong evidence of AABW production from there (Kitade et al. 2014). Density increase with a

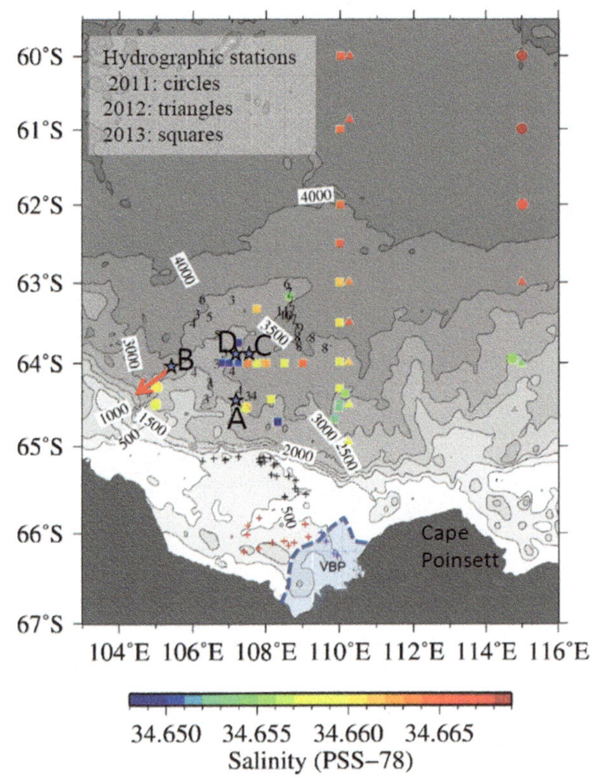

Fig. 10.5 A bathymetry map of the study region. Mooring locations are indicated by stars with A, B, C and D. Red arrow indicates the averaged velocity at 120 m from the bottom at mooring B. Numbers from 5 to 12 indicate months of observation and locations of the ARGO floats. An area surrounded by the dashed line (~8900 km^2) indicates the Vinceness Bay Polynya with sea-ice production >5 m yr^{-1}. CTD data obtained by instrumented elephant seals in February and March are indicated by blue and red crosses, respectively, and obtained near the shelf break region in both months are indicated by black crosses

freshening seen in mid-June and July shows the subduction of the newly formed bottom water in previous winter cooling (Fig. 10.6a–c). Hydrographic stations in Fig. 10.5 were indicated in color coded by values of near-bottom salinity with a manner as follows; salinity averaged in the range of ±25 dbar of the neutral density surface of 28.32 kgm^{-3}. From the figure, it can be seen that the low salinity water

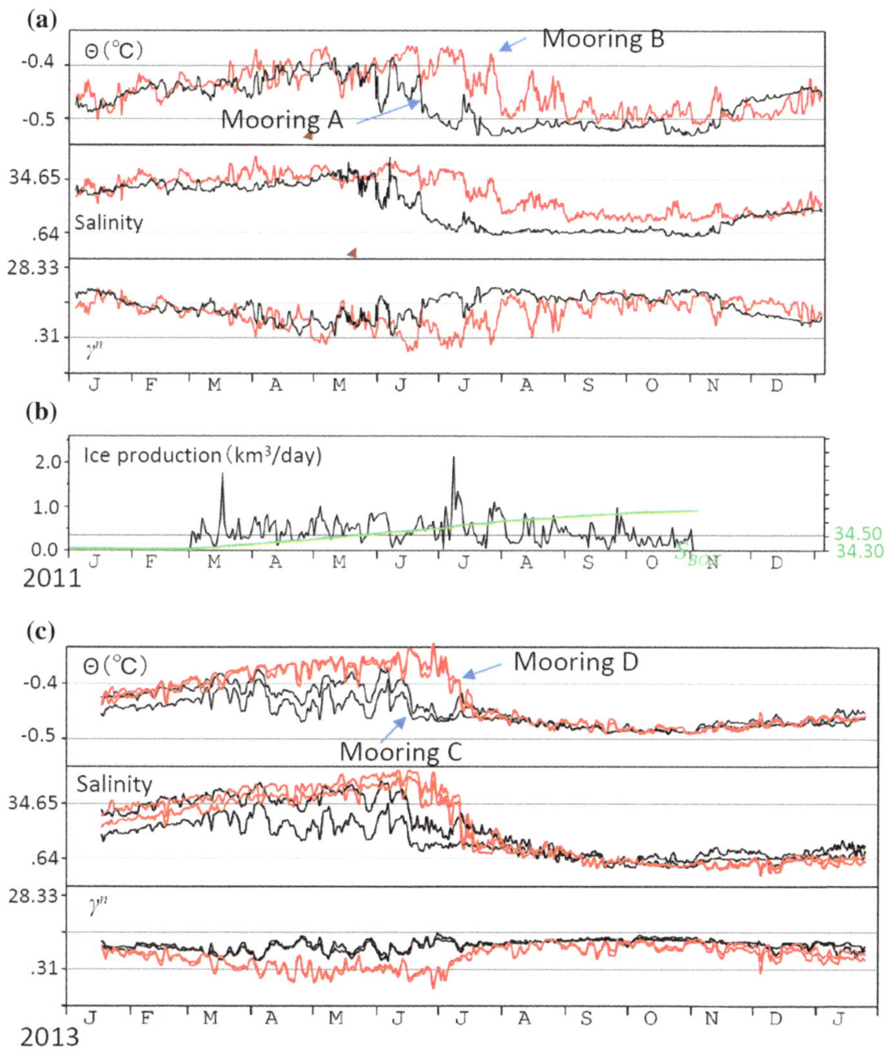

Fig. 10.6 **a** Time series of potential temperature, salinity and neutral density obtained by moorings A (black line) and B (red line). **b** Time series of sea-ice production (black line) estimated from the satellite data. The green line indicates salinity calculated from the salinity flux associated with sea-ice production in VBP. **c** Time series of potential temperature, salinity and neutral density obtained by moorings C (black line) and D (red line)

has spread in the northwestern part of the VBP. This isopycnal salinity contrast indicates that VBP is an independent AABW source from Mertz and Ross Sea.

Analyzing mooring time series (Fig. 10.6a–c) and hydrographic data, together with complementary data from instrumented seals, we identified the presence of newly-formed AABW on the continental slope and link this to DSW forming in and around the VBP. On the continental shelf, Dense Shelf Water (DSW) formation was observed by instrumented seals, in and around the VBP during autumn and we estimated its transport to be 0.16 ± 0.07 ($\times 10^6 m^3 s^{-1}$).

We concluded that the DSW formed in this region, albeit from a modest amount of sea ice production, nonetheless contributes to the upper layer of AABW in Australian–Antarctic Basin. Hereinafter we call the newly found AABW off VBP as the Vincennes Bay Bottom Water (VBBW). Thriving sea-ice production area facing to the Australian–Antarctica Basin exist four polynyas with the exception of the Mertz and VB. For accurate estimate of the flow rate of AABW, the evaluation of the amount of other source of AABW will come become necessary.

10.3 Freshening of Antarctic Bottom Water Off Mertz and Vincennes

Long-term water mass changes during 1994–2012 have been examined from nine repeat hydrographic sections, including data from TR/V Umitaka-Maru, at 140°E where locate just downstream of Mertz polynya (Aoki et al. 2013). They showed significant freshening trends detected within most of the water masses from the bottom to surface. Antarctic bottom water is generally defined as below 0 °C. TS diagrams show that the salinity of the secular in bottom water is reduced (Fig. 10.7). Bottom Water freshened by 0.008–0.009 decade^{-1} below isopycnal surfaces of $\gamma^n \sim 27.3$ (kg/m^3) throughout the study period. The Lower Circumpolar Deep Water on the continental slope underwent freshening at the same rate as the Bottom Water, whereas Modified Shelf Water shows robust freshening at a rate of 0.03 decade^{-1}. Combined with the freshening of near-surface and Bottom Water masses in this region, these data indicate freshening of the entire water column over the continental slope.

Freshening signals have also been clarified in water masses off VBP (Fig. 10.8). In addition to the freshening, warming trend is clearly found in water masses off VBP after 2011. Such widespread freshening is broadly consistent with the enhancement of the global hydrological cycle, together with a possible acceleration of land ice melting as mentioned by Aoki et al. (2013). However, the mechanism which transports such a freshening impact from the sea surface to the seabed is not quantitatively clarified.

Observational result obtained from 2011 to 2013 have shown that the water property of AABW off VBP was mainly influenced by Australian–Antarctic basin AABW (AA-AABW), which is the mixed water of Ross Sea Bottom Water and

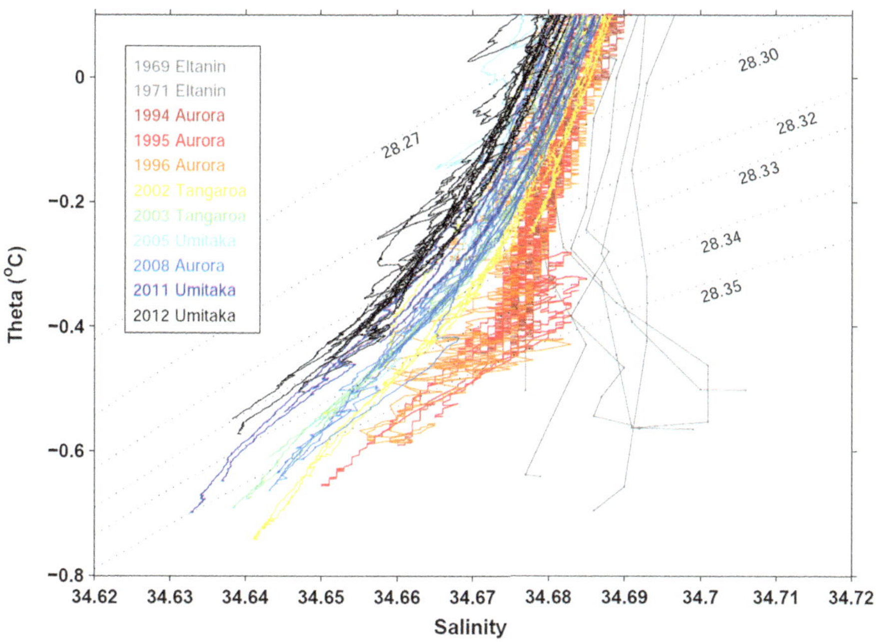

Fig. 10.7 *θ-S* diagram of observations for 60–65S band near 140E (after Aoki et al. [2013]). Legend denotes observation year and vessel. Color lines indicate the CTD observations in the legends. Gray dots denote bottle observations in 1969 and 1971 by R/V Eltanin. Background contours in broken lines are neutral density surfaces

Adélie Land Bottom Water, and the VBBW were distributed over them (Fig. 10.8). However, significant freshening and warming of AA-AABW were observed in the bottom layer off VBP in 2014. This change of water property was considered to increase relative impact of the VBBW on the AABW in the Australian–Antarctic Basin.

Although the mooring observation showed that VBBW properties have remained relatively unchanged in the period of 2011–2014, rapid freshening in AA-AABW of eastern origin implies that VBBW is being denser against AA-AABW (Fig. 10.9). This means that, depending on the year, the newly generated AABW is sometimes sinks below AA-AABW.

In 140E line, there is an inflection point at −0.42 °C by potential temperature and 34.675 by salinity, but it has disappeared since 2000 (Fig. 10.7). Shimada et al. (2012) have reported that freshening of RSBW dominantly affects to the freshening of AABW in AA Basin. Wijk and Rintoul (2014) showed that the amount of high DO water is not clearly changed in the AA Basin, and insisted that total volume transport of RSBW is not changed significantly. Both of these studies indicate that freshening of AA-AABW is significantly influenced by RSBW. Observation made by Umitaka-Maru in 2014 show that relative importance of VBBW was increased by freshening of AABW.

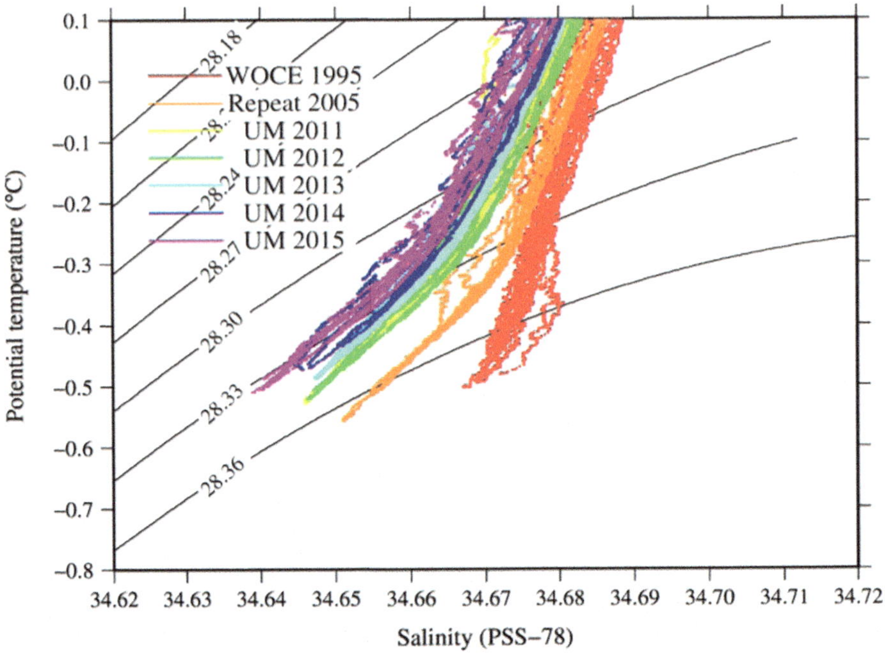

Fig. 10.8 θ-S diagram of observations for 60–65S band along 110E. Legend denotes observation year. Color lines indicate the CTD observations in the legends. Background contours in black lines are neutral density surfaces

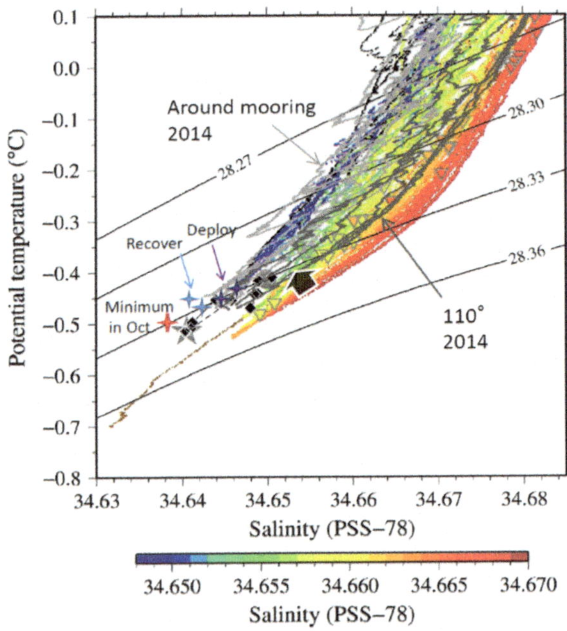

Fig. 10.9 θ-S diagram of bottom water obtained by mooring and hydrographic data during 2011–14 (After Kitade et al. [2014]). Data obtained by Mooring A and B from 2013 Jan. to 2014 Jan. are indicated by crosses. Dark gray line indicate the data along 110°E line, and light gray line indicate the data around mooring stations. Background contours in black lines are neutral density surfaces. The θ-S profile obtained from 2011 to 2013 are color-coded by salinity value, which is averaged in the range of ±25 dbar of neutral density surface of 28.32 kg/m^3

10.4 Problem and Future Project

Vertical fluctuations of salinity signal were observed around 0 °C in recent years as shown in Fig. 10.9. Low salinity signal implies intrusion of water mass of coastal origin, that is, a part of DSW cannot sink to the bottom layer, but can intruding into deep layer (as indicated by VBDW in Fig. 10.10). It is important to understand ventilation in Mid-deep layer along with in the bottom layer, in the evaluation of abyssal ventilation and Meridional Ocean Circulation quantitatively. As these freshening and warming of AA-AABW continues, there is a possibility of change of the deep circulation which significantly affect the climate change.

To grasp freshening process of AABW and to evaluate its global impact, it is necessary to clarify each process described in Fig. 10.10. The mixing process in this study area is considered to be different in seasons and depths. Although it is thought that a process results in formation of DSW holds a key, it is difficult to observe on a continental shelf by Umitaka-Maru which is not an ice breaker. Therefore, we need develop new observation system by new technology and reinforced bio-logging. Thus, in important research in JARE 9th term, construction of a new observation system and the research program which aimed at utilization are formed.

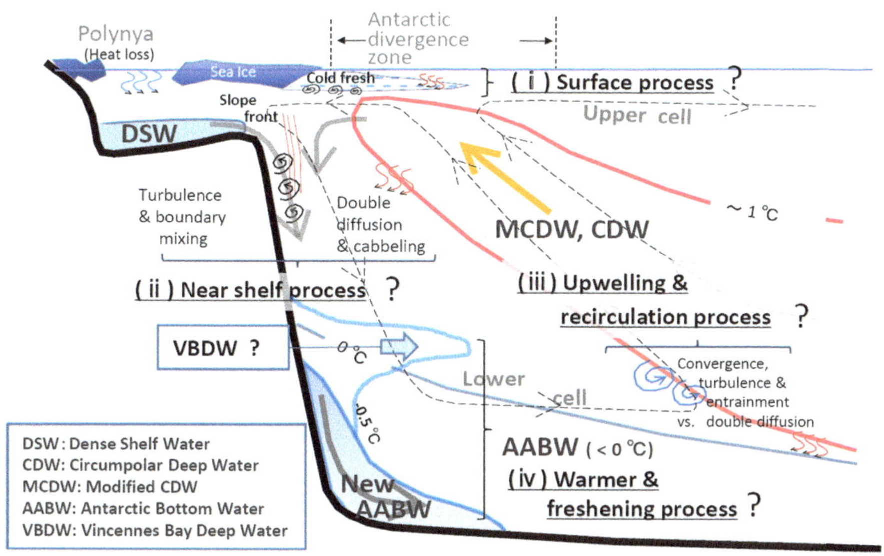

Fig. 10.10 Schematic view of water mass and circulation off VBP. From series of Umitaka-Maru observation, four major processes are considered to be important to clarify the freshening mechanism of AABW

On the other hand, Umitaka-Maru observation group have planned concentrated observation and long term mooring observation in seasonal ice zone. In 2015 and 2016, we will deploy mooring with large numbers of CTD and T sensors covering from bottom to upper part of CDW. In the next few years, we plan to deploy mooring systems to grasp the meridional circulation in Southern Ocean which have never observed directly by humankind. To estimate the upwelling and recirculation in the Southern Ocean, to clarify the importance of recirculation on freshening of AABW, and to evaluate the impact on climate change, it is necessary for us to continue observation in the future.

Acknowledgements We are deeply indebted to the former captains, officers and crew of Umitaka-Maru that has been helped to continue observation in the Antarctic Ocean. We would like to give special thanks to Caps. Kurita, Kitazawa, Noda, and crews who have been involved in the present study. In addition, we would like to thank everyone of the science party, including Drs. Yamaguchi, Ishimaru and Moteki who served as a senior researcher in the Antarctic Ocean observation voyage. This work was supported by JSPS KAKENHI grants 23310003 and 15H01726, MEXT for physical and chemical oceanographic observations under Japanese Antarctic Research Expedition.

References

Aoki S, Kitade Y, Shimada K, Ohshima KI, Tamura T, Bajish CC, Moteki M, Rintoul SR (2013) Widespread freshening in the seasonal ice zone near 140°E off the Adelie land coast, Antarctica, from 1994 to 2012. J Geophys Res Ocean 118:1–18. https://doi.org/10.1002/2013jc009009

Foster TD, Carmack EC (1976) Frontal zonemixing and Antarctic Bottom Water formation in the southern Weddell Sea. Deep-Sea Res 23:301–317

Garabato NAC, McDonagh EL, Stevens DP, Heywood KJ, Sanders RJ (2002) On the export of Antarctic Bottom Water from the Weddell Sea. Deep-Sea Res II 49:4715_4742

Gill AE (1973) Circulation and bottom water production in the Weddell Sea. Deep-Sea Res 20:111–140

Jacobs SS (2004) Bottom water production and its links with the thermohaline circulation. Antarct Sci 16:427–437

Jacobs SS, Amos AF, Bruchausen PM (1970) Ross sea oceanography and Antarctic Bottom Water formation. Deep-Sea Res 17:935–962

Kitade Y, Shimada K, Tamura T, Williams GD, Aoki S, Fukamachi Y, Roquet F, Hindell M, Shio S, Ohshima KI (2014) Antarctic Bottom Water production from the Vincennes Bay Polynya, East Antarctica. Geophys Res Lett 41:3528–3534. https://doi.org/10.1002/2014gl059971

Meredith MP, Watson AJ, van Scoy KA, Haine TWN (2001) Chlorofluorocarbon-derived formation rates of the deep and bottom waters of the Weddell Sea. J Geophys Res 106:28992919

Ohshima KI, Fukamachi Y, Williams GD, Nihashi S, Roquet F, Kitade Y, Tamura T, Hirano D, Herraiz-Borreguero L, Field I, Hindell M, Aoki S. Wakatsuchi M (2013) Antarctic Bottom Water production by intense sea-ice formation in the Cape Darnley polynya. Nat Geosci. https://doi.org/10.1038/ngeo1738

Orsi AH, Johnson GC, Bullister JL (1999) Circulation, mixing, and production of Antarctic Bottom Water. Prog Oceanogr 43:55–109

Rintoul SR (1998) On the origin and influence of Adélie land bottom water, ocean, ice and atmosphere: interactions at Antarctic continental margin. Antarct Res Ser 75:151–171

Shimada K, Aoki S, Ohshima KI, Rintoul SR (2012) Influence of Ross Sea Bottom Water changes on the warming and freshening of the Antarctic Bottom Water in the Australian-Antarctic Basin. Ocean Sci 8:419–432

Tamura T, Ohshima KI, Nihashi S (2008) Mapping of sea-ice production for Antarctic coastal polynyas. Geophys Res Lett 35:L07606

van Wijk EM, Rintoul SR (2014) Freshening drives contraction of Antarctic Bottom Water in the Australian Antarctic Basin. Geophys Res Lett 41:1657–1664. https://doi.org/10.1002/2013gl058921

Chapter 11
Response of Near-Inertial Internal Waves to Various Typhoon-Tracks Around the Tango Peninsula, Japan

Keiichi Yamazaki, Yujiro Kitade, Yosuke Igeta, Yutaka Kumaki and Tatsuro Watanabe

Abstract Response of near-inertial internal waves (NIIWs) to various typhoon-tracks was investigated by analysis of mooring data observed around the Tango Peninsula during eight typhoons. Spectral shapes, energy levels and propagation property of NIIWs were different by typhoon tracks. Amplifications of NIIWs were especially large and the NIIWs propagated with the coast on the right having the characteristic of an internal Kelvin wave along the coast when typhoon have passed northeastward over the Japan Sea. Slab model analysis showed that the amplifications of NIIWs after passage of typhoon were almost induced by rapid temporal variation of local wind, however, duration time and second amplification of NIIWs after passage of typhoons were not explained. To clarify the formation mechanisms of observed properties of NIIWs and response of coastal water to various typhoon-tracks, numerical experiments using multilayer level model with idealized wind stress and realistic topography were performed. Model results showed that the NIIWs energy around the Tango Peninsula were amplified after passage of typhoon by southward propagation of NIIWs generated off the western coast of Noto Peninsula or off Tango Peninsula. In the case which typhoon advanced northward off Kyushu, nevertheless large amplitude NIIWs were generated at wide region in the western side of Oki Island, its energy generated there has

K. Yamazaki (✉) · Y. Kitade
Graduate School of Tokyo University of Marine Science and Technology,
4-5-7 Konan Minato-ku, Tokyo 108-8477, Japan
e-mail: kyamazak@affrc.go.jp

K. Yamazaki
National Research Institute of Fisheries Science,
Japan Fisheries Research and Education Agency, Kanagawa, Japan

Y. Igeta · T. Watanabe
Japan Sea National Research Institute of Fisheries Science,
Japan Fisheries Research and Education Agency, Niigata, Japan

Y. Kumaki
Kyoto Prefectural Agriculture, Forestry and Fisheries Technology Center, Kyoto, Japan

T. Komatsu et al. (eds.), *Oceanography Challenges to Future Earth*,
https://doi.org/10.1007/978-3-030-00138-4_11

not reach around the Tango Peninsula was found. Consequently, the NIIWs energy are greatly strengthen in the wide region around the Tango Peninsula, when the typhoon passed north east ward off the Japan Sea and the NIIWs were widely generated in the eastern side of Noto Peninsula and off the Tango Peninsula. Moreover, slab model and numerical model results showed that amplification of NIIWs can be occur when typhoon is advancing slowly and its radius are large even though the typhoon passed over not Japan Sea but the Honshu.

Keywords Near-inertial internal wave · Typhoon · Coastal strong current Japan sea

11.1 Introduction

Near-inertial period fluctuations (NIPFs) are observed in all over the ocean. Their phase and energy generally propagates toward the equator from their generation area having the characteristics of an internal inertial gravity wave (e.g., Gill 1984; Garrett 2001), so we call them Near Inertial Internal Waves (NIIWs). The Japanese coast forms the southern boundary of the Japan Sea (Fig. 11.1). Therefore, we frequently observe NIIWs associated with the passage of atmospheric disturbances along the Japanese coast. In the northern hemisphere, fluctuation of NIIW is especially amplified along right hand side of trace of typhoon (e.g., D'Asaro 1985). In fact, observational results and numerical experiments have shown that after the passage of atmospheric disturbances, NIIWs were strengthened along the Tango and Noto Peninsulas, which are coastal features of the southern boundary of the Japan Sea (Fig. 11.1) (e.g., Igeta et al. 2009, 2011; Yamazaki et al. 2015). Thus, amplification of the NIIW and its duration period are expected to be changed by the passage track of each typhoon.

Coastal-trapped waves (CTW) accompanied by downwelling or upwelling are also generated by the passage of atmospheric disturbances and propagate along the Japanese coast around the study region (e.g., Isoda et al. 1991; Igeta et al. 2007, 2011). Within the coastal area, both of the NIIWs and the CTWs are frequently confirmed from the observations. In the extreme case, strong currents were observed associated with the atmospheric disturbances, and they were often interpreted by superposition of the large-amplitude NIIWs and the large-amplitude CTWs (Igeta et al. 2007, 2011). There are some discuss about stormy strong current along the coast of Tango Peninsula accompanied by each of the typhoon (Kumaki et al. 2005, 2012; Igeta et al. 2007, 2009). However, they focus on a generation process of observed stormy current to a typhoon. There is not study to clarify responses of the NIIWs to various typhoon-tracks. During a period from June to October in 2004, eight typhoons had gone through around study area along various tracks (Fig. 11.2). Mooring observation at ten locations were carried out near the

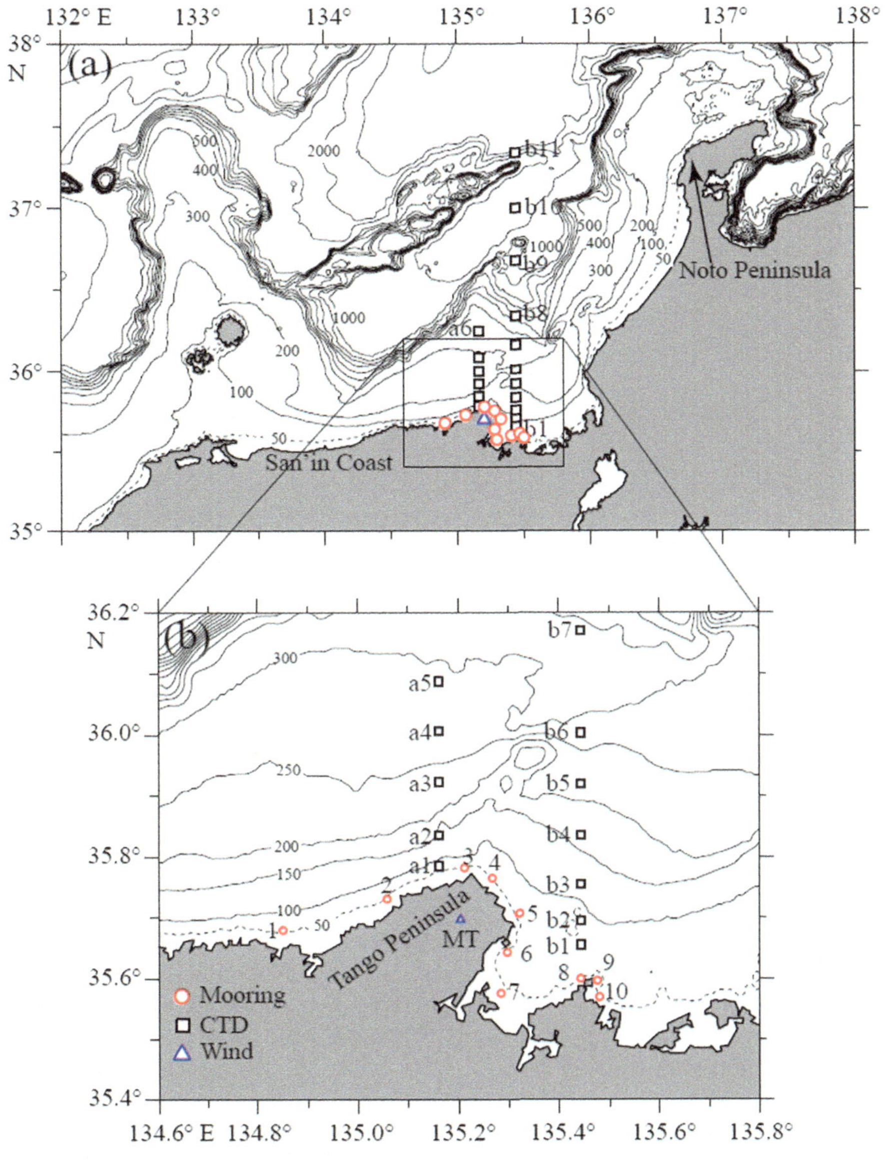

Fig. 11.1 Map showing the locations of **a** CTD (squares) and wind (triangle) stations, and **b** mooring sites (circles) around the Tango Peninsula. Isobaths are shown with numerals in meter

coast of Tango Peninsula during the period. The mooring data showed the distribution of strong current and its duration period for NIIWs. In this study, we analyzed mooring data to clarify properties of NIIWs in the relation with typhoon tracks and discussed their formation mechanisms of by using numerical model.

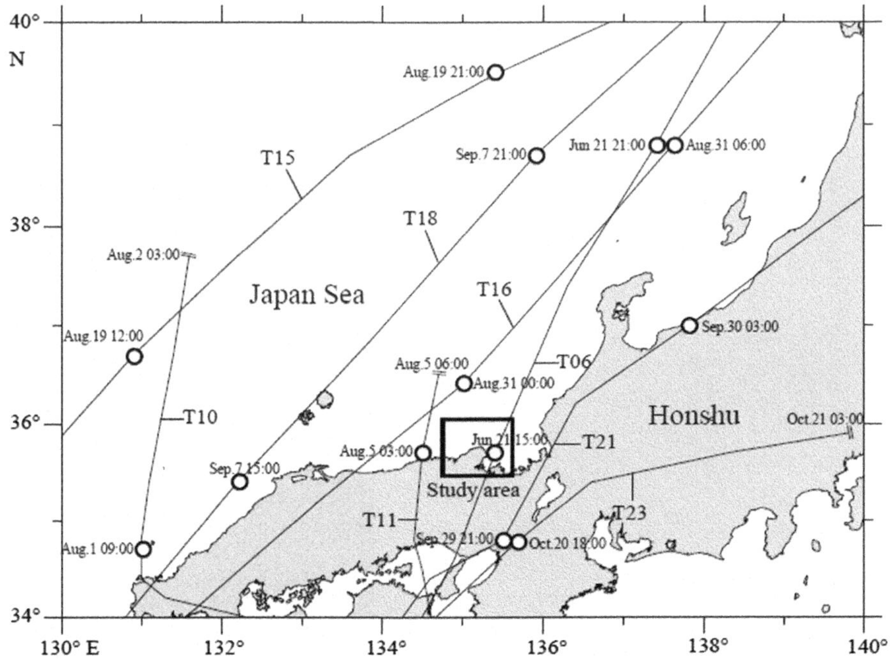

Fig. 11.2 Pathway of typhoons which passed near the study area during period of mooring observation

11.2 Observation and Data

11.2.1 Mooring and Meteorological Data

Mooring observations by using current meters and temperature sensors were performed at ten stations around the Tango Peninsula (Fig. 11.1) from June to October 2004. Electromagnetic current meters with temperature sensor (Compact-EM; JFE Advantech Co., Ltd; accuracy of ± 1.0 cms^{-1}) were deployed at a depth of 15 m, and thermometers with memory storage (MDS-MarkV/T; JFE Advantech Co., Ltd; accuracy of ± 0.05 °C) were deployed at depths of 30 m. The sampling interval was 10 min for all observational instruments.

The vertical profiles of temperature and salinity used in this study were obtained by Conductivity Temperature and Depth profiler (CTD, Falmouth Scientific, Inc.) observations at the stations indicated by black squares (Fig. 11.1) on 3rd–4th September 2004.

To investigate the temporal fluctuation of wind during passage of typhoons, wind velocity data observed at Sta. MT every 10 min were used in this work. The grid point values-mesoscale model (GPV-MSM) wind data sets, which have a horizontal grid size of 0.125° × 0.10° and a temporal resolution of 1 h, provided by the Japan Meteorological Agency (JMA), were used in this work.

11.2.2 Characteristics of Each Typhoon

Position and information of typhoon from JMA are indicated in Fig. 11.2 and Table 11.1. In this study, we describe Typhoon 0406 as T06. Cases of T15, T16, T18, the typhoons were advanced northeastward over the Japan Sea, maximum wind speed and radius of typhoon are higher and larger than the other case. Wind speed and moving speed of typhoon is highest for the case of T18, radius of typhoon is largest for the case of T23. For cases of T10 and T11, which disappeared during passing over the Japan Sea, and case of T21, which advanced over the Honshu, these maximum wind speeds were lower, however maximum wind speed is high for the case of T23. Thus, strong wind area and moving speed of typhoon are different for the case of typhoon.

11.3 Observational Results

11.3.1 Current and Temperature Fluctuations

Throughout the observational period, alongshore current is predominant compared to cross-shore current (Fig. 11.3). Current fluctuations are higher from Sta. 2 to 6. In addition to seasonal variation, observed temperature fluctuates with various periods are higher at 30 m depth than that 15 m depth. Wind fluctuation of

Table 11.1 Characteristic of each typhoon

Typhoon case	Central pressure (hPa)	Max. wind speed (ms^{-1})	Radius of typhoon (km)				[c]Moving speed (kmh^{-1})
			[a]Stormy wind area		[b]Strong wind area		
			Major	Minor	Major	Minor	
T06	970	30	110	60	570	370	63 ± 5
T10	992	23	–	–	370	150	18 ± 7
T11	996	20	–	–	150	50	30 ± 0
T15	970	30	130	130	520	330	55 ± 3
T16	970	30	190	110	650	370	65 ± 9
T18	950	35	170	130	600	410	82 ± 11
T21	990	23	–	–	520	440	55 ± 5
T23	970	30	280	170	750	460	45 ± 9

[a]Average wind speed is greater than 25 ms^{-1}
[b]Average wind speed is greater than 15 ms^{-1}
[c]Average value and standard deviation in the area where shown in Fig. 11.2

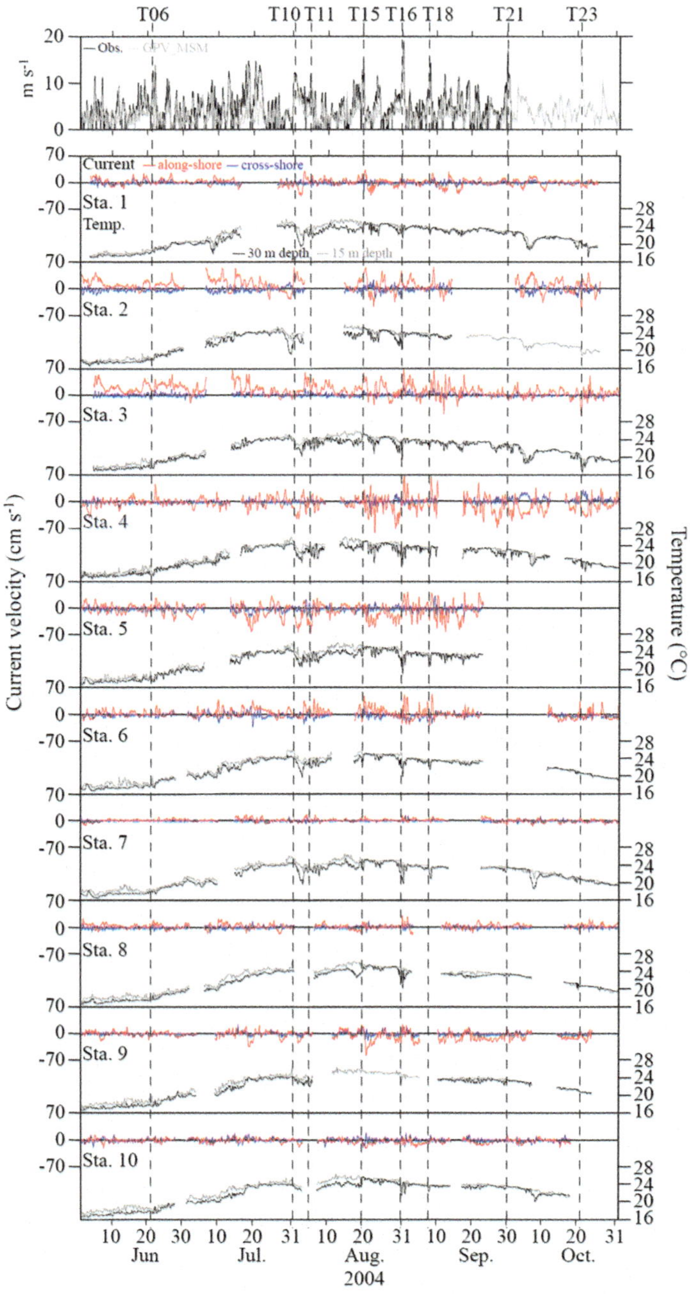

Fig. 11.3 Times series of wind speed at Mt. Taiko (most top panel) and along shore (red) and cross-shore current (blue) and temperature at depths 15 m (gray) and 30 m (black) observed at each station. Vertical dashed lines indicate the time when wind speed reached maximum for each typhoon case

GPV_MSM data give close agreement with that observed at Sta. MT. In this study, we defined the time of maximum wind stress as the typhoons passage time near the Tango Peninsula (vertical dashed lines in Fig. 11.3). Fluctuations of alongshore current and temperature at 30 m depth have amplified associated with passage of typhoons.

Next, we estimated power spectral density of alongshore current and temperature at 30 m depth from 15 July to 31 October to investigate characteristics of phenomenon which dominate around observational site (Fig. 11.4). Conspicuous peaks in alongshore current are found at periods of 25.9 h which is near the O_1 tidal constituent, 20.9 h which is near the local inertial period T_f (=20.4 h) at Sta. 3 and 15.9 and 18.8 h which is shorter than the local inertial period. In the several-day period band, some peaks are found at period from 2 to 10 days. The spectral peaks are also found in temperature fluctuation and its energy density level is highest at 20.9 h. Except for distinct peak at 24 h, relatively small energy spectral peaks are found at 18.7 and 20.9 h in the power spectral density of wind velocity.

Significant correlation, exceeding 0.25 which is the 90% confidence level for coherence-squared, can be found in near inertial period of 18.7 and 20.9 h (Fig. 11.4). Squared coherences of temperature at 30 m depth and northward wind are 0.34 at 20.7 h, significant correlation also found near inertial period. Correlation between temperature at 30 m depth and wind also shows significant value of 0.34 in coherence-squared at 20.7 h period. The relations between wind and current components are not clear at the period around 25.9 h. The 25.9 h period is close to 25.49 h period of O_1 tidal period, indicate the possibility that the O_1 period fluctuations are dominant around the Tango Peninsula. When we extract the NIPFs from observational data, it is necessary to pay attention to these fluctuations.

11.3.2 Distribution Characteristics of NIPFs

To examine the distribution of NIPFs during passage of each typhoon, we estimated the kinetic energy by integration of power spectra from 15.0 to 21.8 h period (Fig. 11.5). Kinetic energy levels of along shore current at Stas. 2–6 are one order in magnitude higher than that at other stations. In the case of T06 and T10, the kinetic energy is low compared to other cases. In the case of T11, the energy is the highest at the Sta. 5. In the case of T15, the energy is high only at Sta. 4 and its value is highest among all cases. In the case of T16, the energy is high at Sta. 3–6, and the highest at the Sta. 3. In the case of T18, the energy is high at from Sta. 3 and 6 (no observation at Sta. 4), and the highest at the Sta. 5. In the case of T21, the energy is high only at Sta. 4. In the case of T23, the energy is high at Stas. 2–4, and its value at Sta. 2 is the highest among all cases. Consequently, magnitudes of kinetic energy significantly depend on their location, and some changes in distribution are also induced by typhoon trace.

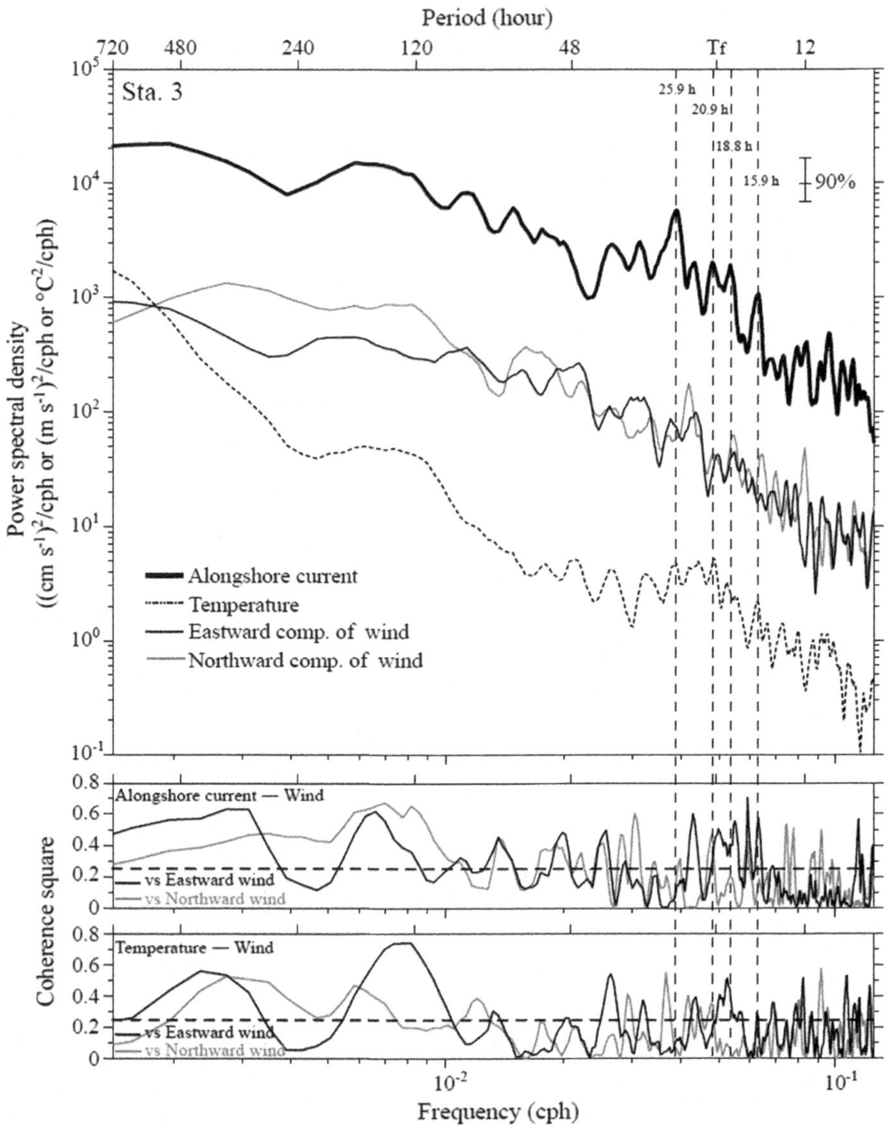

Fig. 11.4 Top: Power spectral density of alongshore current (thick black line), temperature (dotted kine) at Sta. 3 and eastward (thin black line) and northward (thin gray line) wind at Mt Taiko. Vertical dashed line indicates the inertial period at Sta. 3 (20.4 h). Middle: Coherence square between alongshore current and eastward wind (black line) and northward wind (gray line). Low: Coherence square between temperature and eastward wind (black line) and northward wind (gray line)

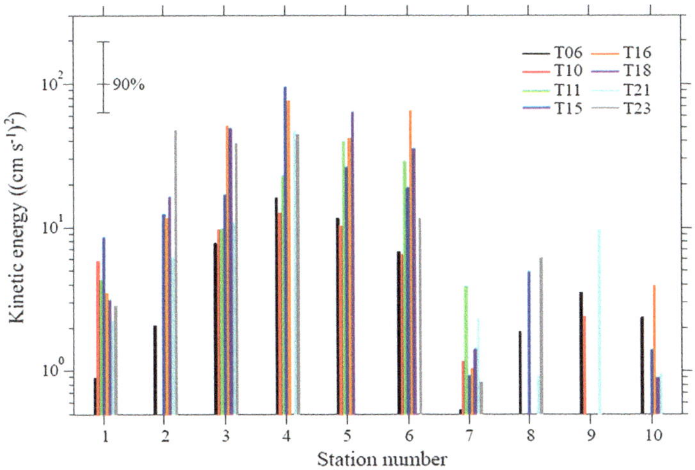

Fig. 11.5 Distributions of kinetic energy of NIPFs observed at each station during passage of each typhoon

11.3.3 Propagation Characteristics of NIPFs

To examine the propagation characteristics of NIPFs induced by each typhoon, we estimated the coherence and phase lag of temperature fluctuations at 30 m depth at all stations in comparison with those at Sta. 3 (Fig. 11.6). Values in Fig. 11.6 are 20 h period component as near-inertial period. Phase lag indicates the delay of the fluctuations to that at Sta. 3. Igeta et al. (2009) pointed out that NIPFs near the coast around the Tango Peninsula propagates the coast on the right as an internal Kelvin waves with near-inertial period. The phase lag which indicates propagation of NIPFs with the coast on the right is found in the cases of T06, T11, T15, and T16, but that is not clear in the other cases. Value of coherence indicates that the NIPFs have propagated from Stas. 3 to 5 in the case of T06 and T15, which have propagated from Sta. 3 to 6 in the case of T11 and T16. We calculated the propagation speed of NIPFs using horizontal distance of adjacent stations and phase lag (Table 11.2).

These values roughly correspond to the value of phase speed of internal Kelvin waves in the 2-layer model (0.81–0.71 ms^{-1}) which were assumed that thickness of upper layer to be 20 m and density difference of upper and lower layer $\Delta\rho/\rho$ to be 0.0027–0.0032. On the other hand, in the results of Igeta et al. (2009) shown as an example, the phase relationship is almost the same results of this study, however, the propagation was just found from Sta. 3 to Sta. 4, the coherences have declined at Sta. 5 drastically. This suggests that the difference of propagation characteristics of NIPFs is due to typhoon tracks and stratification condition.

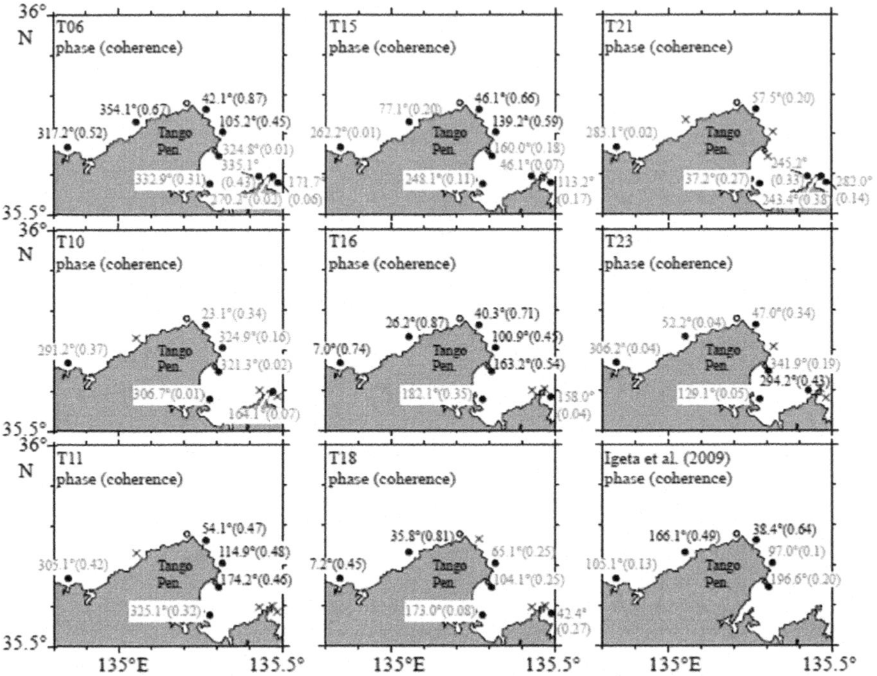

Fig. 11.6 Distributions of phase lags for temperature fluctuations of the 20-h period components. Values in parentheses indicate the amount of coherence square. The phase lags shown by black are indicate that coherence square is larger than 90% confidence limit (0.44)

Table 11.2 Propagation speed of 20-h period fluctuation

Station number	Distance (m)	Propagation speed (ms^{-1})				
		T06	T11	T15	T16	[a]Igeta et al. (2009)
Stas. 3–4	6830	0.81	0.63	0.74	0.61	0.96
Stas. 4–5	8360	0.66	0.68	0.44	2.86	0.77
Stas. 5–6	6350	0.14	0.53	1.52	1.01	0.35

[a]18.5 h period fluctuation

11.3.4 Temporal Variation of Energy of NIPFs

To estimate the temporal variations of kinetic energy density of the NIPFs (KE_I), we investigate the temporal variation of KE_I by following Eq. (11.1) using u_I and v_I which are the 15–22-h band-pass-filtered alongshore and cross-shore currents, respectively.

$$KE_I = \frac{1}{T}\int_{t+\frac{T}{2}}^{t-\frac{T}{2}} \frac{1}{2}\rho\left(u_I^2 + v_I^2\right) dt \tag{11.1}$$

Here, T is the near-inertial period (20 h); ρ is the density (assumed to be 1025.0 kgm^{-3}). Temporal variation of KE_I observed at Sta. 3 which are not exist missing value in all typhoon cases are shown in Fig. 11.7 in conjunction with stick diagram of wind at Sta. MT. The cases that amplification of KE_I is especially high for T16 and T18 (Fig. 11.7e, f), the KE_I increased when wind speed was strengthened with passage of typhoons. The KE_I increased at the same time when the wind speed reached maximum during passage of T16 and that reached the maximum at 30 h later after the wind speed reached maximum during passage of T18. After passage of T18, KE_I is further increased between 11th to 12th on September after amplification of KE_I just after passage of T18. In case of T15 which passed similarly to T18 over the northside of Japan Sea, maximum KE_I is low compared to the case of T18, but its temporal variation is similar (Fig. 11.7d). In the cases of T06 and T21 or T23 that center of typhoons passed eastside of the Tango Peninsula (Fig. 11.7a) and over the Honshu (Fig. 11.7g, h), maximum value of KE_I is low compared to above-mentioned cases. Besides, in the case of T11 which disappeared during moving over the Japan Sea, KEI was not increased (Fig. 11.7c). In the case of T10 which advanced northward off Kyushu, KE_I reached the maximum at 3 days later after the wind speed reached maximum, but its maximum value was low compared to the cases of T16 and T18. Thus, we found that amplitude distribution and duration of NIPFs can be roughly classified by typhoon tracks.

11.4 Response of Coastal Water to Local Wind

We investigated response of NIPFs to local wind using by slab model (D'Asaro 1985). The slab model expresses current fluctuations $Z(Z(t) = u_s + iv_s)$ in the surface mixed layer induced by wind stress according to following basic equation having wind forcing term:

$$\frac{dZ(t)}{dt} + \omega Z(t) = \frac{T(t)}{H} \tag{11.2}$$

where, $T(t) = \frac{\tau_x + i\tau_y}{\rho_w}$, $\omega = r + if$. Here, u_s and v_s are the eastward and northward component of current in the surface mixed layer, respectively, τ_x and τ_y are the eastward and northward wind stress, respectively, H is the thickness of surface mixed layer, ρ_w is water density, r is the damping parameter, and f is the Coriolis parameter. Assuming that integral constant is A, the solution of Eq. (11.2) can be obtained as follows:

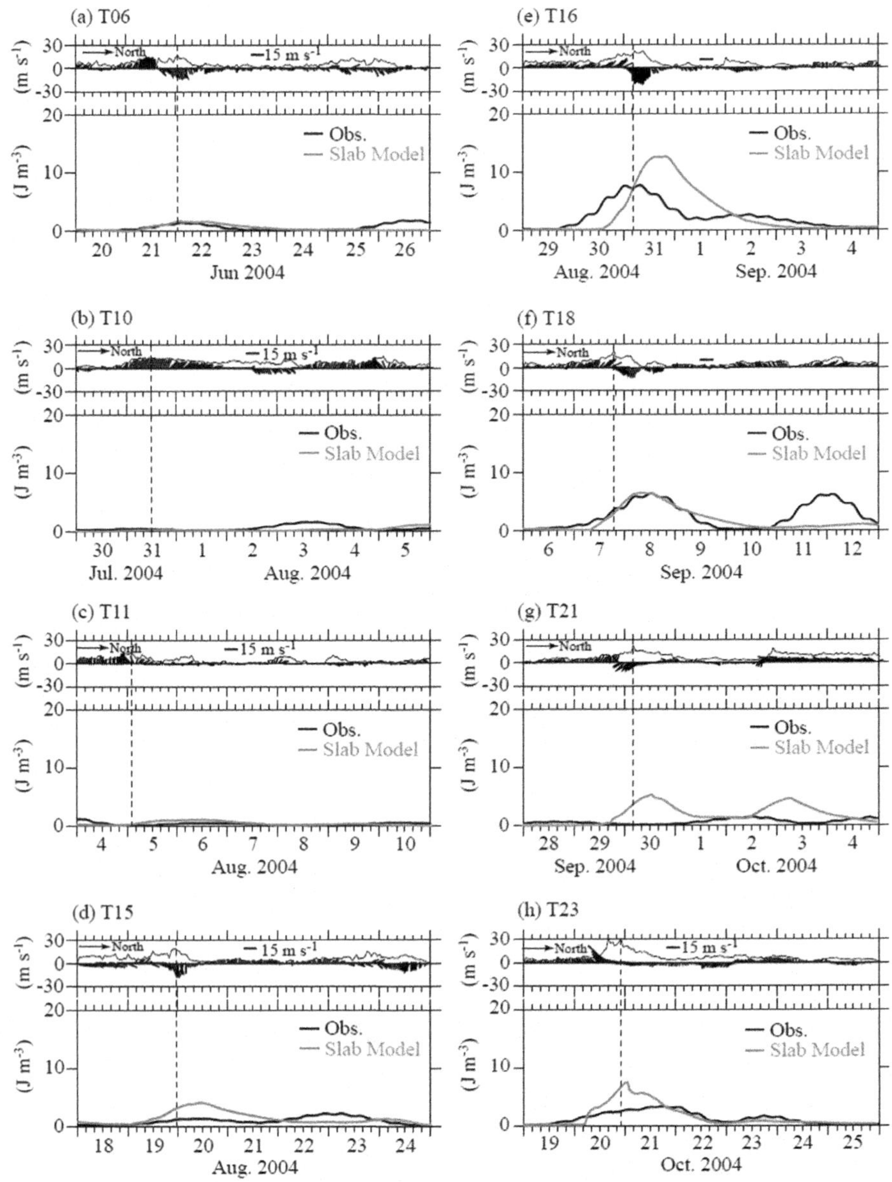

Fig. 11.7 Temporal variations of wind velocity (stick) and wind speed (thin black line) at Mt. Taiko (top panel) kinetic energy density of 20-h period fluctuations at observed at Sta. 3 (black lines) and estimated by using slab model (gray) (low panel) during passage of each typhoon. Vertical dashed lines indicate the time when wind speed was maximum

$$Z(t) = e^{-\omega t}\left\{A - \frac{1}{\omega H}\int \frac{dT}{dt} e^{\omega t} dt\right\} + \frac{T}{\omega H} \tag{11.3}$$

where, the first term and second term on the right side represent inertial oscillation and Ekman transport components of current in the mixed layer, respectively (D'Asaro 1985). In this study, temporal variations of $Z(t)$ can be obtained by applying wind stress calculated from the eastward and northward wind at Sta. MT to Eq. (11.3), assuming constant H = 50 m, $1/r$ = 2 days, ρ = 1025.0 kgm^{-3} and $f = 8.55 \times 10^{-5}$ s^{-1} at 35.78 N and temporal variations of KE_I can be obtained using Eq. (11.1) (gray lines in Fig. 11.7).

Slab model results showed that amplification of KE_I is high as well as observational results in the cases of T16 and T18. From comparison of observation and slab model results, although maximum value of KE_I for the slab model result was over estimated after passage of T18, temporal variation of KE_I during passage of T23 was well reproduced by slab model. These results indicate that NIPFs were induced by local wind in these cases. Although wind direction varied anticlockwise with time during passage of T23, this showed that the KE_I increased by rapid change of wind speed and direction. On the other hand, during passage of T16, whereas the KE_I of observational result was increasing before passage of T16, slab model results showed that the kinetic energy increased after wind velocity rotated clockwise with time. Beside, slab model did not reproduce second amplification of KE_I after passage of T15 and T18. Slab model suggested amplification of energy during passage of T21, however, the amplification did not occur in observational results. These results indicate that the factor other than local wind have given effects on the observed NIPFs.

Slab model results almost reproduced observed NIPFs, however, issue which cannot explain only local wind effect is remained, that is, duration time and second amplification of NIPFs. Next, we investigated response of coastal water to typical typhoon tracks using by numerical model.

11.5 Numerical Experiments

11.5.1 Numerical Model and Conditions

A three-dimensional, multilayer level model having free surface (e.g., Kitade et al. 2011; Yamazaki et al. 2015) was applied to investigate the effect of typhoon tracks. Under the hydrostatic and Boussinesq approximations, equations of motion, continuity, temperature and salinity for an incompressible fluid were differentiated and solved numerically in the model. The parameters used in the model were as follows: coefficient of vertical viscosity $A_v = 1.0 \times 10^{-3}$ m^{2}s^{-1}, coefficient of horizontal diffusivities of temperature and salinity $K_h = 50.0$ m^{2}s^{-1}, coefficient of vertical diffusivities of temperature and salinity $K_v = 1.0 \times 10^{-4}$ m^{2}s^{-1}, and horizontal

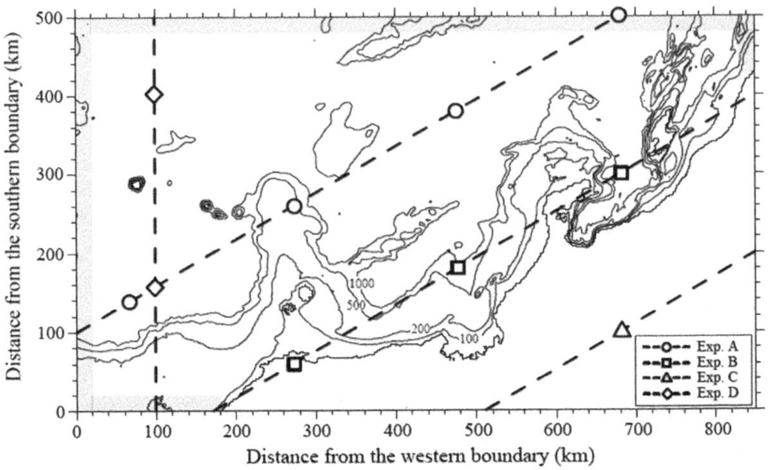

Fig. 11.8 Model domain. Values on the contour lines are in meter. Dashed lines indicate the time variation of position of centor of idealized typhoon

viscosity A_h was calculated with a Smagolinsky eddy parameterization (multiplier 0.1).

Figure 11.8 shows the model domain of 850 km from east to west and 500 km from north to south. JTOPO30 (Marine Information Research Center) is applied for the topographic data and depthson model grid points obtained by linearly interpolation. Depths deeper than 1000 m was set to 1000 m to reduce the calculation load. The grid spacing was 1 km × 1 km horizontally, and the grid points were set at constant depth in z-level, that is, ten vertical levels were set at depths of 5, 15, 30, 50, 80, 140, 230, 580, and 840 m. The clamped condition and the sponge condition were applied along the open boundaries to reduce disturbances nearby, and the non-slip condition was applied along the rigid boundary (e.g., Igeta et al. 2009; Yamazaki et al. 2015).

In this study, we adopted idealized wind field by typhoon in order to explain phenomena clearly occurred in the model. We expressed typhoon's pressure field $P_a(r, t)$ using Fujita's empirical equation

$$P_a = P_\infty - \frac{P_0}{\left\{1 + (r/r_0)^2\right\}^{1/2}}, \tag{11.4}$$

where, $r = (X^2 + Y^2)^{1/2}$ is the distance from the center of typhoon (x_c, y_c) to certain point (x, y), that is, $X = x - x_c$ and $Y = y - y_c$ (Fujita 1952). Here, P_∞ (=1013 hPa) is the peripheral pressure, $P_0(t)$ is the pressure at center of typhoon, and r_0 (=160 km) is the radius of stormy wind area of typhoon. Then, we expressed wind field using following empirical equations:

$$G_x = -(Y\cos\theta + X\sin\theta)\left\{\left(\frac{f^2}{4} + \frac{\rho}{\rho_a} gP_0 \frac{r_0}{\left(r_0^2 + r^2\right)^{3/2}}\right)^{1/2} - \frac{f}{2}\right\},$$

$$G_y = (X\cos\theta - Y\sin\theta)\left\{\left(\frac{f^2}{4} + \frac{\rho}{\rho_a} gP_0 \frac{r_0}{\left(r_0^2 + r^2\right)^{3/2}}\right)^{1/2} - \frac{f}{2}\right\}, \text{ and}$$

$$\mathbf{W} = C_1 C_T \exp(-r\pi/r_i) + C_2\mathbf{G},$$

where $\boldsymbol{G} = (G_x, G_y)$ is gradient wind fields which were inclined θ (=30°)with respect to tangential direction of isobars, ρ and ρ_a are the density of seawater and air, respectively, C_1 and C_2 (=0.95) is the constant, and r_i is the parameter that determines the typhoon shape (e.g., As-salek 1998).

To focus the effect of typhoon track on the amplification of NIPFs around the Tango Peninsula, the same initial stratification condition was applied to all experimental cases, based on the averaged vertical profile of temperature and salinity observed on September 2004 around study area. We assumed that the ocean was initially at rest and only forcing applied was wind stress by moving the center of typhoon with 4 tracks along dashed lines in Fig. 11.8 (Exps. A–D).

11.5.2 Experimental Results

11.5.2.1 Reproducibility of the Model

Model results (Fig. 11.9) showed that kinetic energy of NIPFs was high in the cases of Exps. A and B, and was low in the cases of Exps. C and D. Moreover, NIPFs propagated with the coast on the right in the cases of Exps. A and B and such phase

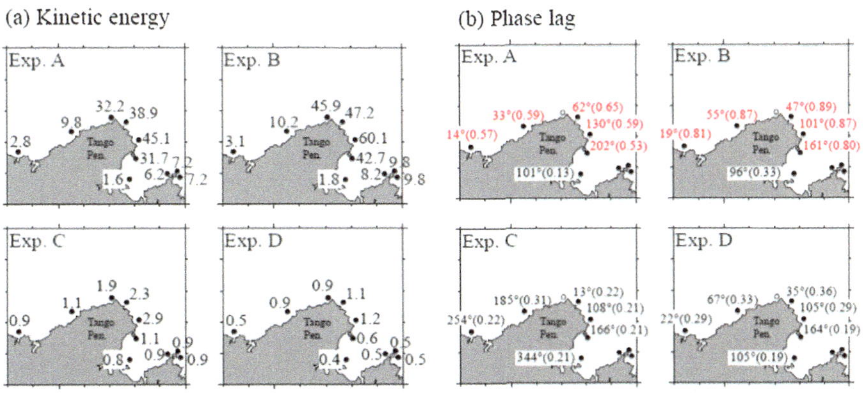

Fig. 11.9 Distributions of **a** kinetic energy and **b** phase lag and coherence of 20-h period fluctuations

propagation characteristics were not seen in the cases of Exps. C and D. These results were good agreement with observational results. This indicates that NIPFs were amplified near the Tango Peninsula and that propagate along the Tango Peninsula when typhoon passed northeastward over the offshore region of Japan Sea. In addition, in the case of T23, although the slab model showed amplification of near-inertial energy, the phase propagation was not seen both observation and model results. This indicates that NIPFs were amplified from east to north coast of Tango Peninsula due to only local wind effect during passage of T23. Therefore, amplification of the NIPFs occurs in the southern area below east coast of the Tango Peninsula when the typhoon has passed through the track, such as Exps. A and B.

11.5.2.2 Generation Area and Propagation Path of Near-Inertial Internal Waves

Figure 11.10 shows the temporal variations of wind and kinetic energy of NIPFs at Sta. 3. In the results of experiments A and B, NIPFs were largely strengthened after passage of typhoons. Wind direction varied clockwise with time in these cases, this variation of wind velocity effectively causes amplifications of inertial oscillations in the northern hemisphere (e.g. D'Asaro 1985). Therefore, amplifications of NIPFs just after passage of typhoon in the cases of Exps. A and B were due to NIIWs which were generated along the typhoon tracks. Second amplifications of kinetic energy density which were not seen in the slab model results, were reproduced in the numerical experimental results. This indicates that propagations of NIIWs generated

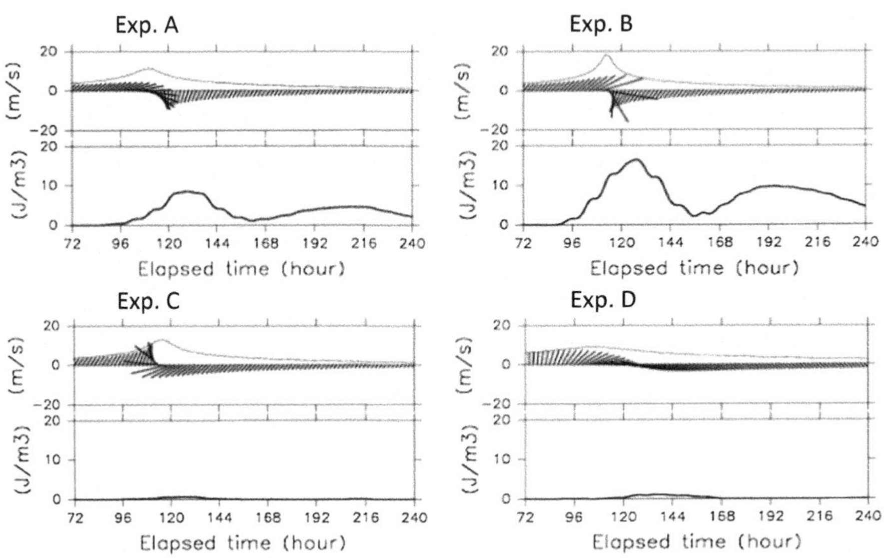

Fig. 11.10 Time series of wind velocity (top) and kinetic energy of NIPFs (bottom) for each model case at monitoring point corresponding to Station 3

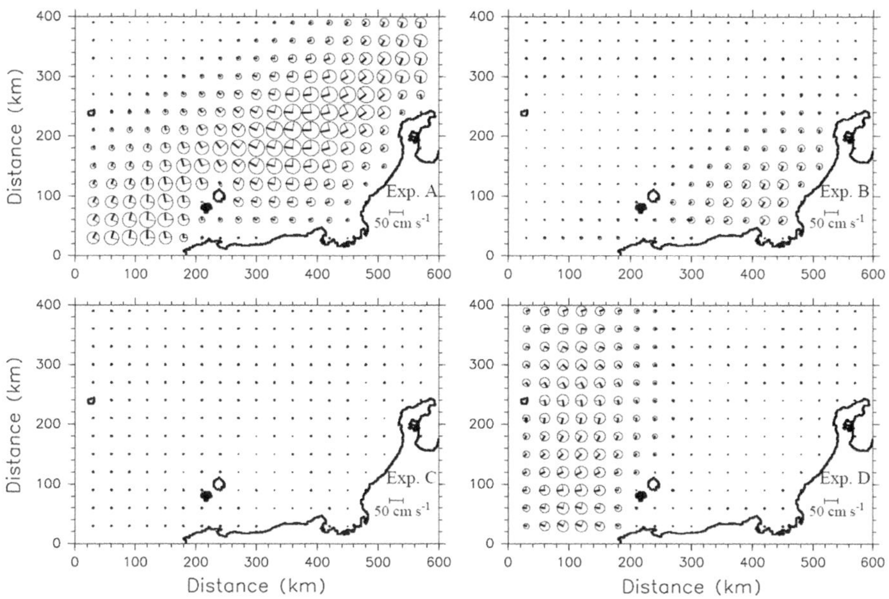

Fig. 11.11 Horizontal distributions of current ellipses of variation component with 20-h period at 15 m depth. Thick (thin) line in the ellipses indicates the current vector at initial phase (5 h after initial phase)

at the offshore region were contributed to amplifications of near-inertial energy after passage of T15 and T18. On the other hand, energy amplifications were low in the cases of Exp. C that wind direction varied clockwise with time and that of Exp. D that the typhoon moved to northward in the western area of model domain.

To investigate the generation area of NIIWs induced by typhoon, current ellipses of 20-h period fluctuations at a depth of 15 m (Fig. 11.11) were estimated by using current amplitude u_0 and phase θ of 20-h period fluctuations ($u(t) = \bar{u} + u_0 \sin(\omega t - \theta)$, where $\bar{u}$ is time mean current, $\omega = 9.18 \times 10^{-5}\ \mathrm{s}^{-1}$).

The amplitude and phase of horizontal current at each model grid were calculated by least square method using 120 h data after passage of typhoons (t = 121–240 h). Thick lines in the ellipses indicate the current vector at initial phase, whereas thin lines indicate the current vector 5 h after initial phase, so we can judge the direction of rotation. In the cases of Exp. A, B and D, the current ellipses in the right side of the typhoon track were almost circle and large, and current vector rotated clockwise with time. Therefore, we found that these regions are the generation area of NIIWs. In Exps. A and D, generation area of NIIWs which were generated in the Japan Sea were wide. Initial phase vectors were delayed for typhoon goes, indicating that propagation direction of NIIWs is the same direction of typhoon goes in its generation area (Gill 1982). On the other hand, in Exp. C, the NIIWs hardly occur in the model. This is because that anticlockwise time variation of wind acted to damp the inertial oscillations.

To explain the propagation process of phenomenon, we investigated the amplitude and phase distributions of temperature at a depth of 30 m for 20-h period fluctuations by the same methods of Fig. 11.11. Here, increasing of phase lag indicates the delay of fluctuations. From phase lag distributions in the cases of Exps. A and D (Fig. 11.12a), the phases are delay moving direction of typhoons around the center of typhoon were found. The distribution of phase lag is very complicated around the Tango Peninsula in Exp. A and in the western side of Oki Island in Exp. D, but phase in these region delay compared to that around the generation area of NIIWs, which indicates that NIIWs propagate to the low latitude area from its generation area. In the case of Exp. B, delay of phase is seen from western side of the Wakasa Bay to eastern coast of the Tango Peninsula and that is continuously delayed toward offshore region. From amplitude distributions (Fig. 11.12b), we found that the amplitudes are especially large along the coast, in the Exp. A, relatively larger amplitudes are distributed around the Oki Island, Tango Peninsula, and Noto Peninsula. In the Exp. B, amplitudes are high at from the north coast of Tango Peninsula to near the coast of Wakasa Bay and the cape of Echizen, but that are low around the Oki Island and western and eastern coast of the Noto Peninsula. In Exp. D, amplitudes are low around the coastal region of Tango and Noto Peninsulas, which are large in the western side of Oki Island.

Next, we investigate the propagation process of NIIWs energy in terms of energy flux. According to Hopkins et al. (2014), we estimated the energy flux of NIIWs $\boldsymbol{F}$ as follows:

$$\mathbf{F} = \frac{1}{T}\int_0^T \int_{-h}^0 p'\mathbf{u}' dz dt,$$

where, h is the water depth and T is near-inertial period ($T = 20$ h), p' is the perturbation pressure calculated from hydrostatic approximation, $\boldsymbol{u}' = (u', v')$is the baroclinic component of horizontal current velocities obtained from the calculation of $(u'(z,t),\ v'(z,t)) = (u(z,t) - \bar{u}(z) - u_0(t),\ v(z,t) - \bar{v}(z) - v_0(t))$ where, $u(z,t)$ and $v(z,t)$ are horizontal current velocities at certain point, $\bar{u}(z)$ and $\bar{v}(z)$ are the vertical profiles time averaged current velocities, and $u_0(t)$ and $v_0(t)$ are the depth averaged current velocities. Each variable used for energy flux estimation are obtained from time series of 20-h period fluctuations reconstructed by amplitudes and phase calculated as the same method of Fig. 11.12. In addition, the total energy of NIIWs TE = KE + PE are also shown in Fig. 11.13b where,

$$KE = \frac{1}{T}\int_0^T \int_{-h}^0 \frac{1}{2}\bar{\rho}\left(u'^2 + v'^2 + w'^2\right) dz dt \text{ and}$$

$$PE = \frac{1}{T}\int_0^T \int_{-h}^0 \frac{1}{2}\frac{g^2\rho'^2}{\bar{\rho}N^2} dz dt$$

From distribution of total energy (Fig. 11.13b), we found that the total energy is high in the generation area of the NIIWs. We can see that NIIWs energy in the Exp. A flows southward off the western coast of Noto Peninsula from its generation

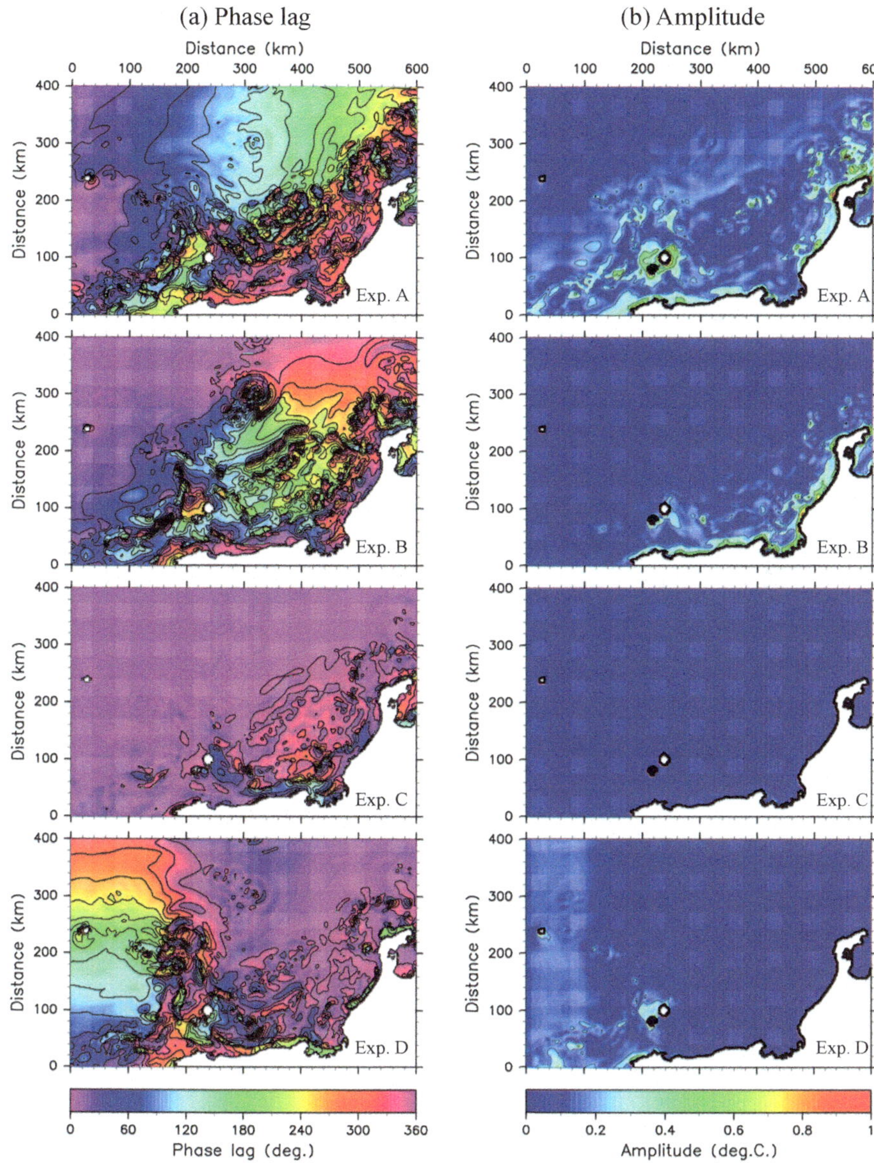

Fig. 11.12 Horizontal distributions of **a** phase lag and **b** amplitude of near-inertial internal waves estimated by using temperature data at a depth of 30 m for each model case

area and flow westward off the Tango Peninsula toward east coast of Oki Islands. In the Exp. B, the NIIWs energy flows southward off the western coast of Noto Peninsula toward the Tango Peninsula and then flows toward offshore from there. In the Exp. C, we cannot see amplification of NIIWs energy. In the Exp. D, NIIWs energy flow in typhoon goes in its generation area and southwestward NIIWs

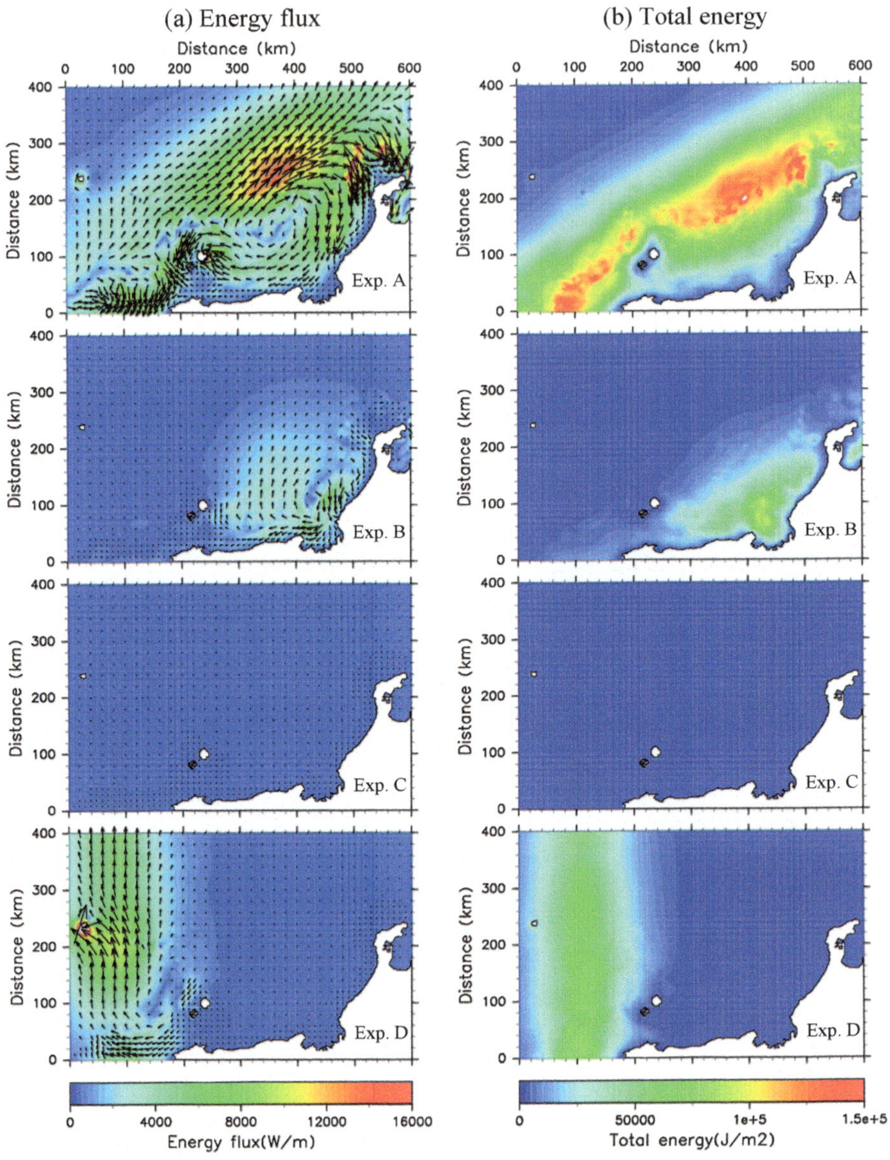

Fig. 11.13 Horizontal distributions of **a** energy flux and **b** total energy of near-inertial internal waves for each model case

energy flux were seen in the western side of Oki Island. In experimental results, the distributions of NIIWs energy flux are corresponding to that of phase lag of NIIWs estimated by harmonic analysis (Fig. 11.13b).

The region where NIIWs are amplified is varied with the different of generation area and propagation process of NIIWs. In the eastern coast of Tango Peninsula, the

NIIWs energy are amplified by the different propagation processes that NIIWs energy propagates southward off the western coast of Noto Peninsula and reached to the Tango Peninsula (Exp. A) and that propagates westward off the western side of Wakasa Bay and reached to the Tango Peninsula (Exp. B). In addition, around the Oki Island, we found that amplifications of NIIWs occur in the eastern coast of Oki Island by that NIIWs generated along the typhoon path propagate eastward off the San'in Coast and reached to eastern side of Oki Island (Exp. A) and that occur in the western coast of Oki Island by eastward propagation of NIIWs generated in the western side of Oki Island (Exp. D).

Finally, moving speed and strong wind area of each typhoon are different as indicated in Table 11.1. Properties of NIPFs vary by difference of moving speed and strong wind area of typhoons, are also considered. Therefore, we performed sensitivity experiments for moving speed and radius of typhoons on the near-inertial fluctuation as follows. Additional cases are varied moving speed of model typhoon every 10 kmh^{-1} from 15 to 95 kmh^{-1}(radius of strong wind area is constant to be 160 km). The other additional cases are varied radius of strong wind area of typhoon every 20 km from 100 to 240 km (moving speed of typhoon is constant to be 55 kmh^{-1}). In both additional cases, typhoon track was the same as Exp. A.

Power spectra density of alongshore current at Sta. 3 in all additional cases shown in Fig. 11.14. The experimental results showed that as typhoon progress speed decreases, the energy level of NIIWs increases (Fig. 11.14a). Similarly, as the radius of typhoon becomes large, the energy level of NIIWs increases (Fig. 11.14b). These results can be considered as a reason why the energy level

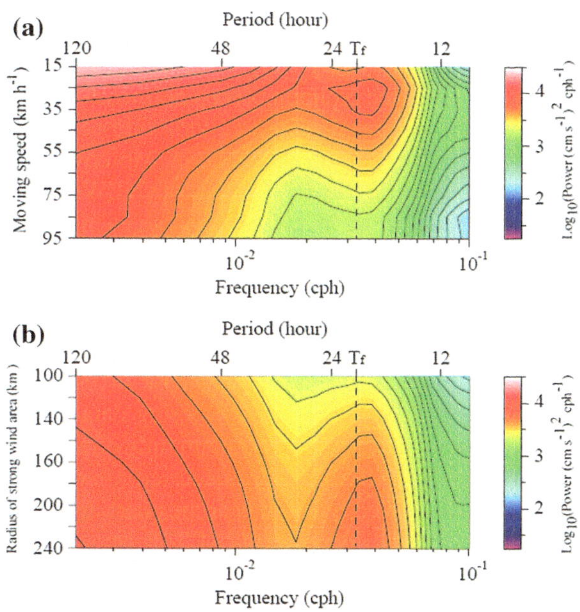

Fig. 11.14 Power spectral density of alongshore current at Station 3

were low in the cases of T10 and T11 that its forcing is small but advanced slowly. In addition, in the cases of T16 and T18 which were similar forcing and tracks, the reason why spectral peak at the inertial period in the case of T16 was clearer than T18 is considered to depend on moving speed of typhoon. From the above results, the region where NIIWs are amplified is varied with radius and moving speed of typhoon in addition to the different of generation area and propagation process of NIIWs was demonstrated.

The case which typhoon passes through the offshore region has been considered as dangerous, because it had excited large near-inertia oscillation in the past. From these result, however, we found that we need pay attention more to a typhoon with large scale which is passing slowly near the target area.

11.6 Conclusion

We showed that energy level and propagation property of NIPFs observed near the coast are varied by difference of typhoon tracks by analysis of mooring observational data. By numerical experiments using idealized typhoons, we found that the NIIWs energy near the Tango Peninsula are amplified by that NIIWs energy generated off the western coast of Noto Peninsula or off Tango Peninsula propagates southward and reaches to around Tango Peninsula and the more generation area of NIIWs are large, the more amplification of NIIW near the coast is large in the case that typhoon advances northeastward over the Japan Sea. In the case which typhoon advances northward off Kyushu, we found that although the NIIWs are generated at wide region in the western side of Oki Island, NIIWs energy generated there do not reach around the Tango Peninsula. Consequently, we concluded that the NIIWs energy are greatly strengthen in the wide region around the Tango Peninsula, when the typhoon passed northeastward off the Japan Sea and the NIIWs were widely generated in the eastern side of Noto Peninsula and off the Tango Peninsula associated with the typhoon. Moreover, slab model and numerical model results showed that amplification of NIIWs can be occur when typhoon is advancing slowly and its radius are large even though the typhoon passed over not Japan Sea but the Honshu. This indicates that we need pay attention more to a typhoon with large scale which is passing slowly near the target area for a forecasting of generation of a strong coastal current due to NIIWs. In this way, we newly organized atmospheric conditions that causes a strong coastal current due to NIIWs around the Tango Peninsula by investigating response of NIIWs on the various-track of typhoons using mooring data and numerical model.

Some issues which cannot explain only effects of typhoon are remained. Second amplification of NIIWs energy occurred after passage of typhoon in the case of Exp. B which moved typhoon near the coast, however, such amplification of NIIWs were not observed after passage of T16 nevertheless T16 advanced in the similar track with Exp. B and its forcing was large. In addition, slab model using realistic wind stress reproduced the amplification of near-inertial period fluctuation, however the amplification has not observed during passage of T21. These results suggested that effects other than typhoon such as sub-inertial currents (e.g. Kunze 1985) influenced the NIPFs around the Tango Peninsula. In the future, there will be needed to investigate the effects of coastal current on the amplifications of NIIWs.

Acknowledgements The authors would like to thank Profs. Jiro Yoshida, Masao Nemoto, Hisayuki Arakawa and Tomoharu Senjyu for their useful discussion. This study was partially supported by the Research Project for Utilizing Advanced Technologies in Agriculture, Forestry, and Fisheries in Japan.

References

As-salek JA (1998) Coastal trapping and funneling effects on storm surges in the Meghna estuary in relation to cyclones hitting Noakhali-Cox's Bazar Coast of Bangladesh. J Phys Ocenogr 28 (2):227–249. https://doi.org/10.1175/1520-0485(1998)028%3c0227:CTAFEO%3e2.0.CO;2

D'Asaro EA (1985) The energy flux from the wind to near-inertial motions in the surface mixed layer. J Phys Oceanogr 15:1043–1059. https://doi.org/10.1175/1520-0485(1985)015%3c1043:TEFFTW%3e2.0.CO;2

Fujita T (1952) Pressure distribution within typhoon. Geophys Mag 23:437–451

Garrett C (2001) What is the "near-inertial" band and why is it different from the rest of the internal wave spectrum? J Phys Oceanogr 31:962–971. https://doi.org/10.1175/1520-0485(2001)031%3c0962:WITNIB%3e2.0.CO;2

Gill AE (1982) Atmosphere-ocean dynamics. Academic, Orlando, Fla., p 662

Gill AE (1984) On the behavior of internal waves in the wakes of storms. J Phys Oceanogr 14:1129–1151. https://doi.org/10.1175/15200485(1984)014%3c1129:OTBOIW%3e2.0.CO;2

Hopkins JE, Stephenson Jr GR, Green JAM, Inall ME, Palmer MR (2014) Storms modify baroclinic energy fluxes in a seasonally stratified shelf sea: inertial-tidal interaction. J Geophys Res. https://doi.org/10.1002/2014jc010011

Igeta Y, Kitade Y, Matsuyama M (2007) Numerical experiment on Kyucho around the Tango Peninsula induced by typhoon 0406. J Oceanogr 63:835–847. https://doi.org/10.1007/s10872-007-0071-0

Igeta Y, Kumaki Y, Kitade Y, Senjyu T, Yamada H, Watanabe T, Katoh O, Matsuyama M (2009) Scattering of near-inertial internal waves along the Japanese coast of the Japan Sea. J Geophys Res 114:C10002. https://doi.org/10.1029/2009JC005305

Igeta Y, Watanabe T, Yamada H, Takayama K, Katoh O (2011) Coastal currents caused by superposition of coastal-trapped waves and near-inertial oscillations observed near the Noto Peninsula, Japan. Cont Shelf Res 31:1739–1749. https://doi.org/10.1016/j.csr.2011.07.014

Isoda Y, Yanagi T, Heung-Jae L (1991) Sea level variations with a several-day period along the southwestern Japan Sea coast. Cont Shelf Res 11:167–182. https://doi.org/10.1016/0278-4343(91)90060-J

Kitade Y, Igeta Y, Fujii R, Ishii M (2011) Amplification of semidiurnal tide observed in the outer part of Tokyo Bay. J Ocenogr 67:613–625

Kumaki Y, Ueno Y, Sobajima N, Matsuyama M (2005) The *Kyucho* current along the Kyoto coast induced by typhoon 0406. Oceanogr Japan. 14:653–664 (in Japanese with English abstract)

Kumaki Y, Kitade Y, Tojima T (2012) *Mae-Kyucho* along the coast of Tango Peninsula. Oceanogr Japan 21:201–217 (in Japanese with English abstract)

Kunze E (1985) Near-inertial wave propagation in gepstrophic shear. J Phys Oceanogr 15:544–565. https://doi.org/10.1175/1520-0485(1985)015%3c0544:NIWPIG%3e2.0.CO;2
Yamazaki K, Kitade Y, Igeta Y, Watanabe T, Senjyu T, Yamada H, Okei N, Ikeda S (2015) Near-inertial internal waves observed near the tip of the northerneast coast of Noto Peninsula, Japan. Oceanogr Japan 24:203–226 (in Japanese with English abstract)

Chapter 12
A High-Resolution Unstructured Grid Finite Volume Model for Currents Around Narrow Straits of Matsushima Bay

Hidekazu Shirai, Ritsuki Kunisato, Shinya Magome, Teruhisa Hattori, Takamasa Takagi, Katsuaki Okabe, Kazufumi Takayanagi and Shigeho Kakehi

Abstract Matsushima Bay is one of the famous cultivation area of oysters. However, it is reported that the seedling collection decline in 2013. It is necessary to clarify this cause in order to take a countermeasure. There is some possibility that strong tidal currents through narrow straits cause oyster larvae in the bay to flow out. It is pointed out that increase of river discharge leads to a large flow of them. This research aims to clarify the influence on the movement of larvae in Matsushima Bay, especially focused on currents and river discharge. In this paper, the numerical model for tidal currents around Matsushima Bay is developed as the first step of the study to clarify their influence on them. The model is based on finite volume method with unstructured grid system and generalized terrain-following coordinate system (FVCOM). The model reproduces the influence on tidal currents by heat flux, precipitation, evaporation, surface wind stress and river discharge. The results are compared with some observation data to verify the validity of the model. The characteristics of currents around Matsushima Bay are also investigated from the result reproduced by the model, focused on the influence of fresh water from rivers.

Keywords Numerical simulation of tidal currents · Matsushima bay
Narrow straits seedling of oysters

H. Shirai (✉) · R. Kunisato · S. Magome · T. Hattori · T. Takagi · K. Okabe · K. Takayanagi
Sanyo Techno Marine, Inc, Osaka, Japan
e-mail: shirai@stm.co.jp

S. Kakehi
Tohoku National Fisheries Research Institute, Shiogama, Japan

T. Komatsu et al. (eds.), *Oceanography Challenges to Future Earth*,
https://doi.org/10.1007/978-3-030-00138-4_12

12.1 Introduction

Matsushima Bay is enclosed in several islands. It is provided with oceanic water through some islands from Open Ocean and with fresh water from rivers. It is good growing environment for oyster. Cultivating oysters is flourishing in this area. In order to maintain them, it is important to comprehend the environment around Matsushima Bay.

There are some researches focused on seawater around Matsushima Bay. Seawater exchange in Matsushima Bay through the several straits is investigated by Watanabe et al. (1972). It is suggested that tidal current is dominant in the bay. The relation between oyster larvae and less saline water are discussed by Sato et al. (1960). In their study, it is confirmed that less saline water flow into Matsushima Bay through Sabusawa strait and Wanigabuchi strait from the outside of the bay. It is pointed out that fresh water from Naruse River is much larger than Takagi River and it flows into Matsushima Bay (Kakehi et al. 2016). Moreover, they indicate the possibility that the topographical changes (around Katsugiura strait) lead to increase the influence of freshwater from the Naruse River.

As mentioned above, there is some possibility that the current around the bay have much influence on the environment around Matsushima Bay. However, it is not much clear how the current, which temporally and spatially changes, influences on the environment in the bay. It is difficult to investigate the temporal and spatial variation of the tidal currents around the bay. Thus, the numerical model for tidal and residual currents around the Matsushima Bay is developed to clarify their influence on environment in the bay in this paper. In order to verify the validity of the model, results of numerical model are compared with some observed data which are conducted in 2014. The characteristics of currents around Matsushima Bay are also investigated from the result reproduced by the model, focused on the influence of fresh water from rivers.

12.2 Numerical Model

There are many islands of various size in Matsushima Bay, which surround the inside bay. Narrow straits along them link inside and outside of the bay, of which the boundary are complicated. Water depth is made deeper toward outside of the bay and changeful around the straits as shown in Fig. 12.1.

In this study, FVCOM (Chen et al. 2003, 2004) is applied to consider the topographical characteristics as mentioned above. It is a three dimensional model based on finite volume method with unstructured grid system and generalized terrain-following coordinate system.

Unstructured grid system has advantage to maintain accurate reproduction of complicated boundary. Figure 12.2 shows the computational domain. This grid system enables the boundary of computational domain to be set along complicated

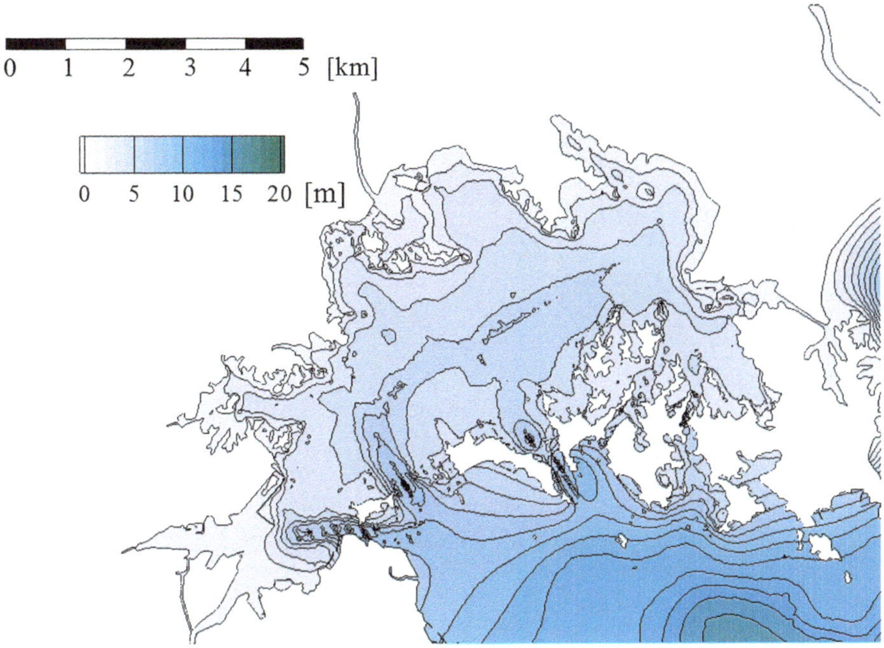

Fig. 12.1 Bathymetry around the Matsushima Bay

straits as shown in the figure. Using the terrain-following coordinate system, an irregular bottom shape bathymetry can be well expressed. Vertical axis is specified as hybrid coordinate system with 8 layers in this study.

The tidal level and velocity components are calculated by continuity and momentum equations. The Smagorinsky eddy parameterization method (Smagorinsky 1963) is used to calculate horizontal diffusion. Mellor and Yamada (1982) turbulent closure model is applied to evaluate the vertical eddy viscosity and vertical thermal diffusion coefficient.

Water temperature and salinity are calculated by temperature and salinity equations which consist of convection term, horizontal and vertical eddy diffusion term and thermal and salt diffusion terms. Water temperature and salinity influence on the currents by the change of density which is calculated by density equation.

The model calculates the tidal and residual currents with consideration of the influence of heat flux, precipitation, evaporation, wind and river discharge by this model. Heat flux are calculated by COARE 2.6 (Fairall et al. 1996) based on bulk method (Liu et al. 1979). Calculated surface net heat flux and shortwave flux are used to give surface boundary condition for water temperature. Surface boundary condition for water temperature is specified by surface net heat flux and shortwave flux. Surface boundary condition for salinity is influenced on by precipitation and evaporation. The absorption of downward irradiance is considered in temperature equation (Paulson and Simpson 1977; Simpson and Dickey 1981). The effect of

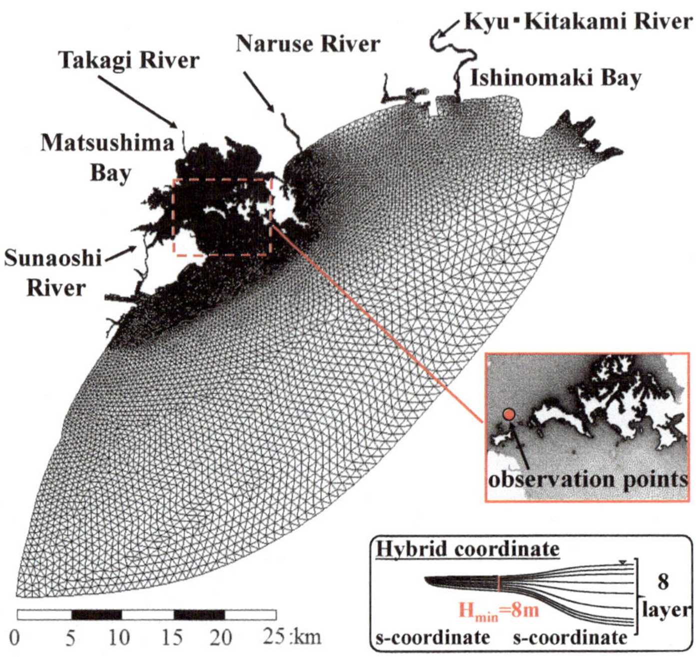

Fig. 12.2 Computational domain

wind is considered as the surface wind stress in momentum equation which is expressed by drag coefficient and wind speed components. The effect of fresh water flow from rivers is also taken into account in the model.

As mentioned above, the tidal level, velocity, water temperature and salinity are calculated with consideration of the influence of heat flux, precipitation, evaporation, wind and river discharge by this model.

12.3 Computational Conditions

Several data are collected to simulate the tidal currents around Matsushima Bay: tidal level, water temperature and salinity at open boundary.

Tidal levels at Sendai and Ayukawa port are used as the open boundary conditions. Figure 12.3 shows tidal levels at Sendai and Ayukawa port calculated from nine tidal components (M2, K1, O1, S2, Sa, P1, K2, N2, Q1). Tidal levels at open boundary are spatially interpolated by these tidal levels.

Observation data around the bay is provided by Miyagi Prefecture Government (www.pref.miyagi.jp/soshiki/mtsc/kankyoutyousa.html). The data is used to set boundary conditions of water temperature and salinity. The observation is conducted at several points around Matsushima Bay about once a month.

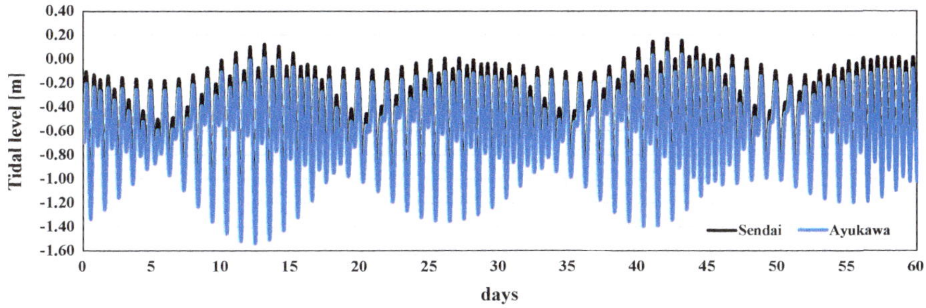

Fig. 12.3 Tidal levels at Sendai and Ayukawa port

Moreover, FRA-ROMS (http://fm.dc.affrc.go.jp/fra-roms/index.html) isalso obtained to consider the temporal changes of temperature and salinity well. FRA-ROMS is forecasting data using ocean modeling system. They provide the data of water temperature and salinity at daily intervals.

In this study, it is considered that heat flux, precipitation, evaporation and wind have influence on current. GPV data (Index of/arch/jmadata/data/gpv/original, http://database.rish.kyoto-u.ac.jp) are obtained to estimate these effects. GPV data is forecasting data relative to weather and climate: air temperature, relative, humidity, air pressure, wind velocity, cloudiness and rainfall and so on. The averaged values of them are used to estimate the effect of heat flux, precipitation, evaporation and wind shear stress. The averaged values are calculated from the data inside of the computational domain.

It is important to consider influence of fresh water from rivers on temporal variation and spatial distribution of salinity around the bay. Takagi River, Naruse River and Kyu-Kitakami River are taken into account of in this study. Takagi River is one of the representative rivers inside of Matsushima Bay. Naruse River and Kyu-Kitakami River are outside of the bay. These discharges are estimated from rainfall in each river basin area.

The tidal and residual currents around Matsushima Bay are simulated in the period between July 1 2014 and August 31 2014, using the data mentioned above.

12.4 Results and Discussion

Numerical model is developed and applied to currents around Matsushima Bay under the condition mentioned above. Tidal and residual currents are simulated in the period between the beginning of July 2014 and the end of August. The results of numerical simulation are compared with observation data to verify the validity of the model. Observation point passes through the main strait as shown in Fig. 12.2. Tidal level, velocity, water temperature and salinity were observed from June to September, 2014.

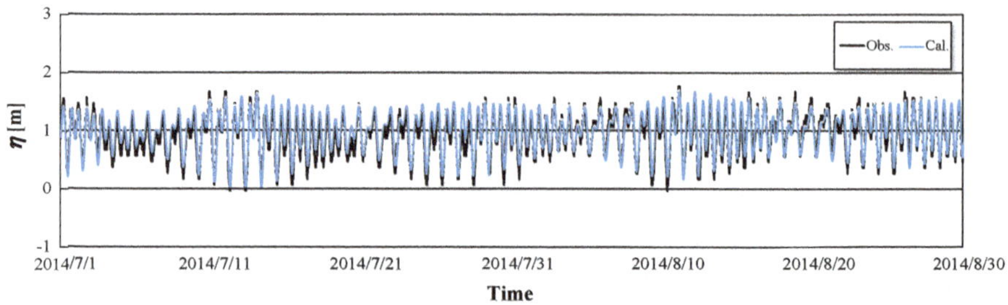

Fig. 12.4 Free surface elevation at St.1

Figure 12.4 shows temporal changes of tidal level. The blue line is the result of numerical simulation. The black line indicates observation. The tidal level of simulation well agrees with the observations.

However, tidal current ellipses (M2 component) do not agree well with each other as shown in Fig. 12.5. Numerical results of velocity components are overestimated as shown in the figure.

One of the reasons is that the spatial distribution of velocity is varied around this area as shown in Fig. 12.6. It is also possible that the spatial uniformity of wind speed components caused the overestimation of the flow velocity. So further investigation is needed to enhance the reproducibility of velocity in the model.

The water temperature at observation point simulated by the model is compared with observation in Fig. 12.7. Blue line and black line indicate the results of numerical simulation and observation. Numerical results of water temperature are generally 1–2 degree Celsius lower than observed data at 1 m depth. However, the temporal variation of water temperature at 1 m depth is reproduced well in the numerical model. Moreover, numerical result of water temperature at 1 m above the bottom are comparatively good agreement with observed value.

Figure 12.8 shows the comparison of salinity of simulation result and observation. The numerical model reproduces 1–2 higher than observation. But the temporal variation of salinity at both surface and bottom layers are reproduced well by the model. This figure also shows that they have things in common, compared with salinity of simulation result and observation. The both salinity of simulation and observation comparatively largely decrease during floods in Naruse River (18 Jul 2014–22 Jul 2014, 6 Aug 2014–11 Aug 2014).

Figure 12.9 show the spatial distributions of salinity. They are horizontal distribution of salinity at surface layer and cross-sectional distribution of salinity during the flood. Then, salinity at observation point is largely decreased. This figure indicates that less saline water from Naruse River spreads along the coastal line and flow into Matsushima Bay through Katsugiura strait and the other straits. It results in large decrease of salinity in Matsushima bay. Thus, it is found that the fresh water from Naruse River outside of the bay have much influence on the inside of

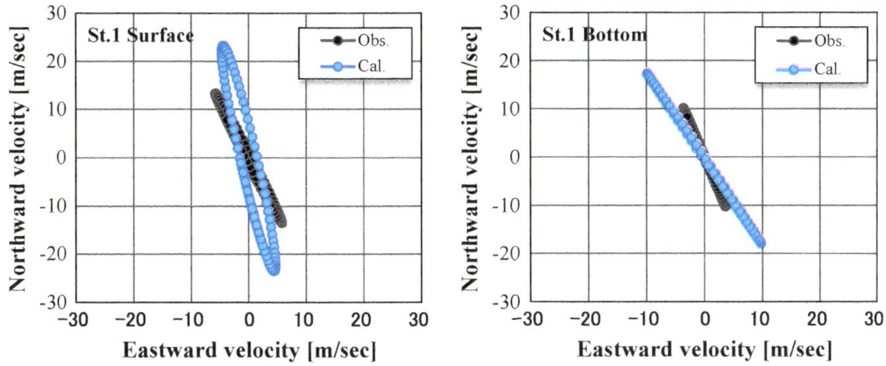

Fig. 12.5 Tidal current ellipses at 1 m depth and at 1 m above the bottom (St.1)

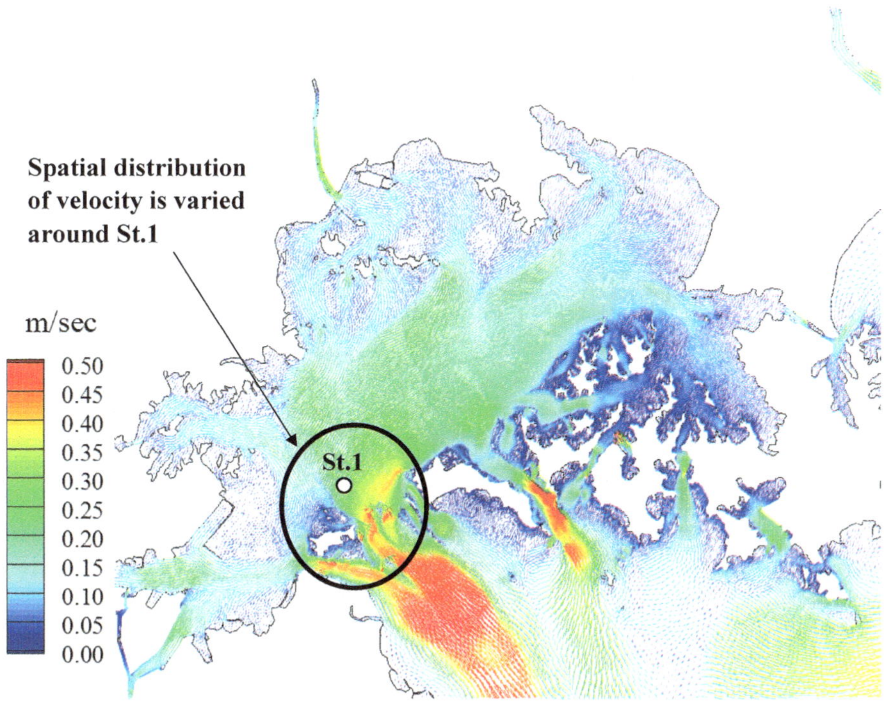

Fig. 12.6 Spatial distribution of velocity around Matsushima Bay

the bay after flood. This result agrees with the measurement results by Sato et al. (1960) and akehi et al. (2016). It is concluded that the model can reproduce the characteristics of the tidal and residual currents around the Matsushima bay.

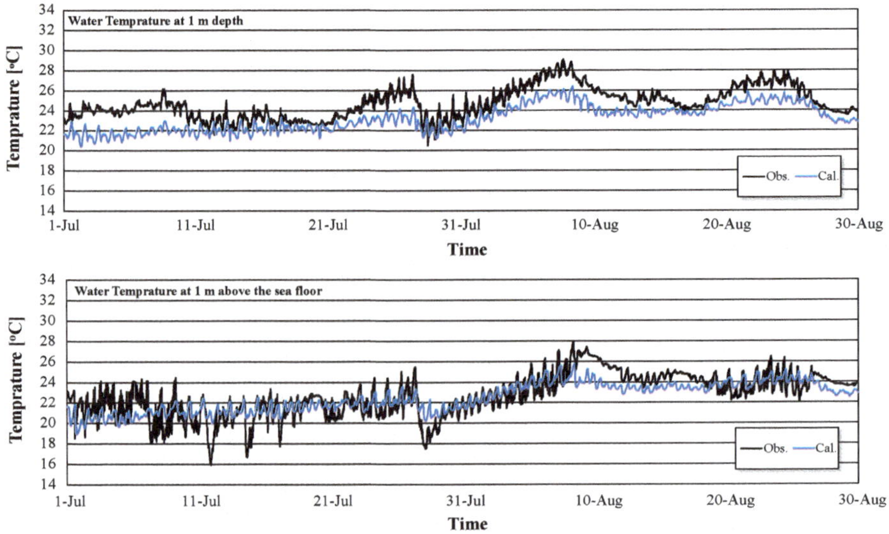

Fig. 12.7 Water Temperature at 1 m depth and at 1 m above the bottom (St.1)

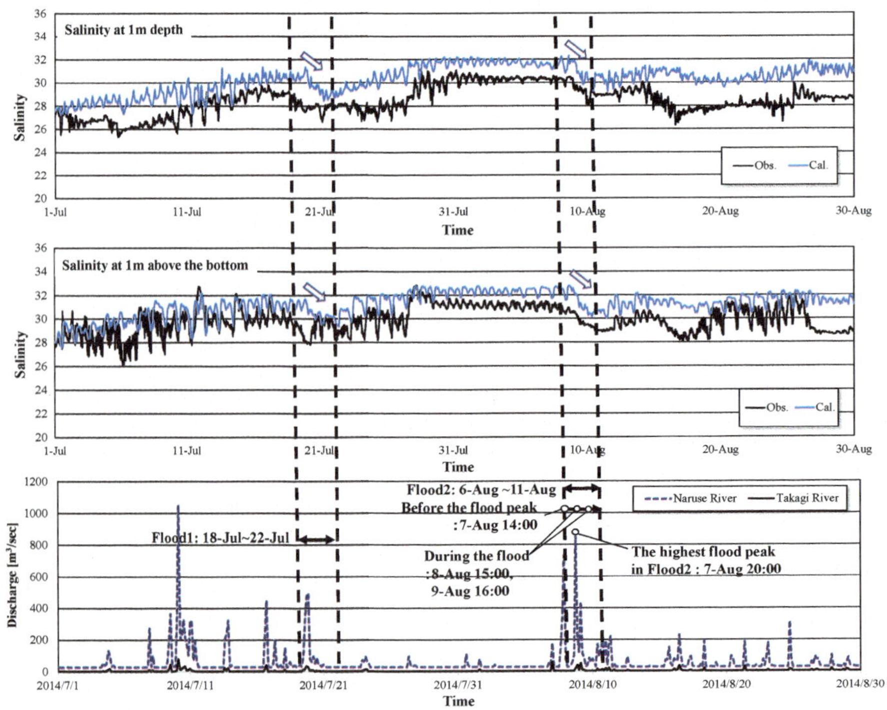

Fig. 12.8 Salinity at 1 m depth and at 1 m above the bottom (St.1)

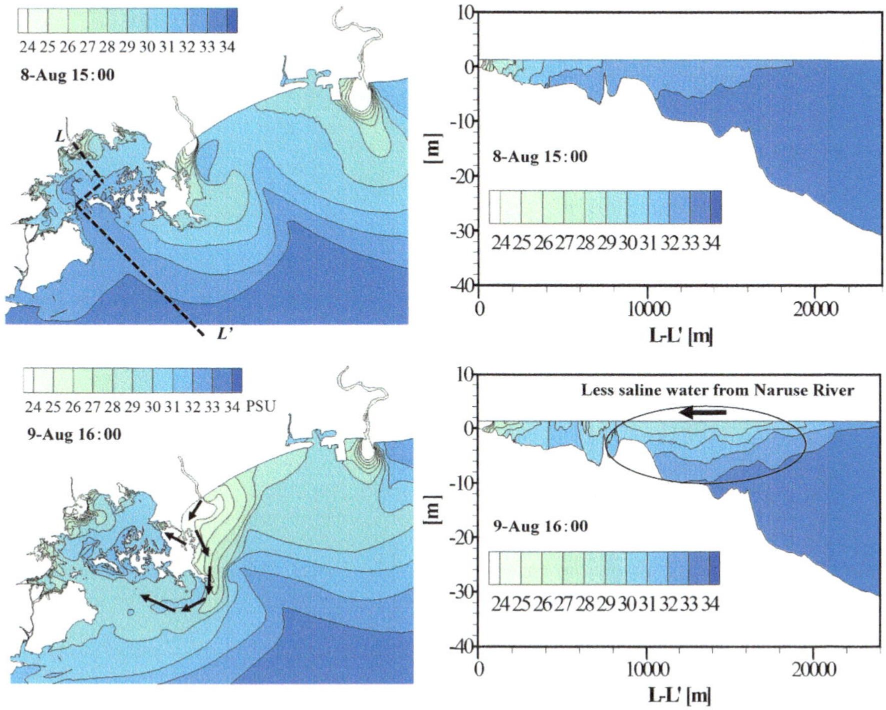

Fig. 12.9 Horizontal and cross-sectional distributions of salinity

12.5 Conclusion

In this paper, the numerical model for tidal currents around narrow straits of Matsushima Bay was developed. The influence on tidal currents by heat flux, precipitation, evaporation, surface wind stress and river were considered in the model. To verify the validity of the model, the result of numerical simulation was compared with observation data which were carried out between June and September 2014.

It was confirmed that the model could calculate the tidal and residual currents around the narrow straits in details and reproduce the temporal variation of water temperature and salinity to some degree. The model reproduces the influence on Matsushima Bay by the river outside of the bay, Naruse River.

Further investigations should be conducted to improve the reproduction of the model by considering the spatial distribution of heat flux, precipitation, evaporation, surface wind stress. For the future, the tracer advection movements on the tidal and residual currents will be conducted by the improved model to predict the movement of oyster larva and to clarify the mechanism of them.

Acknowledgements This project is founded by Ministry of Agriculture, Forestry and Fisheries. In this study, FVCOM3.2.2 are applied with Prof. Changsheng Chen (University of Massachusetts – Dartmouth)'s consent. FRA-ROMS data are provided by Dr. Takashi Setou (National Research Institute of Fisheries Science). GPV data is provided by Japan Meteorological Agency. The data of water temperature and salinity in Matsushima Bay provided Miyagi Prefecture Government are used through the simulations. We are deeply grateful for them.

References

Chen C, Liu H, Beardsley RC (2003) An unstructured, finite volume, three dimensional, primitive equation ocean model: application to coastal ocean and estuaries. J Atmos Ocean Technol 20:159–186

Chen C, Cowles G, Beardsley RC (2004) An unstructured grid, finite-volume coastal ocean model: FVCOM User Manual. UMASS-Dartmouth Technical Report-04-0601. University of Massachusetts, School of marine Science and Technology, New Bedford, MA, p 183

Fairall CW, Bradley EF, Rogers DP, Edson JB, Young GS (1996) Bulk parameterization of air-sea fluxes for tropical ocean-global atmosphere coupled-ocean atmosphere response experiment. J Geophys Res 101:3747–3764

Kakehi S, Kamiyama T, Abe H, Hanawa S, Oota H, Matsuura R, Oshino A (2016) Mechanisms leading to the decline in Pasific oyster Crassostrea gigas seedling in Matsushima Bay, Japan. Fish Sci 82:499–508

Liu WT, Katsaros KB, Businger JA (1979) Bulk parameterization of air–sea exchanges of heat and water vapor including the molecular constraints at the interface. J Atmos Sci 36:1722–1735

Mellor GL, Yamada T (1982) Development of a turbulence closure model for geophysical fluid problem. Rev Geophys Space Phys 20:851–875

Paulson CA, Simpson JJ (1977) Irradiance measurements in the upper ocean. J Phys Oceanogr 7:952–956

Sato S, Kanno H, Kikuchi S (1960) On the low-salinity green water and oyster larvae during the seed-oyster-collecting season in Matsushima Bay. Bull Tohoku Region Fish Res Lab 16:87–117 (In Japanese)

Simpson JJ, Dickey TD (1981) The relationship between downward irradiance and upper ocean structure. J Phys Oceanogr 11:309–323

Smagorinsky J (1963) General circulation experiments with the primitive equations, I. The basic experiment. Mon Weather Rev 91:99–164

Watanabe K, Abe K, Sato K, Suzuki K (1972) Current structures in Matsushima Bay -mainly about volume transport and current velocity. Bull Miyagi Pref Fish Inst 6:89–138 (in Japanese)

Chapter 13
Observation of Near-Bottom Current on the Continental Shelf Off Sanriku

Daigo Yanagimoto, Kiyoshi Tanaka, Shinzou Fujio, Hajime Nishigaki and Miho Ishizu

Abstract We measured current velocity sections including near-bottom current within 10-m range above the bottom about 100-m depth off Otsuchi Bay with using a subsurface towfish mounting an acoustic Doppler current profiler. Four-time navigating observations showed the transition of the current distribution through flood tide. The near-bottom current has smaller spatial scales than the upper currents and its temporal variation is not in phase with tidal current which has a baroclinic structure in the upper layers. We did not find organized near-bottom current widely perhaps due to absence of the bottom boundary layer.

Keywords Near-bottom current · Towed acoustic doppler current profiler Continental shelf · Otsuchi bay

13.1 Introduction

The giant tsunami caused by the earthquake on 11 March 2011 heavily changed marine environments in Otsuchi Bay, a ria bay in Sanriku coast. Clarification of water exchange between outer and coastal oceans is very helpful for reconstruction of aquaculture industry based on the marine environments. Bottom boundary layer on the continental shelf is thought to play an important role in the water exchange across the front between outer and coastal waters. Over Middle Atlantic Bight off the east coast of North America, where coastline extends north to south and coastal current flows southward like those of Sanriku coast, offshore bottom boundary current across the shelf break front is indicated due to the effect of bottom friction under a vertically uniform along-shelf flow near the shelf break (Gawarkiewicz and Chapman 1991). On the other hand, massive cold water intrusion, which is thought

D. Yanagimoto (✉) · K. Tanaka · S. Fujio · M. Ishizu
Atmosphere and Ocean Research Institute, University of Tokyo, Tokyo, Japan
e-mail: daigo@aori.u-tokyo.ac.jp

H. Nishigaki
Faculty of Education and Welfare Science, Oita University, Oita, Japan

T. Komatsu et al. (eds.), *Oceanography Challenges to Future Earth*,
https://doi.org/10.1007/978-3-030-00138-4_13

to originate in the bottom layer over the shelf, often occurs in Otsuchi Bay (Furuya et al. 1993). It is not known how the deep outer water runs aground the shelf break and comes inshore, and whether the bottom boundary current interact with the cold water intrusion or not and how if so. In order to clarify these issues, we first need to investigate near-bottom current distribution widely over the continental shelf regardless of whether the near-bottom current is bottom boundary one.

Many current measurements have been done with using mooring arrays for clarifying dynamics of bottom boundary current (Williams et al. 1987, and so on). For investigating current distribution in a wide area, navigating observations with shipboard acoustic Doppler current profilers (ADCPs) are more suitable than mooring observations. Tanaka et al. (2015) clearly showed a baroclinic tidal current structure with a node at about 30-m depth at the mouth of Otsuchi Bay in summer with using shipboard ADCPs. However, it is difficult to measure near-bottom current by shipboard ADCPs. ADCPs generally have three or four transducers emitting sound beams which tilt at 10–25° from vertical direction and have side lobes. Even if downward beams reach the bottom, we cannot measure currents precisely within the lowest 10% range from the seafloor to the instrument due to contamination by bottom reflection of the side lobes. That is, shipboard ADCPs have data blanks within 10-m range above the seafloor at 100-m depth for example. Because the bottom boundary layer typically exists within this range, we try to use a subsurface towfish mounting an ADCP close to the bottom.

We aim to establish the method of direct, wide-area measurement of near-bottom current within 10-m range above the seafloor and to grasp the spatial distribution of near-bottom current speed and its temporal variation on the continental shelf off Otsuchi Bay.

13.2 Methods

We use a towfish "V-fin" mounting a 500-kHz ADCP "ADP 500" manufactured by Xylem, whose pressure rating is 500 m and nominal profile range is 70–120 m. Because ADP 500 has only three transducers, echo intensity and signal-to-noise ratio (SNR) are only parameters for validation of current measurement. According to Xylem's manual, the minimum among three SNR values should be higher than 3 dB in order to obtain significant signal. NOAA Technical Report also mentions that the lower limit of SNR is 3–5 dB for good profile measurement by ADP while SNR over 15 dB is recommended (Shih et al. 2001).

We must monitor the depth of V-fin in real time so that it is surely towed at 50-m height from the seafloor or lower in order to measure near-bottom current. However, V-fin is towed with not communication cable but just wire rope. We adopted a very handy acoustic telemetry system of Vemco, Canada, which is generally used in bio-logging field. A transmitter "V16TP" equipped with pressure sensor with a weight of about 20 g was tied to a shackle connecting the V-fin to wire rope. We keep receiving telemetry signals from the transmitter by an omnidirectional hydrophone "VH165", which is hung down to the sea and is connected to a receiver "VR100" by a cable.

On 30 July 2015, we boarded R/V Yayoi (12 ton, 20 persons), a vessel of International Coastal Research Center, Atmosphere and Ocean Research Institute, and towed the V-fin about 50-m depth for about 2 nautical miles between C1 and C2 along 100-m isobaths off Otsuchi Bay (Fig. 13.1).

We set ADP 500 mounted at the bottom of V-fin to measure current profiles downward with cell sizes of two meters by intervals of 10 s. As summarizing in Table 13.1, we towed it four times: 09:05–10:07 (observation line L1), 10:25–10:59 (L2), 11:06–11:40 (L3), and 13:47–14:19 (L4).

All these lines were done during flood tide at Kamaishi and Miyako, the south and north neighbouring port respectively, from 9:00 to 16:00 (Fig. 13.2). The observation lines L1 and L2 were done at the early period of flood tide, L3 at its middle period, and L4 at its late period. We also used a ship-mounted ADCP (RDI Workhorse 300 kHz) of R/V Yayoi which can measure current profiles with nominally 120-m range, and made CTD/O_2 observations at C1 and C2 (not shown in this article).

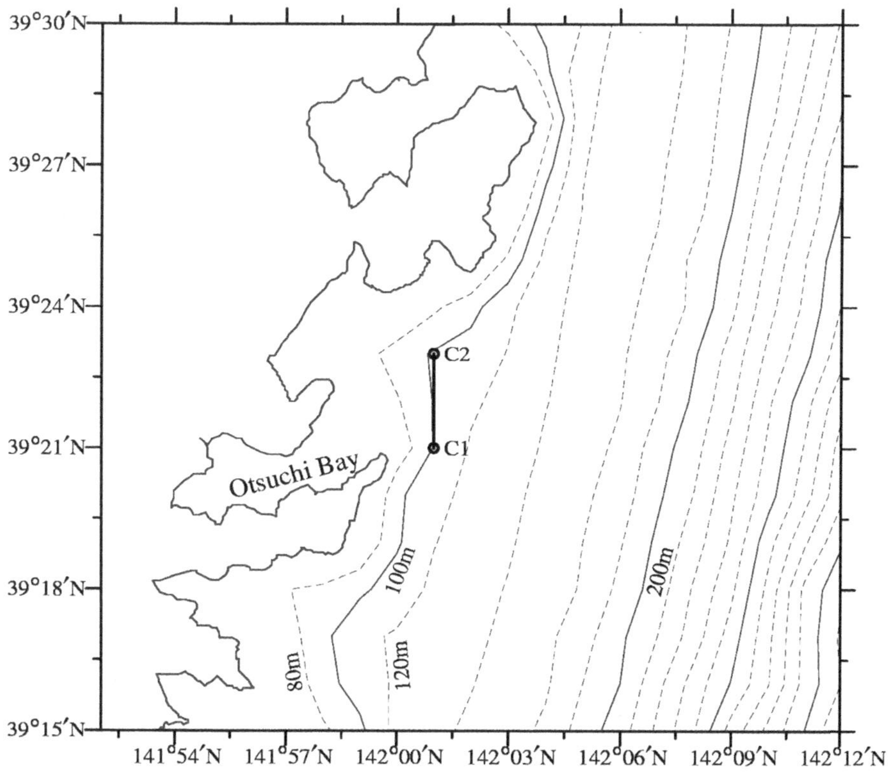

Fig. 13.1 Observation area off Otsuchi Bay, Sanriku coast. Navigating observation with ship-mounted and towed ADCPs was operated between C1 and C2. Isobaths are based on 1 Arc-Minute Global Relief Model (ETOPO 1)

Table 13.1 Observation lines along 100-m isobaths off Otsuchi Bay on 30 July 2015. V-fin was towed with speed of 2 knot along L1, otherwise 4 knot

Line no.	Start				End			
	Time (JST)	Latitude	Longitude	Depth (m)	Time (JST)	Latitude	Longitude	Depth (m)
L1	09:05	39°22.926′N	142°01.087′E	105	10:07	39°20.990′N	142°01.002′E	107
L2	10:25	39°21.002′N	142°01.003′E	107	10:59	39°23.015′N	142°00.997′E	103
L3	11:06	39°23.015′N	142°01.026′E	104	11:40	39°20.983′N	142°00.996′E	103
L4	13:47	39°21.005′N	142°01.000′E	108	14:19	39°23.007′N	142°01.002′E	104

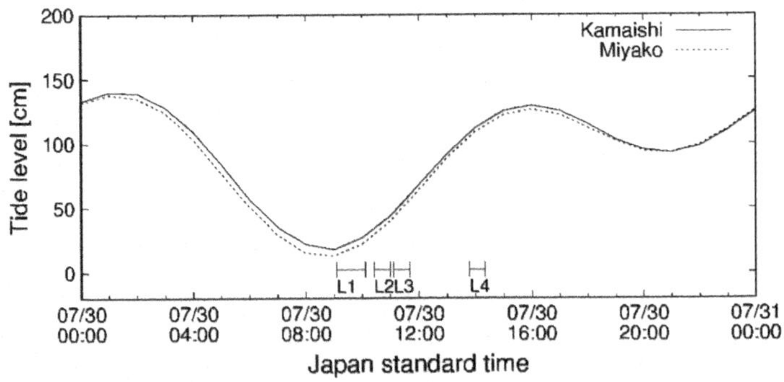

Fig. 13.2 Astronomical tidal levels at Kamaishi and Miyako on 30 July 2015. Otsuchi Bay locates between them. We used tide levels published by Japan Meteorological Agency

13.3 Results

13.3.1 Measurement Quality of ADCP

In order to remove contamination of the acoustic reflection from the bottom, we simply cut out data within the lowest 10% range of the vertical distance between V-fin and the bottom. In the remaining V-fin data, any SNR is higher than 5 dB through four sections (Fig. 13.3).

They are rather high values for determining that the current velocity measured by V-fin is significant in the whole section at every line. Because SNR lower than 15 dB distributes widely over the bottom as shown by black points in Fig. 13.3, we need to pay attention to statistical velocity error which should be decreased by some average operating. Echo intensity is higher than 12 dB through four sections. On the other hand, the ship-mounted RDI 300 kHz ADCP obtained enough high quality data with the sum of percent goods 1 and 4 over 75% through the whole sections.

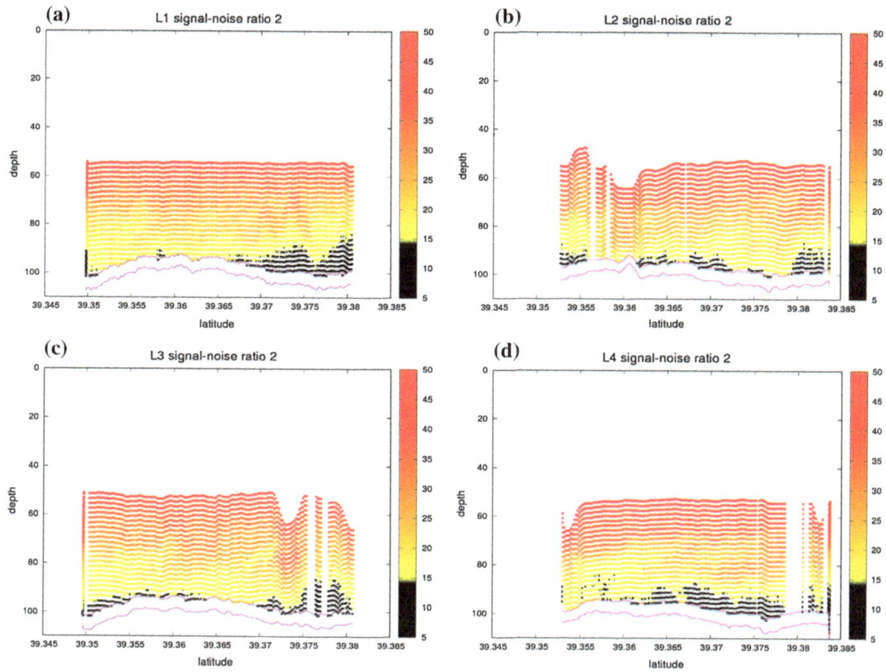

Fig. 13.3 Sections of signal-to-noise ratio (dB), SNR, by V-fin along L1 (**a**), L2 (**b**), L3 (**c**) and L4 (**d**) (from top left to bottom right). The bottom and the upper limit of the blank are shown by curves. Black points show SNR lower than 15 dB

13.3.2 Velocity Field

The ship-mounted ADCP of R/V Yayoi and the towed ADCP of V-fin enabled us to obtain current profiles from 9-m depth to 5-m height above the bottom. We combined these datasets with matching the V-fin current with the ship-mounted ADCP current at 70-m depth. Figure 13.4 show the sections of eastward (offshore) velocity components (U) gridded by 0.001° horizontally and 2.5 m vertically. Except for L1 when flood tide just began, there are two U cores opposite to each other above 60-m depth from L2 to L4. Average U is 2.8–3.3 cms^{-1} in the offshore core around 20-m depth and −2.6–−8.2 cms^{-1} in the inshore core around 50-m depth. These two cores are robust during flood tide.

The inshore-U core around 50-m depth sometimes reaches the bottom especially in the middle portion of the line; around 39.37°N and 39.38°N at L2 and around 39.37°N at L4.

However, offshore U often becomes dominant with a relatively strong core around 80-m depth especially in the southern portion of the line. Average U at 80-m depth south of 39.36°N is 10.9 cms^{-1} at L3 and 6.5 cms^{-1} at L4. Offshore currents

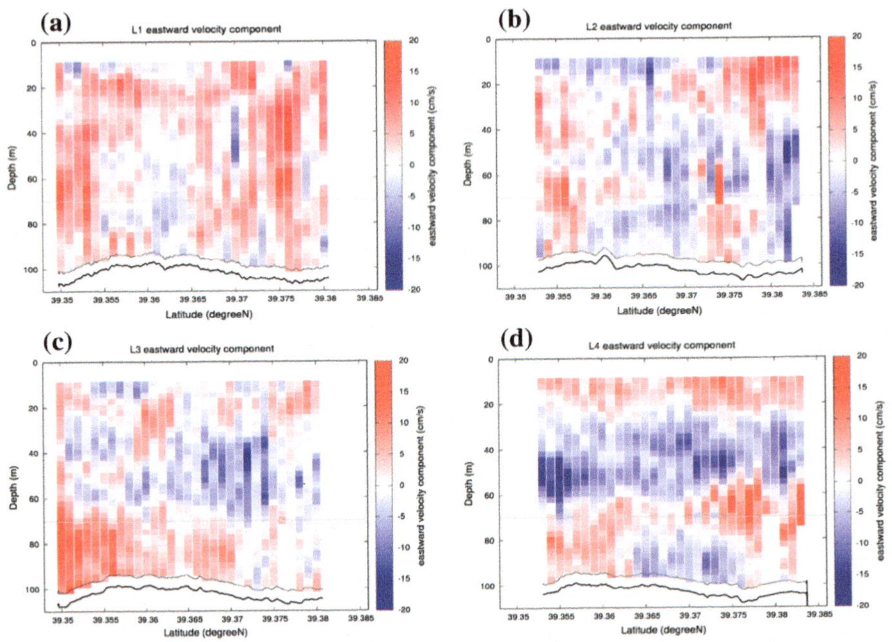

Fig. 13.4 Sections of eastward current velocity component at L1 (**a**), L2 (**b**), L3 (**c**) and L4 (**d**) (from top left to bottom right). Current velocity measured by the ship-board ADCP above 70-m depth and by the V-fin below 70-m depth are combined and flattened into 0.001° × 2.5 m grids by Gaussian filter. The bottom and the upper limit of the blank are shown by curves. The depth of 70 m is shown by grey lines

appear also in the other places below 60-m depth and most of them reach the bottom.

We flatten the near-bottom current vectors furthermore with vertical mean under 10-m height above the bottomand temporal 5-min running mean along the bottom (Fig. 13.5).

They are also not apparently coherent. Southeastward (offshore) current seems to be dominant mainly in the southern portions of L1, L3 and L4. In the other places, current direction changes by observation; northward (alongshore) at L1 and L2, eastward (offshore) at L3, and westward (inshore) at L4. These do not occur in phase with the tide observed in the upper layers.

13.4 Discussion

Current velocity sections over the continental shelf off Otsuchi Bay were observed from the upper layer at 10-m depth to the lowest layer within 10-m range above the bottom about 100-m depth with a ship-mounted ADCP and a towfish mounting an

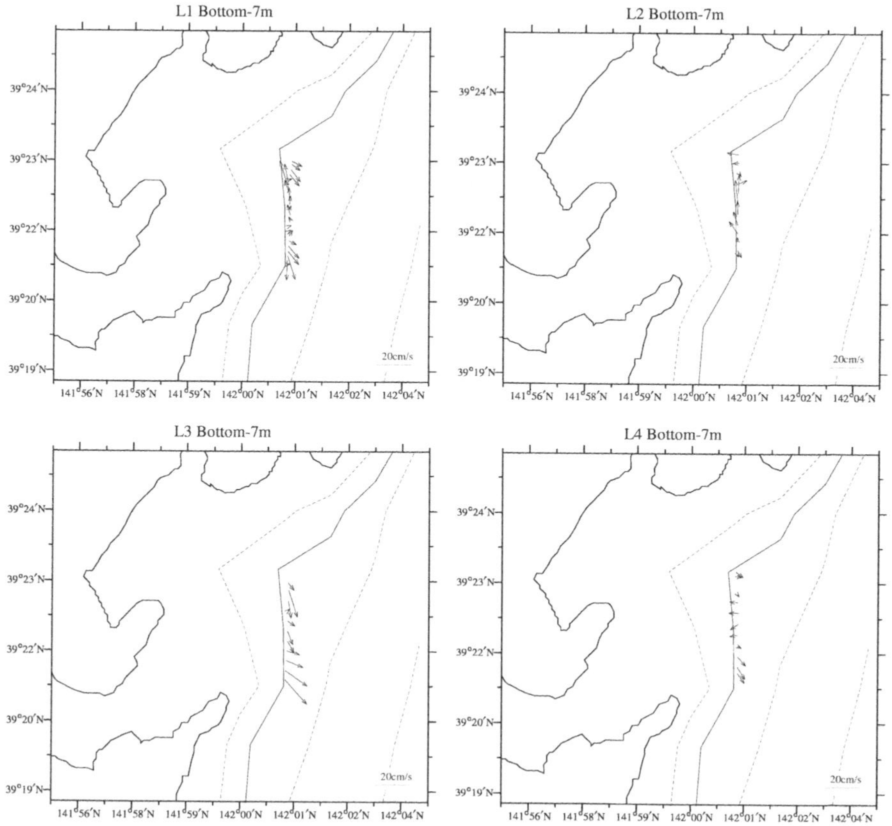

Fig. 13.5 Current distributions near the bottom by V-fin along L1, L2, L3 and L4 (from top left to bottom right). Current velocity components are averaged vertically within 10-m height above the seafloor and flattened by temporal 5-min running mean

ADCP. We obtained the transition of spatial current distribution from the start to the end of flood tide. In the upper layers, we found that zonal current velocity components are offshore around 20-m depth and inshore around 50-m depth during flood tide. Baroclinic tidal current structure, which is clearly observed at the mouth of Otsuchi Bay by Tanaka et al. (2015), is indicated also over the continental shelf. On the other hand, offshore near-bottom velocity component which was dominant especially in the southern portion of the observation line suggests more complex current structure. It is not apparent whether tide causes such a complex current structure. We need to observe near-bottom current also during ebb tide.

CTD/O_2 observations operated at the both end points of ADCP lines did not indicate the bottom boundary layer unfortunately. If the bottom boundary layer develops, we expect that organized bottom boundary current would be found widely. We need to pile up current data by further V-fin surveys in various conditions such as in winter time, under a strong southward coastal current, and so on.

Also we need to try V-fin surveys at the shelf break, where the bottom boundary layer may play the most important role in water exchange between the outer water and the coastal water.

Acknowledgements The authors would like to thank Captain M. Hirano and his crew of R/V Yayoi. The authors would also thank the Tohoku Ecosystem-Associated Marine Sciences for providing financial support to this project.

References

Gawarkiewicz G, Chapman DC (1991) Formation and maintenance of shelfbreak fronts in an unstratified flow. J Phys Oceanogr 21:1225–1239
Furuya K, Takashi K, Iizumi H (1993) Wind-dependent formation of phytoplankton spring bloomin Otsuchi Bay, a Ria in Sanriku, Japan. J Oceanogr 49:459–475
Shih HH, Sprenke J, Payton C, Mero T (2001) Towing basin speed verification of acoustic doppler current profiling instruments. NOAA Technical Report NOS CO-OPS 033, 53 p
Tanaka K, Komatsu K, Itoh S, Yanagimoto D, Ishizu M, Hasumi H, Sakamoto TT, Urakawa S, Michida Y (2015) Baroclinic circulation and its high frequency variability in Otsuchi Bay on the Sanriku ria coast, Japan. J Oceanogr Special issue: Oceanographic observations after the 2011 earthquake off the Pacific coast of Tohoku. https://doi.org/10.1007/s10872-015-0338-9
Williams AJ 3rd, Tochko JS, Koehler RL, Grant WD, Gross TF, Dunn CVR (1987) Measurement of turbulence in oceanic bottom boundary layer with an acoustic current meter array. J Atmos Ocean Technol 4:312–327

Part IV
Innovative Research

Chapter 14
Control of Pressure-Driven Microdroplet Formation and Optimum Encapsulation in Microfluidic System

Mathias Girault, Akihiro Hattori, Hyonchol Kim, Kenji Matsuura, Masao Odaka, Hideyuki Terazono and Kenji Yasuda

Abstract Formation of stable micro-droplets in multiphase flow is an important step to perform numerous microfluidic applications such as sorting experiments. We herein investigate the conditions of formation of stable micro-droplets using a flow focusing microfluidic device. Two single phases and four different multiphase flow regimes were observed depending on the pressures of fluids. By tuning sample stream pressure against fixed lower oil stream pressure, stable droplet regime can create microenvironment with a diameter ranged from 30 μm to 140 μm. Results obtained show that the formation of strictly size controlled droplets can encapsulate single cell-sized bead into droplet. Moreover, the limit between unstable and stable droplet regimes was the most suitable to efficiently encapsulate cell-sized bead in droplet sorting application. This limit can be precisely monitored by using the change of the droplet speed found at the threshold between these two regimes.

Keywords Microfluidics · Encapsulation · Flow control

M. Girault · A. Hattori · H. Kim · K. Matsuura · M. Odaka
Kanagawa Academy of Science and Technology, KSP East 310, 3-2-1 Sakado, Takatsu, Kawasaki 213-0012, Japan

H. Kim · M. Odaka · H. Terazono
Institute of Biomaterials and Bioengineering, Tokyo Medical and Dental University, Chiyoda, Tokyo 101-0062, Japan

K. Yasuda (✉)
Department of Physics, Faculty of Science and Engineering, Waseda University, 3-4-1 Okubo, Shinjuku-ku, Tokyo 169-8555, Japan
e-mail: yasuda@yasuda-lab.jp

K. Yasuda
Waseda Bioscience Research Institute in Singapore (WABIOS), 11 Biopolis Way #05-01/02, Helios, Singapore 138667, Singapore

K. Yasuda
CREST, Japan Science and Technology Agency, Kawaguchi, Japan

T. Komatsu et al. (eds.), *Oceanography Challenges to Future Earth*,
https://doi.org/10.1007/978-3-030-00138-4_14

14.1 Introduction

Most of microfluidic cell sorter systems need to master two techniques: (i) create stable droplets suitable for biological assays and (ii) control the path of each droplet in the channel. Except specific object ordering methods, common studies mainly used random cell encapsulation to capture cells in droplet (Edd et al. 2008; Köster et al. 2008). According to the Poisson's distribution, the encapsulation can be tuned by modifying the concentration of cells in the sample and the volume of each droplet. However, in random case the number of droplets containing a single cell inside can reach only 36.7% of the total number of droplets (Mazutis et al. 2013). This theoretical maximum percentage is difficult to reach in the low range of flow rates due to the swimming speed of some planktons and sedimentation of cells in the channel. Sedimentation of cells is scarcely investigated in the literature by comparison to other parameters such as droplet speeds or sorting ability. However, development of an integrate lab-on-a-chip design needs to better characterize the range of flow rates suitable to efficiently encapsulate cell.

In sorting microfluidic applications, the formation of droplets was usually performed by two major methods named: (i) flow rate controlled and (ii) pressure controlled systems (Gu et al. 2011; Mazutis et al. 2013). In a flow rate controlled system, the syringe pump maintains a specified volume flow rate by adjusting the force applied to the syringe. In contrast, the pressure controlled system is a static pressure approach, where the flow rate can vary during the droplet formation. Although no difference between these two methods was found in a single-phase flow, significant variations in the droplet regime were reported in the multiphase flow (Ward et al. 2005). Droplet formation in a multiphase flow is usually interpreted by a competition between interfacial, inertia and viscous forces of fluids (Zhao et al. 2006). When the Bond (Bo), Reynolds (Re) and Weber numbers (We) are lower than 1 in most microfluidic applications, the viscous force and interfacial forces are the dominant forces to control the flow pattern (Gu et al. 2011). However, calculation of these forces is particularly difficult because they depend on several factors such as the wetting properties of the materials as well as channel geometry of the chip (Salim et al. 2008). Consequently, no general scaling law can precisely predict the transition between each droplet regimes (Gu et al. 2011). In this context, we investigated a method using a pressure-controlled system in order to create an emulsion of homogenous micro-droplets containing objects like-cell with a diameter size up to 20 µm. The herein study also describes the different flow regimes with a particular focus on the generation of stable droplet flow needed to develop sorting microfluidic applications.

14.2 Materials and Methods

14.2.1 Microchannel Design

Rectangular microchannel fabrication was achieved by soft lithography using a poly (dimethylsiloxane) (PDMS) product (Duffy et al. 1998). The height of channels was 25 μm. The width of oil channels was 100 μm. Sample channel width was 100 μm from the sample inlet to the distance of 200 μm before the flow focussing area. From these distance, the width of sample channel became narrower and reached 50 μm at the flow focussing area. Narrow sample channel was designed to limit backflow of oil into the sample channel upstream when the pressure ratio P_{sample}/P_{oil} was lower than 0.5. The width of the downstream channel was 140 μm. A water repellent reagent [5v/v% of 1H, 1H, 2H, 2H—(Heptadecafluorodecyl) trichlorosilane (Wako)] in pure hydrofluoroether HFE-7300 (Novec) was used as an hydrophobic solution to coat the microchannel (Tice et al. 2003; Mazutis et al. 2013).

14.2.2 Droplets Formation and Optical Setup

A dual pumps system composed of two independent compact air cylinders (Misumi, MSCCN50-50) was used to generate droplets by flow focusing an aqueous stream with two streams of fluorinated oil (FC-40) containing a surfactant (Dolomite, Pico-Surf I; 2 *v/v%*) (Anna et al. 2003; Baroud et al. 2010). Pressure of each cylinder of the dual pumps system was controlled by using a set of two air pressure sensors (Keyence, AP-C30, detection resolution 0.1 kPa). In this present study, pressures are reported relative to the atmosphere. An inverted microscope IX71 (Olympus) equipped with a high speed and high sensitivity color complementarity metal-oxide-semiconductor (CMOS) camera (Thorlabs GmbH, DCC3240C) was used to record videos. The CMOS camera was connected to the computer where the software Lab VIEW 2013 was installed. Ten minutes after the stabilization of the pressures in the inlets, the different multiphase flows were recorded for video analysis. In this present study, stable droplet flow pattern is defined as a single droplet flow centred in the microchannel and moving linearly in the channel.

14.2.3 Bead Encapsulation

Encapsulation experiments were performed using 20 μm Duke standard beads (Thermo scientific) in deionised water. This bead was selected to be in the range of the nanoplankton size and slightly lower than the channel height (25 μm). To obtain a maximum of single bead per droplet, concentration of beads was optimized according to the Poisson's distribution (Mazutis et al. 2013; Wu et al. 2013). The percentage of bead encapsulation depending on the flow regimes was investigated by maintaining the oil pressure at two pressures (5 kPa or 10 kPa) and stepwise increased the sample pressure. At each change of sample pressure, the number of encapsulation and the number of beads stacked in the sample channel were counted. The result was expressed in percentage as follow:

$$Encapsulation\,(\%) = \frac{(Number\ of\ beads\ encapsulated)}{(Nb\ Total)} \times 100 \qquad (14.1)$$

where, *Nb Total* is the sum of the beads encapsulated and the unmoving beads in the channel after maintaining the oil and the sample pressures during 20 min.

14.3 Results and Discussion

14.3.1 A Competition Between the Inertial and the Viscous Forces in a Fixed Geometry

Droplet and single phase flows were investigated in order to better understand the main factors controlling their formations. By using a dual pump system, combinations of 400 different pressures were measured and associated flow patterns were recorded. Flow patterns were discriminated into six different cases characterizing either droplet or single phase flow patterns (Fig. 14.1).

Droplet diameter varied from 30 μm to 133 μm in the stable droplet flow and from 71 to 131 μm in the unstable droplet flow (Figs. 14.2 and 14.3).

Among the different flow regimes, the minimum droplet size (30 μm) of the chip was slightly smaller than the flow focusing contraction width (50 μm) and highlighted the geometry dependence in the pressure-controlled system. Figure 14.4 tends to confirm the geometry dependence between the orifice size of the flow focusing area and both the droplet and plug flow regimes ($y = 2(P_{sample}/P_{oil})^{0.92}$, $R^2 = 0.94$).

Moreover, results obtained showed that droplet diameter was limited to 71 μm (around half of the channel width 140 μm) in the stable droplet regime flow at the frequency higher than 2 Hz (Fig. 14.3). However, at a frequency lower than 2 Hz, higher droplet diameter can be created in the stable droplet pattern. This shift in droplet diameters between low frequency (<2 Hz) and higher frequency could be linked to the change in dominant force between the inertial and the viscous forces. In other words, the role of the inertia of the droplet may be more important in the unstable flow pattern than in the case of stable flow regime. In this context, calculations of the Reynolds (Re) numbers were performed according to the geometry of the microfluidic devices in view to verify if any possible increase of the Re

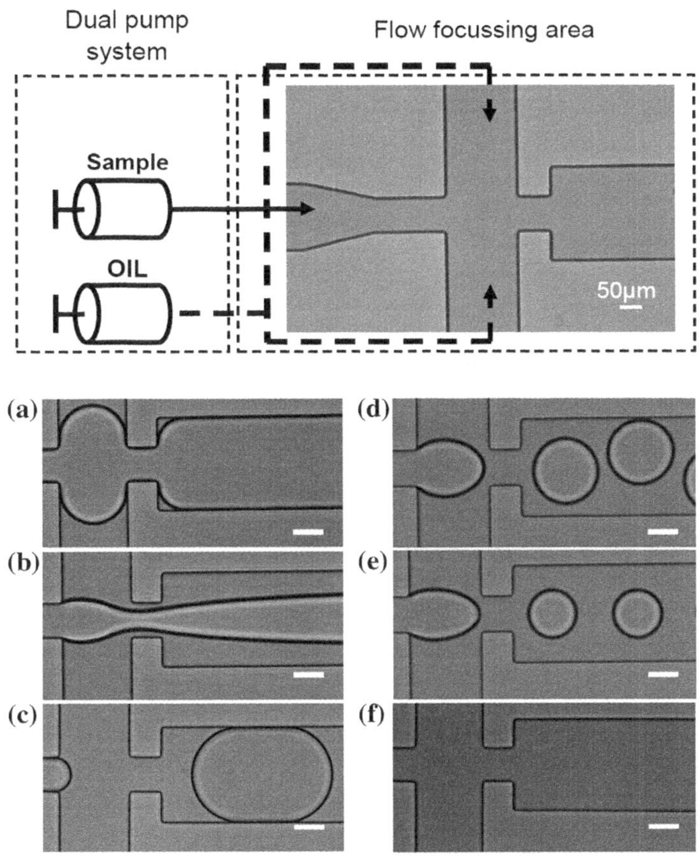

Fig. 14.1 Schematic drawing of the experimental devices and pictures of the six different flow regimes (scale bar 50 μm). **a** is the micrograph of the channel filled with sample (sample only), **b** an annular flow pattern, **c** the droplet with diameter size higher than the channel width (plug pattern), **d** the unstable droplet pattern where the droplets are not centred in the channel, **e** the droplet centred in the channel (stable droplet pattern), **f** shows the channel only fill with oil (oil only)

numbers can describe any change in the flow regimes. By using the dual pump system, the Re number varied from 0.28 to 0.42 in the range of pressures tested. This result suggested that the flow should be laminar in the channel and droplets are expected to move straightly in the channel. Re numbers lower than 1 were in agreement with the most common microfluidic devices and means that any induced vortices should die away in the channel (Tice et al. 2004; Baroud et al. 2010; Pan and Arratia 2013). Although the laminar flow can be observed in some specific cases at high Reynolds numbers (>2000), presence of vortexes under low Reynolds number such as in common microfluidic devices was usually reported to be impossible (Brody et al. 1996). However, both theoretical and experimental

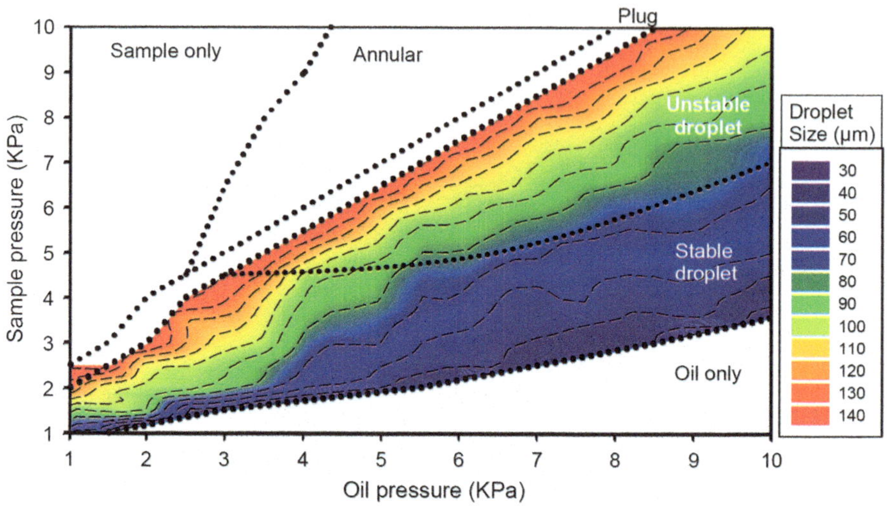

Fig. 14.2 Droplet size (μm) and flow regimes depending on the pressures (KPa) applied on the sample and oil inlets

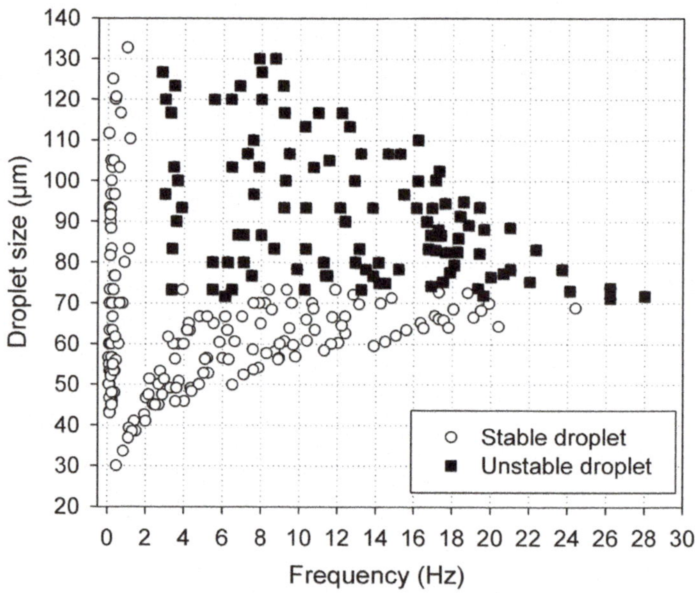

Fig. 14.3 Diameter of the droplet of the stable and unstable droplet regimes depending on the frequency of the droplet formation (Hz)

approaches demonstrated that specific geometric structures such as small diameter microchannels, wavy wall structures as well as wall roughness can modify the stability of the fluid despites a Reynolds number lower than 1000 (Kandlikar 2005;

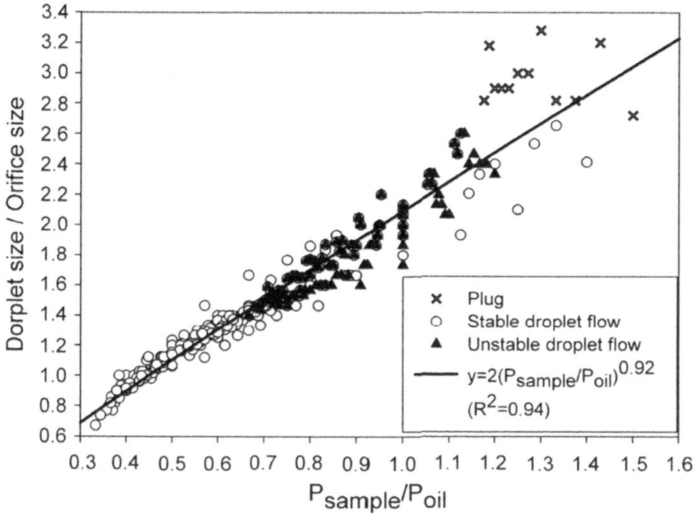

Fig. 14.4 Relationship between the droplet diameters normalized to the orifice size of the sample channel and the sample to oil pressures. The cross symbols are the plug flow regime (Fig. 14.1c), the white circles are the stable droplet pattern (Fig. 14.1d) and the black triangles show the unstable droplet pattern (Fig. 14.1e). The black line is the power regression curve calculated using the entire data set (n = 382)

Szumbarski et al. 2007). Wang et al. (2014) also reported turbulent flow in microfluidic devices with a Re number equal at 1 and suggested that turbulent flow could be cautiously investigated with regard to the viscosity and geometric design.

14.3.2 Limit Between the Continuous and the Multiphase Flows

Detection of the limit between the continuous and the multiphase flows is particularly important in order to find the lower pressure suitable to create small round shape droplet containing cell. Figure 14.2 shows that the limit between the oil only and the stable droplet can be discriminated into two phases: (i) a first phase characterized by the low oil and sample pressures (<5.5 kPa and 2 kPa, respectively) and (ii) a second phase where the ratio of P_{sample}/P_{oil} is almost constant (0.33 to 0.36; Fig. 14.2). At these threshold values, the diameter of the droplets was the smallest observed and the distances between two consecutive droplets were maximal. Formation of the droplet smaller than the orifice width was reported as a dripping mode and was characterized by a higher difference between the flow rates of the sample and the carrier fluid (Zhou et al. 2006). The difference in flow rates led to an increase of the viscous stresses and an elongation of cylindrical neck narrower than the orifice (50 μm). In this condition, droplets were created due to a combination of the end-pinching and capillarity wave instabilities (Lee et al. 2009). According to the stable flow regime and small diameter of the droplets, results may suggest that limit between the oil-only and the stable droplet flows could be suitable for microfluidic applications such as on chip sorting devices. However, the droplet frequency was particularly low and the encapsulation of the beads in the droplet was not optimum despite a high speeds of the droplet found (Fig. 14.5).

Association of the low frequency and the low volume the droplet ($\sim$15 pL) which is characterized by the low P_{sample}/P_{oil} ratio tend to increase the sedimentation in the channel and lead to a low efficiency of the bead encapsulation (Fig. 14.5). At low P_{sample}/P_{oil} ratio, role of surface tension is also highlighted when no droplet was created despite a positive pressure in the inlet of the dispersed phase.

In this condition, the normal and shear stresses on the liquid-liquid interface exerted by the external flow are greater that the force exerted by the internal fluid, hence a static spherical cap can be observed in the sample channel. Capillarity

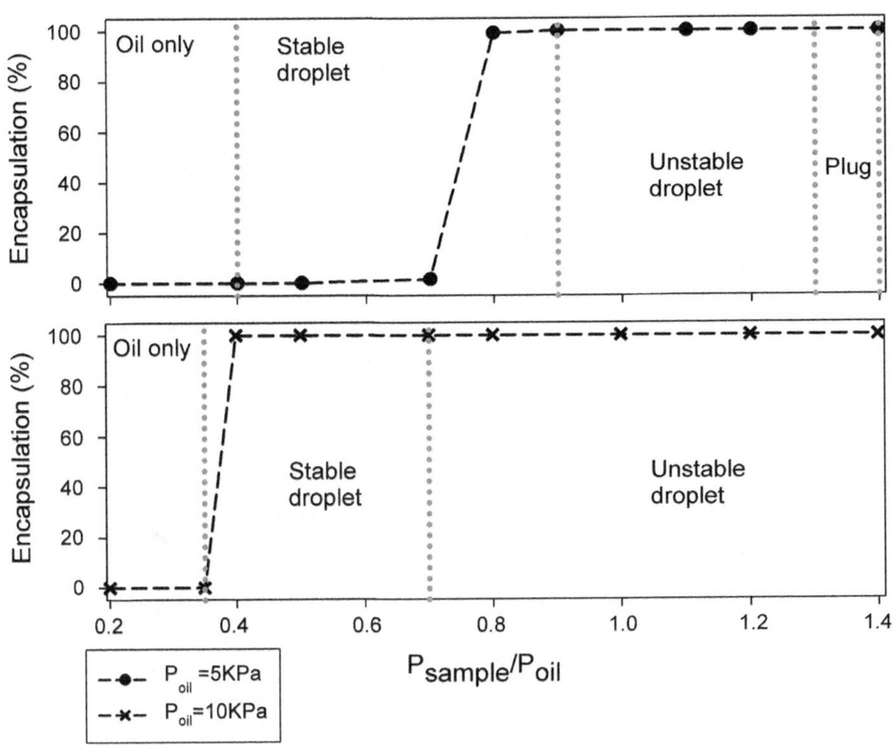

Fig. 14.5 Encapsulation of 20 μm standard beads (%) depending on the ratio P_{sample}/P_{oil}. The upper (black dots) and the bottom panels (black crosses) are the encapsulation of beads when P_{oil} was fixed to 5 kPa and 10 kPa, respectively. The dotted gray lines show the limit between the different flow regimes

number (Ca < 1) estimated using our microfluidic devices confirmed that the round the major role of the interfacial tension to create the round shape droplet in the range of stable droplet flow pressures.

14.3.3 Limit Between the Stable and Unstable Droplet Flows

Abrupt change between the stable and the unstable flows is difficult to measure in the microfluidic channels because of the lack of velocimeter capable to measure turbulence at this scale without addition of beads (Li et al. 2005; Avila et al. 2011). In this present study, discrimination between the stable and unstable droplets seems not to result in the change of the wet condition of the channel because droplet regimes were routinely observed in the same pressure condition and at each lab on a chip tested. Therefore, limit between the stable and the unstable flow regimes can only depend on the flow properties and chip geometry. Among the flow properties, distance between two consecutive droplets can discriminate two types of flow regime (Fig. 14.6).

However, by using only this parameter, discrimination was not accurate especially when the P_{sample}/P_{oil} ratio varied from 0.6 to 0.8.

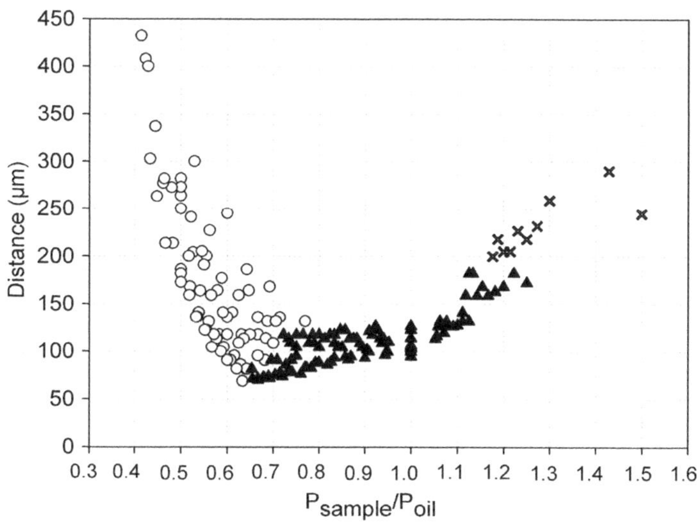

Fig. 14.6 Distance between the center of two droplets (μm) depending on the ratio of sample to oil pressures and the two-phase flows regime pattern. The open circles, triangles and crosses are the stable droplet flow (Fig. 14.1e), the unstable droplet flow (Fig. 14.1d) and plug pattern (Fig. 14.1c), respectively

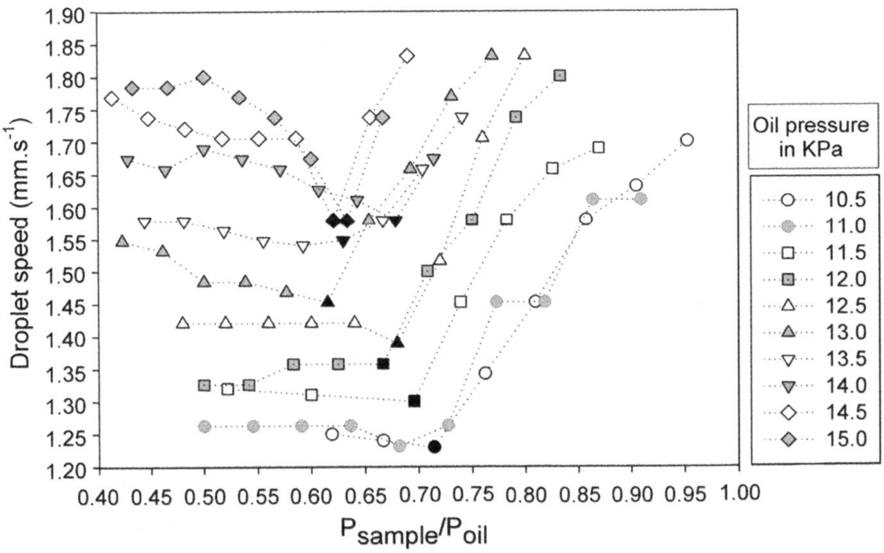

Fig. 14.7 Droplet speed (mm s^{-1}) at different constant oil pressures (KPa) depending on the pressure ratio P_{sample}/P_{oil}. The symbols filled in black show the transition between stable and unstable droplet flows

To enhance the detection of the change of droplet regime, we observed that a propagation of instability from the end of microfluidic channel to the droplet formation area occurred in these ranges of P_{oil}/P_{sample} values. According to these observations the limit between stable and unstable droplet regimes was studied by measuring the speed of the droplets in the channels (Fig. 14.7).

By fixing the oil pressure and stepwise increasing the dispersed phase fluid, the hydrodynamic of the multiphase flow was characterized by two trends (i) a nearly constant or decrease of droplet speed and (ii) a sudden increase of droplet speed. The change of droplet speed corresponded precisely to the transition between the stable and the unstable droplet flow regimes. The critical value of the minimum of the droplet speed also suggested that the increase of pressure on dispersed phase not necessary led to an increase of droplet speed in the channel. This counter intuitive observation pointed out the complex relationship between the droplet size and the speed of the carrier flow, especially in rectangular channel, where carrier fluid and droplet speed were reported to differ in some occasion (Dreyfus et al. 2003; Jakiela et al. 2011; Wu et al. 2013). To control the role of frequency (f) and the distance between two consecutive droplets (d) on the speed (U), a dimensionless graph were also plotted by converting to an equivalent Re number (Fig. 14.8) as reported in the study of Ward et al. (2005).

Based on the value of the droplet speed, results obtained suggested that the plot is nearly constant and independent of the flow control parameter (e.g. pressures on the inlet tubes, Fig. 14.8). The dispersion of the observations along the linear regression curve reflected in part the experimental error in the measurement of the

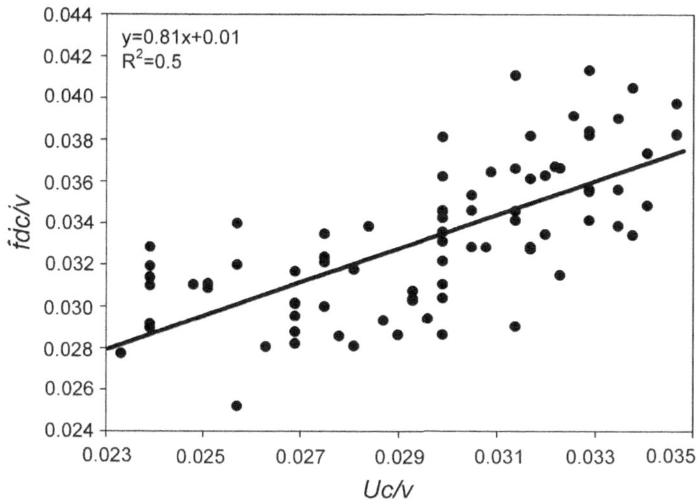

Fig. 14.8 Frequency of formation of droplet (f) the distance between the centers of two consecutive droplets (d) depending on the speed of droplet measured in the microchannel (U). Both parameters are made dimensionless by scaling with the viscosity (v) and half of the width channel (c = a/2)

space between two consecutive droplets and their speeds. However, this result tended to confirm that the viscous effect should play a key role by comparison with the inertial effects in the dynamical process.

14.3.4 The Suitable Flow Regime for Encapsulation

Figure 14.5 shows the percentage of encapsulation of the beads depending on the pressures ratio P_{sample}/P_{oil}. The percentage of encapsulation is characterized by a drastic increase (0–100%) in the stable droplet phase. This increase was found in the range of pressure lower than 0.1 kPa which is the detection resolution of the pressure sensor. No encapsulation of beads was observed at the limit between the oil only flow and the stable droplet regime. In contrast, the limit between the stable and unstable droplet regimes was characterized by the encapsulation of all the beads in the droplets These results indicated that the most robust encapsulation of the beads 20 μm (high frequency and the speed of droplets) was found in the higher P_{sample}/P_{oil} ratio and suggested that the plug or unstable droplet flow regimes could be used to correctly encapsulate objects into the droplet flows (Fig. 14.5). However, in some applications such as sorting microfluidic systems, both diameter of the droplet and the space between two consecutives droplets are particularly important to control. For example, Mazutis et al. (2013) reported that a distance of ten droplet diameters between two consecutive droplets is needed to efficiently sort a targeted droplet in the microfluidic channel. According to the results obtained, the short distance between two droplets in the unstable flow regime will hamper the droplet sorting. Size of the droplets is also a key parameter to maintain alive the cell encapsulated and created micro batch cultures. In this condition, plug type can be used to reach a large volume and a high encapsulation rate but also increases the possibility to encapsulate more than one cell by plug according to the Poisson's distribution. Moreover, classical active sorting system based on the fluorescence usually need droplet type flow to deviate and sort droplet. Consequently, only the higher P_{sample}/P_{oil} ratio of the stable droplet regime located near the limit stable-unstable droplet flow regimes appeared to be the most suitable range of pressures for the sorting experiment using our microfluidic device.

14.4 Conclusions

By using a pressure-controlled flow focusing device two single phases and four different multiphase flow regimes were observed and their formations were examined. Although the Reynolds number was always lower than 1 in the droplet formation area, unstable droplet regime flow were observed and suggested that in addition to the viscosity, the channel geometry in small micro fluidics is important and should integrate the entire chip design either than a single section. Results obtained also highlighted that the limit between the unstable and stable droplet regimes was the most suitable for sorting application due to the compromise between the droplet size, frequency, distance between two consecutive droplets and encapsulation of beads. The limit between stable and unstable droplet flows can be detected using the drastic change of the droplet speed found at the threshold between the two regimes. Finally, results obtained show that the increase of the pressure in the sample inlet not necessary increase the droplet speed underlying the difference in speed between the carrier fluid and the droplet flow.

Acknowledgements This work was supported by Kanagawa Prefector's local basic science funding for the On-chip Cellomics Project at the Kanagawa Academy of Science and Technology. This research was also supported by JST, CREST. We gratefully thank Ms. H. Mikami and Ms. M. Hasegawa for their helps during the experiments.

References

Anna SL, Bontoux N, Stone HA (2003) Formation of dispersions using "flow focusing" in microchannels. Apply Phys Lett 82:364–366

Avila K, Moxey D, De Lozar A, Avila M, Barkley D, Hof B (2011) The onset of turbulence in pipe flow. Science 333:192–196

Baroud CN, Gallaire F, Dangla R (2010) Dynamics of microfluidic droplets. Lab Chip 10:2032–2045

Brody JP, Yager P, Goldstein RE, Austin RH (1996) Biotechnology at low Reynolds numbers. Biophys J 71:3430–3441

Dreyfus R, Tabeling P, Willaime H (2003) Ordered and disordered patterns in two-phase flows in microchannel. Phys Rev Lett 90:144505

Duffy DC, McDonald JC, Schueller OJA, Whitesides GM (1998) Rapid prototyping of microfluidics system in poly(dimethylsiloxane). Anal Chem 70:4974–4984

Edd JF, Di Carlo D, Humphry KJ, Köster S, Irimia D, Weitz DA, Toner M (2008) Controlled encpsulation of single-cells into monodiperse picolitre drop. Lab Chip 8:1262–1264

Gu H, Duits MHG, Mugele F (2011) Droplets formation and merging in two-phase flow microfluidics. Int J Mol Sci 12:2572–2597

Jakiela S, Makulska S, Korczyk PM, Garstecki P (2011) Speed of flow of individual droplets in microfluidic channels as a function of the capillary number, volume of droplets and contrast of viscosities. Lab Chip 11:3603–3608

Kandlikar SG (2005) Roughness effects at microscale – reassessing Nikuradse's experiments on liquid flow in rough tubes. Bull Pol Ac Tech 53:343–349

Köster S, Angilè FE, Duan H, Agresti JJ, Wintner A, Schmitz C, Rowat AC, Merten CA, Pisignano D, Griffiths AD, Weitz DA (2008) Drop-based microfluidic devices for encapsulation of single cells. Lab Chip 8:1110–1115

Lee W, Walker LM, Anna SL (2009) Role of geometry and fluid properties in droplet and thread formation processes in planar flow focusing. Phys Fluid 21:032103. https://doi.org/10.1063/1.3081407

Li H, Ewoldt R, Olsen MG (2005) Turbulent and transitional velocity measurements in a rectangular microchannel using microscopic particle image velocimetry. Exp Therm Fluid Sci 29:435–446

Mazutis L, Gilbert J, Urig WL, Weitz DA, Griffiths AD, Heyman JA (2013) Single-cell analysis and sorting using droplet-based microfluidics. Nat Protoc 8:870–891

Pan L, Arratia PE (2013) A high-shear, low Reynolds number microfluidic rheometer. Microfluid Nanofluid 14:885–894

Salim A, Fourar M, Pironon J, Sausse J (2008) Oil–water two-phase flow in microchannels: Flow patterns and pressure drop measurements. Can J Chem Eng 86:978–988

Szumbarski J (2007) Instability of viscous incompressible flow in a channel with transversely corrugated walls. J Theor App Mech Pol 45:659–683

Tice JD, Lyon AD, Ismagilov RF (2004) Effects of viscosity on droplet formation and mixing in microfluidic channels. Anal Chim Acta 507:73–77

Tice JD, Song H, Lyon AD, Ismagilov RF (2003) Formation of droplets and mixing in multiphase microfluidics at low values of the Reynolds and the capillary numbers. Langmuir 19:9127–9133

Wang GR, Yang F, Zhao W (2014) There can be turbulence in microfluidics at low Reynolds number. Lab Chip 14:1452–4958

Ward T, Faivre M, Abkarian M, Stone HA (2005) Microfluidic flow focusing: Drop size and scaling in pressure versus flow-rate-driven pumping. Electrophoresis 26:3716–3724

Wu L, Chen P, Dong Y, Feng X, Liu BF (2013) Encapsulation of single cells on a microfluidic device integrating droplet generation with fluorescence-activated droplet sorting. Biomed Microdevices 15:553–560

Zhao Y, Chen G, Yuan Q (2006) Liquid-liquid two-phase flow patterns in a rectangular microchannel. AlChE J 52:4052–4060

Zhou CF, Yue PT, Feng JJ (2006) Formation of simple and compound drops in microfluidic devices. Phys Fluids 18:092105. https://doi.org/10.1063/1.2353116

Chapter 15
Development of a De-oiling System for Seabed Sediments

Yoshichika Ikeda, Motohiro Miki, Hisayuki Arakawa and Mitsuru Izumi

Abstract In northeastern Japan, a large number of fuel-oil tanks were destroyed by the tsunami that followed the March 2011 earthquake, with the spilled oil settling on the seabed in coastal areas. To remove the oil from the seabed, a de-oiling system was developed using superheated steam, and the efficacy and cost of the process assessed. The system was designed to process seabed sediments at a speed of 20 kg/h and supply steam at a temperature of <700 °C. Superheated steam at >300 °C was capable of removing contaminated oil from seabed sediments. The de-oiling capacity of the system was maximized at a processing speed 30 kg/h of seabed sediment. The system proved effective in the de-oiling of contaminated seabed sediments, with estimated operating costs of ~20,000 JPN/m^3 at a steam temperature of 300 °C and processing speed of 30 kg/h.

Keywords De-oiling superheated steam · Seabed sediments · Spilled oil Tsunami

15.1 Introduction

Numerous fuel-oil tanks on the coast of northeastern Japan were destroyed by the tsunami that followed the March 2011 earthquake. In Miyagi Prefecture, approximately11 kilotonnes (kt) of fuel oil were washed into Kesennuma Bay (Yamamoto et al. 2012; Sakai 2013; Yokoyama et al. 2014). Most of the spilled oil was transported outside the bay by back-waves from the tsunami where the oil was mixed with sand/suspended particles and deposited on the seabed. The massive volume of oil deposited on the seabed raised concerns regarding environmental damage and its effect on marine organisms (Suchanek 1993; Carls et al. 1999; Morales-Caselles et al. 2006). Therefore, the removal of contaminated sediments was imperative.

Y. Ikeda (✉) · M. Miki · H. Arakawa · M. Izumi
Tokyo University of Marine Science and Technology, 5-7, Konan-4, Minato, Tokyo 108-8477, Japan
e-mail: link-yi@kaiyodai.ac.jp

T. Komatsu et al. (eds.), *Oceanography Challenges to Future Earth*,
https://doi.org/10.1007/978-3-030-00138-4_15

A number of methods are employed to deal with major oil spills, such as chemical oil dispersants (Lessard and Demarco 2000; Prevot et al. 2001) or biodegradation and bioremediation by microorganisms (Tsutsumi et al. 2000; Flayyih and Jawhari 2014), and physical techniques, such as adsorption/collection (Toyoda et al. 1999), dredging (Islam and Parks 2014) and using superheated steam (Smith 2002), to remove the dispersed oil.

Chemical methods using dispersants can remove oil in seabed sediments cost effectively and over a relatively wide area. However, the chemicals are known to adversely affect many marine organisms (Lessard and Demarco 2000; Tenjin and Akimoto 1978). Biological methods are safer because they utilize microbial activity, but a prolonged period of action is necessary to remove large quantities of oil (Inada 2009). Physical methods, such as the use of adsorptive materials and barrages, are suitable for removal of oil from the sea surface but not applicable to de-oiling seabed sediments.

In this study, a de-oiling system using superheated steam for seabed sediments was developed, its effectiveness evaluated and a cost–benefit analysis performed based on the results.

15.2 Materials and Methods

15.2.1 Superheated Steam de-Oiling

A mechanical device was developed that removes oil from sediment using superheated steam and a rotary kiln capable of agitating the sedimentary particles. Steam temperature was maintained at 300–700 °C with a sediment throughput of 20 kg/h (Fig. 15.1, Table 15.1).

15.2.2 Operation of the System

Seabed sediments were collected from Kesennuma Bay on August 2014 using a vacuum aspirator that caused minimal turbidity. The collected sediments were brought back to the laboratory and dried to ~10% water content in a drying oven. Temperature in the retort (combustion chamber) and gas/electricity consumption during operation of the system are shown in Fig. 15.2.

Combustion chamber temperature reached the target value after ~120 min from commencement of heating, and remained stable. After 210 min, when the temperature of the sediment was sufficiently high, the rotary kiln became operational and started the de-oiling process.

Properties of the exhaust gas during the system operation are shown in Table 15.2. The quantity of exhaust particles reached 1729 g/h.

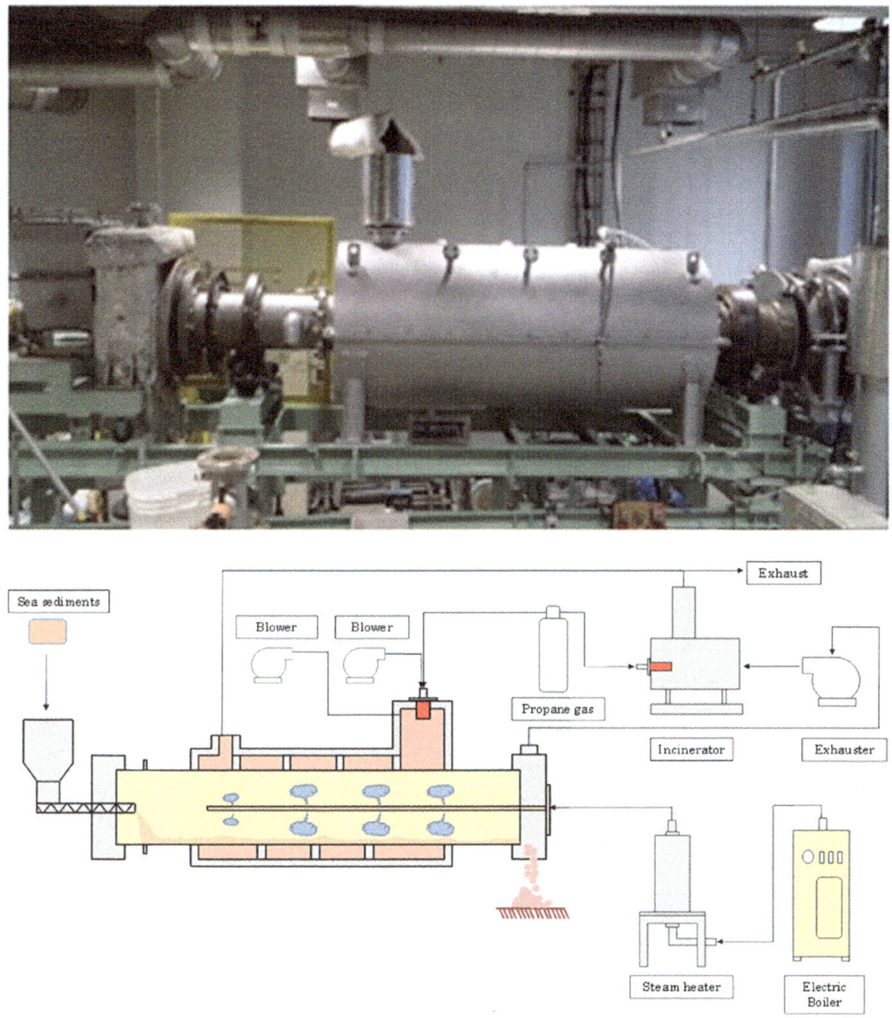

Fig. 15.1 The superheated steam de-oiling system

Table 15.1 Specifications of superheated steam system

Item	Specifications
Model	External heat-type rotary kiln
Object of processing	Oil contaminated soil
Particle size of object	Less than 15 mm
Water content of object	Less than 15.0%
Processing speed	20 kg/h

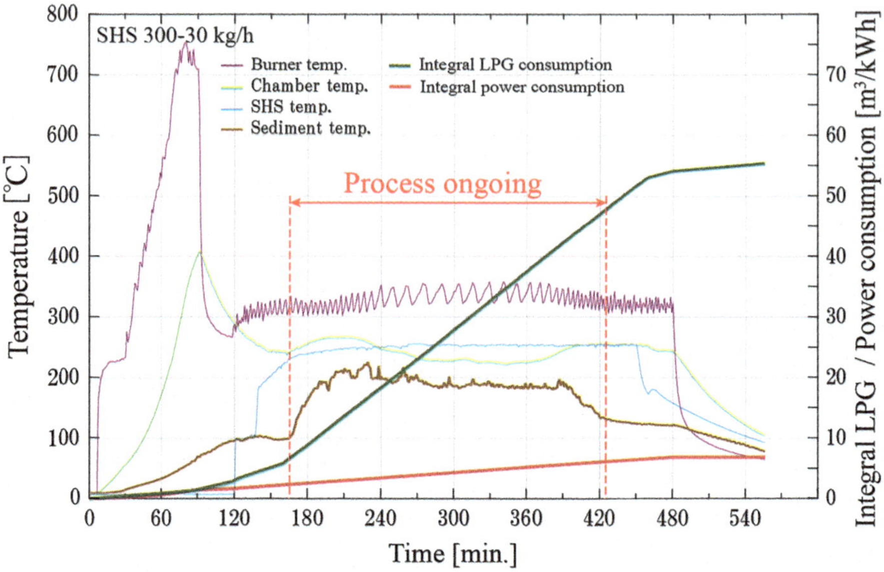

Fig. 15.2 Time variations of temperature of steam and sediments in the operation

Table 15.2 Properties of exhaust gas during the operation

Items		Mean	SD
Quantity of particulate matter	g/h	1729	287
Exhaust gas inflow (Wet)	m^3N/h	26.5	5.1
Exhaust gas inflow (Dry)	m^3N/h	12.5	5.2
Exhaust gas speed	m/s	2.9	0.6
Exhaust gas temperature	°C	175	8.9
Water content of exhaust gas	vol.%	53.9	11.0
Oxygen concentration	vol.%	16.1	1.1
Carbon dioxide concentration	vol.%	3.8	1.1
Carbon monoxide concentration	vol.%	<0.2	
Nitrogen concentration	vol.%	80.1	0.15

The efficiency of the de-oiling process was compared under the following experimental conditions: (1) at temperatures of 400–500 °C in the combustion chamber without the use of steam, (2) at temperatures of 300–500 °C in the combustion chamber with the use of steam, and (3) at sediment processing speeds between 20 and 40 kg/h.

15.3 Results and Discussion

15.3.1 De-Oiling Conditions

Normal hexane extract (NHE) and total petroleum hydrocarbon (TPH) levels in seabed sediments collected in Kesennuma Bay were 3500 and 600 mg/kg, respectively. NHE and TPH were not detected in sediments at temperatures >450 °C without the use of steam. At 400 °C, NHE was below the limit of detection but TPH was 200 mg/kg. However, with steam at 400°C, TPH was below the detection limit, indicating that the de-oiling process was more effective with steam (Table 15.3).

The effects of low temperatures and processing speeds were also examined. At a sediments processing speed of 20 kg/h, both NHE and TPH were undetectable in sediments after de-oiling at 300 °C with steam. When the processing speed was increased to 30 kg/h, NHE and TPH were undetected. However, at 40 kg/h, NHE and TPH levels were 500 and 300 mg/kg, respectively (Table 15.4).

From the results, operating the system at 300 °C with steam and a 30 kg/h processing speed yielded optimal de-oiling conditions.

Table 15.3 Effect of steam temperature on de-oiling

Condition	Processing speed (kg/h)	Concentration of n-hexane extracts (mg/kg)	Concentration of TPH (mg/kg)
Before treatment	20	3500	600
400 °C without steam	20	<100	200
450 °C without steam	20	<100	<100
500 °C without steam	20	<100	<100
300 °C with steam	20	<100	<100
350 °C with steam	20	<100	<100
400 °C with steam	20	<100	<100
450 °C with steam	20	<100	<100
500 °C with steam	20	<100	<100

Table 15.4 Effect of processing speed of seabed sediments

Condition	Processing speed	Concentration of n-hexane extracts (mg/kg)	Concentration of TPH (mg/kg)
Before treatment		2700	1500
300 °C with steam	30 kg/h	<100	<100
300 °C with steam	40 kg/h	500	300

System stability was compromised by dew condensation, unstable retort temperature and fine dust particles. Therefore, to maximize operating conditions, a residual heat jacket, heaters and a dust collector was installed in the system.

15.3.2 Estimation of Operational Costs

In a cost–benefit analysis, gas and electricity consumption was monitored for de-oiling contaminated sediments at 300 °C with steam.

A time series of the temperature of various elements, quantity of gas used and electricity consumption is shown in Fig. 15.2. Gas and electricity consumption between 210 and 460 min from the start of operation, when the temperature of sediment samples was stable, was calculated as 3.532 m^3 and 41.69 kWh, respectively, where gas flow-meter measurements were conducted at 0 °C atmospheric temperature and 1 atm atmospheric pressure, and the density of propane gas was 2.01 kg/m^3. Thus, the quantity of the gas consumed was 7.1 kg. Therefore, the quantity of gas and units of electricity required to treat 1 tonne dry weight (DW) of oil-contaminated sediment was calculated as 92.6 kg/tonne and 543 kWh/tonne, respectively.

The unit price rates of propane gas and electricity were, respectively, 334,000 JPY/tonne in 2014[1] and 19.5 JPY/kWh.[2] In the case of dried sediment, the unit prices per unit dry weight of processed sediment were 30,900 JPY/tonne for gas and 10,600 JPY/tonne for electricity; a total of 41,500 JPY/tonne DW.

From the density and water content of bottom sediment in Kesennuma Bay, the following relationship was obtained:

$$\rho_M = 0.059 \times u^2 - 9.338 \times u + 995.5 \tag{15.1}$$

where ρ_M and u indicate soil density(kg/m^3) and water content (%), respectively.

The water content of seabed sediment in Kesennuma Bay was 55.1%. Then, from Eq. (15.1), the dry soil included in 1 m^3 of seabed sediment was 660.1 kg.

$$C_v = C_w \times 660.1 \tag{15.2}$$

where C_v and C_w denote the processing cost per unit volume (JPY/m^3) and the cost per unit weight (JPY/kg), respectively.

Therefore the processing price of wet sediment per unit volume was 27,400 JPY/m^3.

[1]Japan LP Gas Association, http://www.j-lpgas.gr.jp/stat/kakaku/, accessed 14 March 2016.

[2]TEPCO, http://www.tepco.co.jp/e-rates/individual/data/chargelist/chargelist01-j.html, accessed 14 March 2016.

At a processing speed of 40 kg/h, low levels of both NHE and TPH were detected in sediment after the de-oiling operation while, at 30 kg/h, NHE and TPH were undetectable (Table 15.4). Therefore, the most cost-effective operation was $\sim$20,000 JPY/m^3 at a processing speed of 30 kg/h.

From the above, the operational cost for processing 1 m^3 of sediment was 20,000 JPY. Therefore, to de-oil a 1 km^2 area (thickness: 0.1 m) of highly contaminated sediment in Kesennuma Bay using the superheated steam method would cost 2 billion JPY.

The cost of drying and collecting the processed sediments was not included in this estimate; thus, the actual operating costs would be much higher.

15.4 Conclusion

A superheated steam system was developed capable of removing oil from seabed sediments. Optimal operation conditions of the system were at a steam temperature of 300 °C and processing speed of 30 kg/h. The cost for the de-oiling operation was estimated at 20,000 JPN/m^3.

Acknowledgements This study was supported by "Tohoku Ecosystem-Associated Marine Science" research project funds from the Ministry of Education, Culture, Sports, Science and Technology of Japan. The cooperation of Mr. Umakakeba of Aoki-asunaro Construction Co. Ltd. and his staff for collecting sediment from the seabed in Kesennuma Bay is greatly appreciated.

References

Carls MG, Rice SD, Hose JE (1999) Sensitivity of fish embryos to weathered crude oil: Part 1. Low-level exposure during incubation causes malformation, genetic damage, and mortality in larval pacific herring (*Clupea pallasi*). Environ Toxic Chem 18:481–493

Flayyih I, Jawhari HAI (2014) Ability of some soil fungi in biodegradation of petroleum hydrocarbon. J App Environ Microbiol 2:46–52

Inada T (2009) Possibility of contaminated soil purification and bottom mud purify cation with the superheated steam. Plan Constr, 47–52. (In Japanese)

Islam S, Parks J (2014) Economic assessment of dredging operations in Mobile Bay, USA and some additional notes on sustainable management of dredged material. J Coast Conserv 18:157–165

Lessard RR, Demarco G (2000) The significance of oil spill dispersants. Spill Sci Technol Bull 6:59–68

Morales-Caselles C, Jimenez-Tenorio N, Gonzalez de Canales ML, Sarasquete C (2006) Ecotoxicity of sediments contaminated by the oil spill associated with the tanker "Prestige" using juveniles of the fish *Sparus aurata*. Arch Environ Contam Toxicol 51:652–660

Prevot AB, Gulmini M, Zelano V, Pramauro E (2001) Microwave-assisted extraction of polycyclic aromatic hydrocarbons from marine sediments using non-ionic surfactant solutions. Anal Chem 73:3790–3795

Sakai K (2013) The fisheries damage of the East Japan great earthquake disaster and correspondence of the fisheries experimental station. Nippon Suisan Gakkaishi 79:93–94 (in Japanese)

Smith RM (2002) Extractions with superheated water. J Chromatogr A 975:31–46

Suchanek TH (1993) Oil impacts on marine invertebrate populations and communities. Amer Zool 33:510–523

Tenjin A, Akimoto Y (1978) Effects of oil and oil dispersant on the abalone and sea urchin. Bull Fukushima Pref Fish Exp Stat 5:89–97

Toyoda M, Moriya K, Aizawa J, Inagaki M (1999) Sorption and recovery of heavy oil by using exfoliated graphite. J Chem Soc Jpn 3:193–198

Tsutsumi H, Kono M, Takai K, Manabe T, Hraguchi M, Yamamoto I, Oppenheimer C (2000) Bioremediation on the shore after an oil spill from the Nakhodka in the sea of Japan. Field tests of a bioremediation agent with microbiological cultures for the treatment of an oil spill. Marine Pollut Bull 40:320–324

Yamamoto M, Yokoyama K, Yoshinaga I (2012) Change of heavy metals and oil spilled in Kesennuma Bay and Moune Bay after the disaster 3.11. Aquabiology 34(6):538–544. (in Japanese)

Yokoyama K, Hara R, Kohno M, Shimada K, Yoshinaga I (2014) Impact of tsunami disaster on bottom sediments of the Kesennuma Bay. J.JSCE, Ser. B2. Coast Eng 70:496–500

Chapter 16
Development of an Optical Detection System of Fuel Oil on Seabed Sediments

Akira Matsumoto, Kazuki Toguchi, Yoshichika Ikeda and Hisayuki Arakawa

Abstract Many oil tanks around coastal northeastern Japan were destroyed by the tsunami caused by the March 2011 earthquake. In Kesennuma Bay, Miyagi Prefecture, roughly 11,000 kL of fuel oil were flushed into the bay. Much of the oil was deposited on the seabed and mixed with sediment particles. Effective removal of the oil from seabed sediments requires a detailed understanding of its distribution throughout the seabed. Using ultraviolet (UV) fluorescence, we developed a new system of detecting oil in sediments and examined the effectiveness of the system. The in situ seabed sediments of Kesennuma Bay emitted fluorescence at a dominant wavelength of 550 nm under UV irradiation. The fluorescence was stronger in seabed sediments with higher concentrations of oil. The detection limit of oil in Kesennuma Bay occurred at a water depth of 6–7 m. Thus, the high turbidity and deep seabed in Kesennuma Bay prevent effective survey with the proposed system. In clear waters, however, the detection limit depth is estimated at 20–30 m, indicating that the system may be used to understand the distribution of oil in seabed sediments.

Keywords Fuel oil pollution · Optical detection system · Fluorescence Kesennuma bay

16.1 Introduction

The tsunami that occurred in East Japan after the earthquake on March 11, 2011 destroyed 21 of 23 oil tanks on the coast of Kesennuma Bay, Miyagi, and more than 11,000 kL of oil were spilled into the sea (Yamamoto et al. 2012; Sakai 2013).

Spilled oil generally drifts to shore or offshore and forms an emulsion with the seawater in the area of the spill (Mori and Toyama 1994). In the 2011 spill, a large

---

A. Matsumoto · K. Toguchi · Y. Ikeda · H. Arakawa (✉)
Tokyo University of Marine Science and Technology, 5-7, Konan-4, Minato, Tokyo 108-8477, Japan
e-mail: mattun_s15@yahoo.co.jp

T. Komatsu et al. (eds.), *Oceanography Challenges to Future Earth*,
https://doi.org/10.1007/978-3-030-00138-4_16

proportion of the spilled oil flowed out of Kesennuma Bay with the dilatational wave of the tsunami, while some oil adhered to mud and sand particles suspended by the tsunami and were subsequently deposited on the seabed (Sakai 2013).

The occurrence of oil on bottom sediments can cover the gills and cell membranes of aquatic organisms, causing death by suffocation or altered functional activity. Some organisms also take up oil in the sediment, and the lingering smell of oil can cause serious damage to various fisheries (Blumer et al. 1970, Blumer 1971; Ogata and Fujisawa 1991; 1999). After the earthquake, a large area of the bay contained concentrations of normal hexane extraction material (n-hexane extracts) 1,000 mg/kg over normal levels (Yamamoto et al. 2012; Arakawa and Nakamura 2016). There is concern for the influence this material will have on the coastal fisheries. Thus, it is essential to determine the distribution of oil in Kesennuma Bay.

Optical methods of measuring oil concentrations on the sea surface or in the water column include infrared (Fornaca et al. 1995), fluorescence (Hover and Plourde 1995; Brown and Fingas 2003; Sasano et al. 2008; Kim et al. 2010) and visible light methods (Morinaga et al. 2003). Oil on the seabed, however, can only be measured by sampling the sediment. These point observations cannot give detailed distributions for a space-time interval.

One property of oil is that it emits fluorescence under ultraviolet (UV) irradiation. We developed a new system for detecting oil on sediments using UV-excited fluorescence and examined the effectiveness of the system.

16.2 Materials and Methods

16.2.1 Fluorescence of Oil in Seabed Sediments

Seabed sediments were sampled and the water column beam attenuation coefficient was measured at 11 stations in Kesennuma Bay, Miyagi in December 2012 and July 2013 (Fig. 16.1). Sediments were collected by Ekman Berge. After collection, sediment samples were moved to an airtight container and immediately frozen. Samples were divided for analysis of n-hexane extracts concentration (Ministry of the Environment of Japan 2012) and fluorescence.

The beam attenuation coefficient (C'_λ) of seawater in Kesennuma Bay was measured with AC-S and C-Star beam transmissometers (Wet-labs Inc., Oregon; measurement wavelengths of 400–750 nm and 370 nm, respectively). Using a CTD, both transmissometers continuously measured the transmittance of seawater from the surface to the sea bottom.

Sediment fluorescence was measured with a Jaz spectroradiometer (Ocean optics Inc., Florida). Sediment samples from each station were defrosted and moved to cylindrical plastic containers in a dark room. Both the light receptor of the spectroradiometer and the excitation light were set 1 cm above the sediment surface and

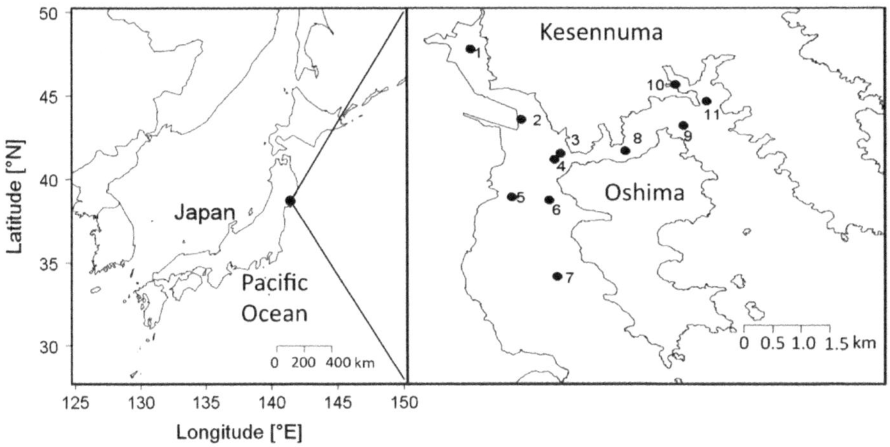

Fig. 16.1 Study area and sampling stations in Kesennuma Bay

were used to measure the fluorescence intensity of the sediments at each wavelength. Ten measurements were taken at random points above the sediment.

16.2.2 Oil Detection in Sediments Using Ultraviolet-Excited Fluorescence

To provide a continuous description of the distribution of oil on the sediment over a short time, a detection system using UV-excited fluorescence was developed with reference to the existing LIDAR system (Sasano et al. 2016). The system consisted of a detector that emits a laser and receives the fluorescence and a control structure in which the fluorescence is recorded as an electrical signal (Fig. 16.2).

The detector sends a UV laser (wavelength: 355 nm; output: 12 mJ/pulse) from the sea surface to the bottom. The fluorescence emitted by the oil on the seabed is then received in a photomultiplier tube and a telescope on the sea surface, where the fluorescence intensity is converted into an electrical signal. In the control structure, the electrical signal is sent to an oscilloscope and a personal computer. GPS coordinates are simultaneously recorded on the personal computer. The type of laser, photomultiplier tube and telescope utilized are shown in Table 16.1.

The UV laser and telescope were installed on a canoe (336 cm × 76 cm, length × width) with a clear bottom of polycarbonate. A glass window was set in the clear bottom of the canoe to transmit the UV laser. Floats on either side of the canoe reduced rolling of the hull.

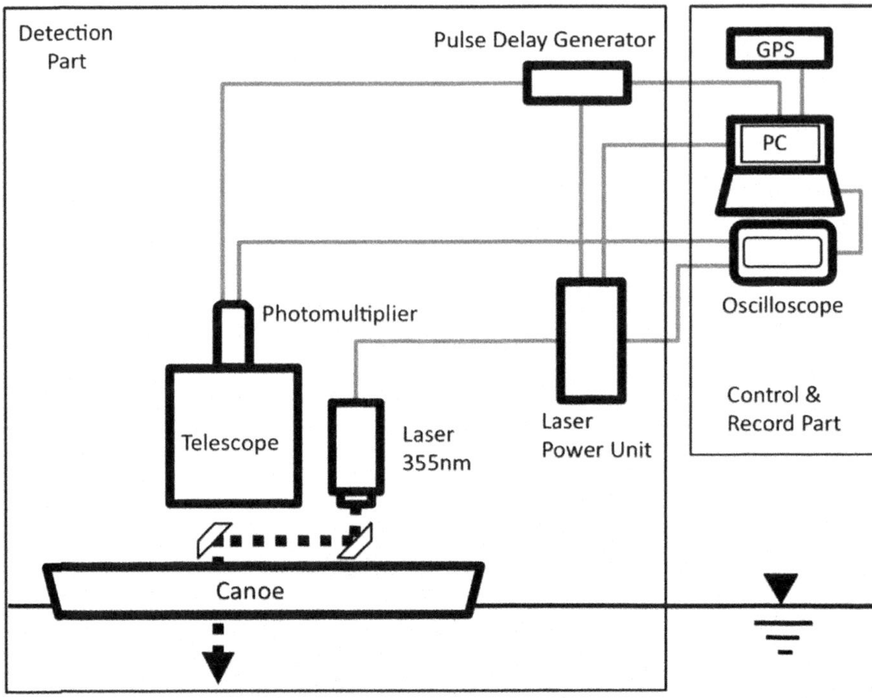

Fig. 16.2 System of detecting oil concentrations in sediment using an ultraviolet laser

16.2.3 Fluorescence Intensity of Oil on Sediments

Knowledge of the fluorescence intensity of oil in the laser output (output to an oscilloscope) is essential for estimating the detection limit depth of oil in the field. The detection system described above was used to measure the fluorescence intensity of oil on sediments.

The fluorescence detector was set 1.5 m above the floor in a dark room. Fuel oil A (marine diesel oil; kinematic viscosity 20 mm^2/S at 50 °C) or C (marine fuel oil; kinematic viscosity 167 mm^2/S at 50 °C) was poured into a glass Petri dish (91 × 21 mm) to a depth of 15 mm. The Petri dish was installed directly beneath the laser window. Laser intensity was regulated with flash lamp energy of 3.0–4.5 J. After irradiation with each intensity of the laser, the fluorescence intensity of the oil was measured and transmitted to the oscilloscope as a voltage (V) reading. The measurements were repeated 100 times at constant flash lamp energy and the average value was calculated.

The relationship between the oil concentration in the sediments and the fluorescence intensity was also measured. Fuel oil A or C was mixed with bottom sediments taken from Kesennuma Bay to create samples of differing oil concentrations (1–800 g/kg). Oils already occurring in the Kesennuma Bay sediments

Table 16.1 Specifications of the oil detection system on seabed sediments

Device	Specification	
Laser	Quantel ultra 50	Nd:YAG(3)
	Wavelength	355 nm
	Power source	12 mJ/pulse
	Pulse width	10 ns
	Repetition	20 Hz (max)
Gated PMT	Hamamatsu photonics	GaAsP, usual OFF
	Gain	$2*10^6$
	Gate time	>100 ns
Telescope	Vixen VMC200 l	
	Collection mirror diameter	200 mm
	Observation Wavelength	500 nm
Pulse delay generator	Hamamatsu photonics	TTL
	Channels	3 ch
	Pulse width	10 ns (min)
Oscilloscope	View Go 2	
	Channels	2 ch
	Frequency band	200 MHz

were removed by steaming at 400 °C for 60 min prior to mixing. The fluorescence detector was set 1.5 m above the floor in the dark room, and Petri dishes of the created samples were installed directly beneath the laser window. The fluorescence intensity of each sample was measure dafter irradiating it with the UV laser. Measurements were repeated 100 times on each sample.

16.3 Results and Discussion

16.3.1 Oil Fluorescence in Kesennuma Bay Sediments

Irradiation of a vinyl bag containing fuel oil A and C with UV light resulted in strong fluorescence of fuel oil A, although the fluorescence of fuel oil C was not clearly visible.

A spectroradiometer was used to determine the spectral distribution of the fluorescence of (1) the in situ bottom sediments at each station in Kesennuma Bay, (2) the bottom sediments from which the oil was removed and (3) fuel oil A and C (Fig. 16.3). Fuel oil A fluorescence peaked at both 430 nm and 490 nm, while fuel oil C peaked only weakly at 500–600 nm (Fig. 16.3a). The in situ bottom sediments showed a weak peak in fluorescence at 500–600 nm that was smaller than that of fuel oil C. Although the majority of the oil spilled into Kesennuma Bay was fuel oil A (Sakai 2013), the fluorescence of the bottom sediments was more similar in wavelength distribution to the fluorescence of fuel oil C. This change in the composition of the oil may be caused by loss of the volatile component during fire after the outflow or over time on the seabed.

The bottom sediment fluorescence intensity at 550 nm was strongest at Stas. 4 and 7 in December 2012 (Fig. 16.3b) and at Stas. 2, 6 and 7 on July 2013 (Fig. 16.3c).

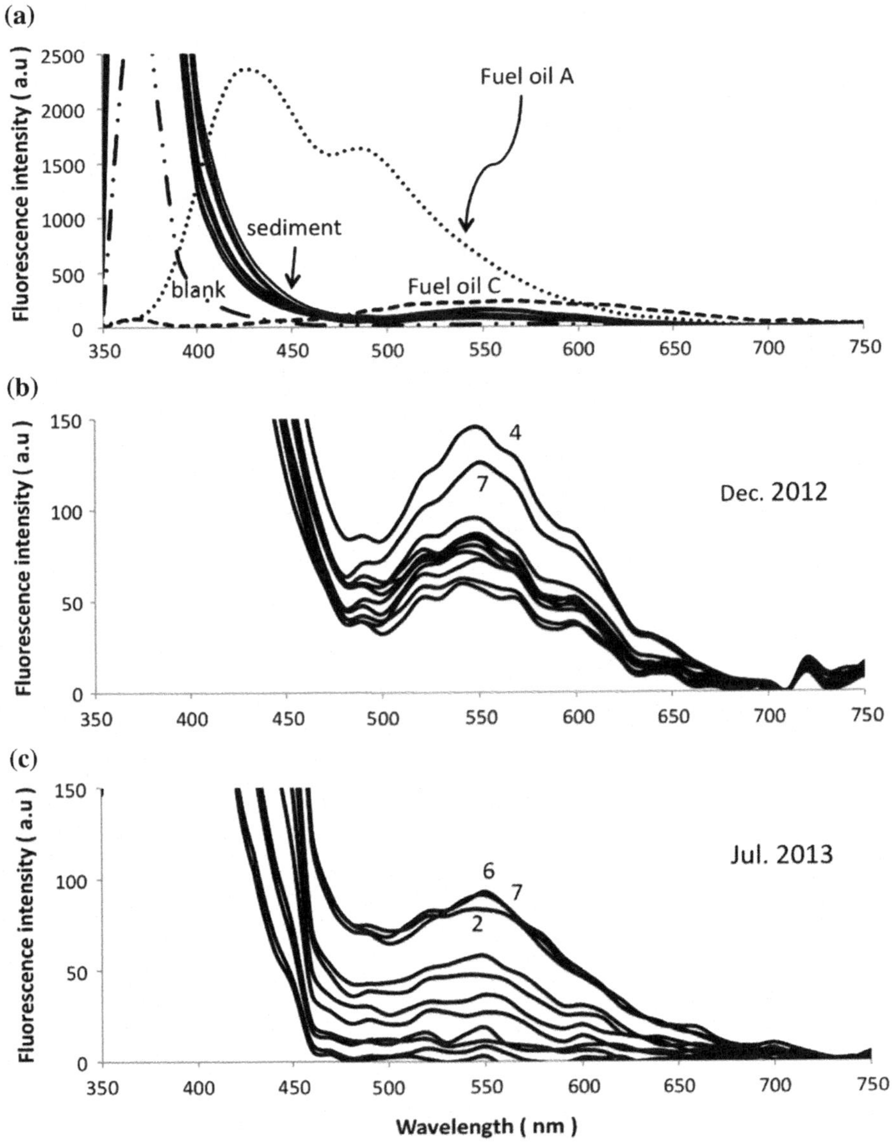

Fig. 16.3 Spectral distributions of oil fluorescence on seabed sediments of Kesennuma Bay **a** over the entire sampling period, and in **b** December 2012 and **c** July 2013

The concentration of n-hexane extracts in the bottom sediment of Kesennuma Bay was used as an index of oil concentration (Fig. 16.4). Concentrations of n-hexane extracts were high in the inner bay and in the channel, particularly at Stas. 4 and 3 in December 2012 (8000 and 5900 mg/kg, respectively). One of the oil tanks submerged and damaged in the tsunami leaked oil near Stas. 4 and 3. At all stations except Sta. 7, the concentration of n-hexane extracts was found to be 1000 mg/kg above the water quality standard for fisheries (Japan fisheries resource conservation association 2013).

The relationship between sediment fluorescence intensity and content of n-hexane extracts is shown in Fig. 16.5. Peak areas were calculated using image analysis (Image J). Ingeneral, fluorescence of the bottom sediments was stronger where n-hexane extract content was higher (Fig. 16.5). In rare cases, however,

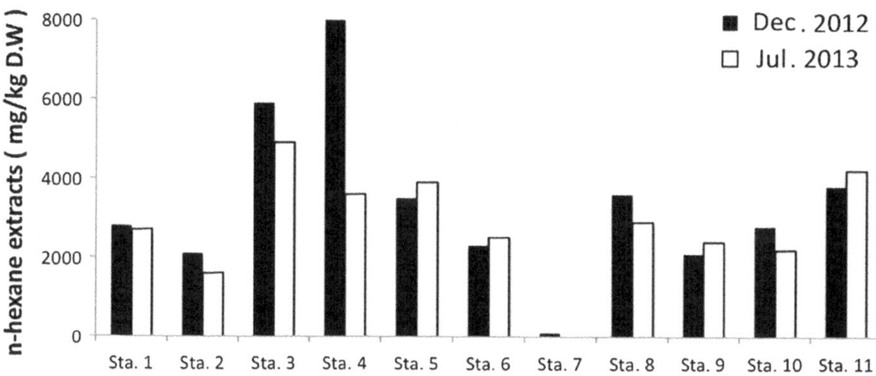

Fig. 16.4 Concentrations (mg/kg) of n-hexane extracts in sediments of Kesennuma Bay

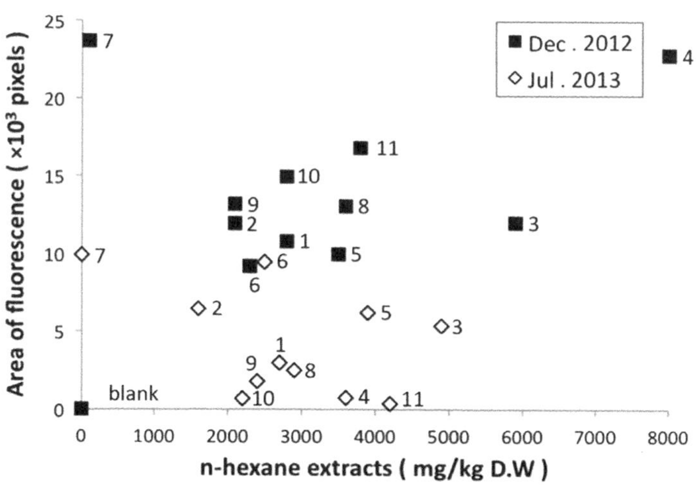

Fig. 16.5 Relationship between oil florescence of sediments and n-hexane extracts of seabed sediments

fluorescence was high where the concentration of n-hexane extracts was low (Sta. 7 on December 2012 and July 2013), or fluorescence was undetectable where n-hexane extract content was high (Sta. 4 and 11 on Jul. 2013; Fig. 16.5). Strong fluorescence at Sta. 7 despite low content of n-hexane extracts was likely caused by a large quantity of shell pieces in the sediments. Shells such as those of scallop have been shown to emit fluorescence (Shimono et al. 2004).

Overall, these results demonstrate that measuring the UV-excited fluorescence of the seabed can reveal the presence of oil on the sediment as well as the composition and the state of the oil (Fig. 16.3).

16.3.2 Operative Components of the Detection System

The system of oil detection relies on the measurement of four items: (1) The attenuation of the UV laser from the sea surface to the bottom; (2) The relationship between the intensity of the UV laser and the fluorescence intensity of the oil; (3) The relationship between the concentration of oil on the sediments and the fluorescence intensity; and (4) The attenuation of fluorescence intensity in seawater. Processes (1) and (4) can be described by measuring the beam attenuation coefficient in the water column. The relationships (2) and (3) were determined with a laboratory experiment.

The fluorescence of fuel oil A increased linearly with increasing laser strength (Fig. 16.6) according to the following equation:

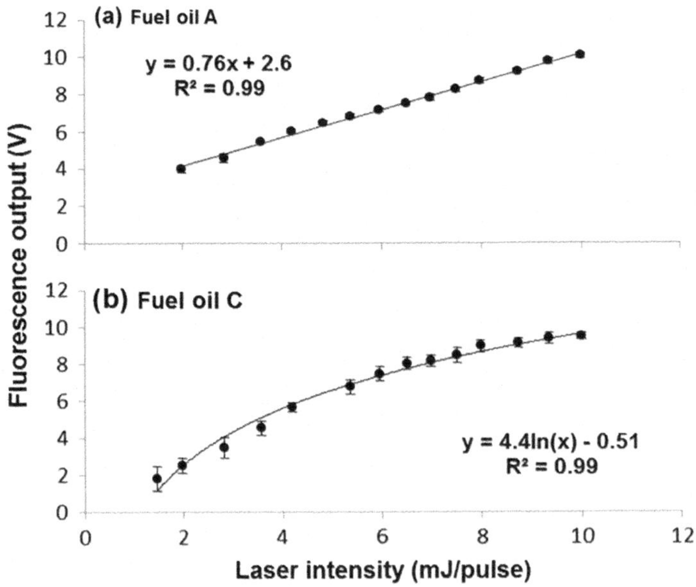

Fig. 16.6 Relationship between laser intensity and oil fluorescence

$$y = 0.76x + 2.6\left(R^2 = 0.99\right), \tag{16.1}$$

where y is the oscilloscope output (V) and x is the laser intensity (mJ/pulse). The maximum fluorescence output was 10.1 V.

The fluorescence of fuel oil C increased logarithmically with increasing laser intensity according to the following equation:

$$y = 4.4\ln(x) + 0.51\left(R^2 = 0.99\right), \tag{16.2}$$

where y is the oscilloscope (V) and x is the laser intensity (mJ/pulse). The maximum fluorescence was also 10.1 V.

The fluorescence of fuel oil A also increased linearly with increasing oil concentration up to 475 g/kg, but showed a marked increase at 500 g/kg oil (Fig. 16.7).

Fluorescence saturated at an oil concentration of 600 g/kg (Fig. 16.7a). Between concentrations of 0–475 g/kg, the linear regression of oil concentration (x) and the fluorescence output (y) was:

$$y = 0.0044x + 3.2\left(R^2 = 0.83\right), \tag{16.3}$$

In contrast, the fluorescence of fuel oil C increased slowly with oil concentration, saturating at 700 g/kg oil (Fig. 16.5b). Logistic regression described the following relationship:

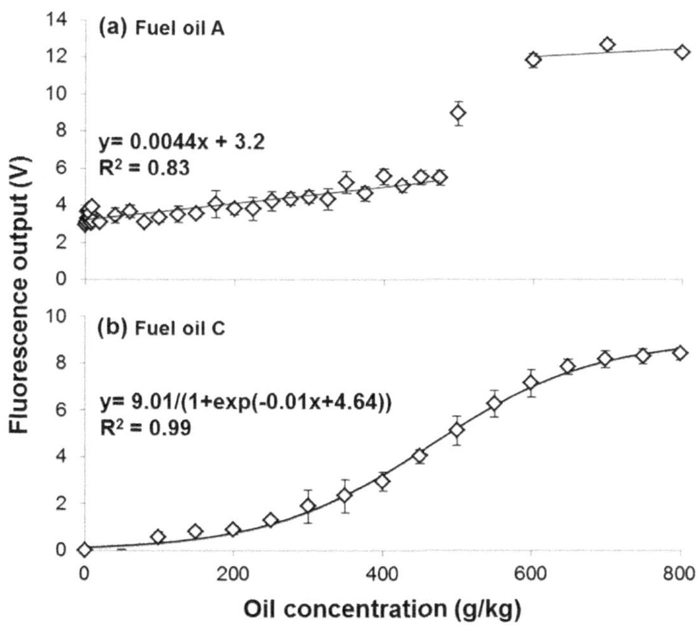

Fig. 16.7 Relationship between oil concentrations in the sediments and fluorescence output (V)

$$y = 0.91/(1 + exp(-0.01x - 4.64))\left(R^2 = 0.99\right), \quad (16.4)$$

The attenuation of the laser (item 1) and the oil fluorescence from the sediments (item 4) in Kesennuma Bay were determined by measuring the beam attenuation coefficient in winter (December) and summer (July). The mean beam attenuation coefficient was calculated for wavelengths of 370 nm and 550 nm. In winter, the beam attenuation coefficient was uniform from the surface to the seabed. In summer, however, the beam attenuation coefficient of the surface layer (<5 m) was extremely high; thus the mean coefficient was calculated separately for the layers above and below 5 m depth (Table 16.2). In the Table 16.2, *C* at 370 nm and 550 nm in wavelength added *C*' measured by the transmissometer to *C* of pure water at each wavelength (Smith and Baker 1981).

Table 16.2 Detection limits depth of oil on seabed sediments in Kesennuma Bay

Dec. 2012	$C_{370\ nm}$ (m^{-1})		$C_{550\ nm}$ (m^{-1})		Detection limit depth (m)	
					Fuel oil A	Fuel oil C
Sta. 1	0.84 ± 0.18		1.27 ± 0.18		2.7	2.3
Sta. 3	0.22 ± 0.04		0.49 ± 0.05		7.6	7.2
Sta. 5	0.24 ± 0.07		0.46 ± 0.31		8.0	7.2
Sta. 6	0.26 ± 0.03		0.59 ± 0.06		6.4	6.0
Sta. 8	0.21 ± 0.03		0.58 ± 0.13		6.7	6.5
Sta. 9	0.20 ± 0.03		0.48 ± 0.08		7.8	7.5
Sta. 10	0.21 ± 0.05		0.45 ± 0.07		8.3	7.8
Sta. 11	0.21 ± 0.03		0.47 ± 0.08		8.0	7.5
Average	0.30		0.60		6.1	5.6
Jul. 2013	$C_{370\ nm}$ (m^{-1})		$C_{550\ nm}$ (m^{-1})		Detection limit depth (m)	
	<5 m	≧5 m	<5 m	≧5 m	Fuel oil A	Fuel oil C
Sta. 1	0.95 ± 0.79	0.47 ± 0.28	1.16 ± 0.69	0.50 ± 0.16	2.9	2.1
Sta. 2	0.71 ± 0.39	0.31 ± 0.17	1.05 ± 0.41	0.42 ± 0.09	3.3	2.7
Sta. 3	0.53 ± 0.29	0.24 ± 0.04	0.61 ± 0.13	0.39 ± 0.06	5.6	3.9
Sta. 4	0.74 ± 0.64	0.31 ± 0.11	0.73 ± 0.29	0.47 ± 0.13	4.6	2.9
Sta. 5	0.53 ± 0.36	0.31 ± 0.23	0.72 ± 0.35	0.45 ± 0.12	4.8	3.8
Sta. 6	0.58 ± 0.46	0.28 ± 0.07	0.68 ± 0.31	0.52 ± 0.15	5.0	3.6
Sta. 7	0.74 ± 0.56	0.39 ± 0.12	0.87 ± 0.30	0.63 ± 0.18	3.9	2.8
Sta. 8	0.51 ± 0.17	0.27 ± 0.03	0.79 ± 0.27	0.47 ± 0.07	4.5	3.7
Sta. 9	0.38 ± 0.16	0.33 ± 0.11	0.55 ± 0.11	0.53 ± 0.18	6.4	5.1
Sta. 10	0.36 ± 0.17	0.33 ± 0.13	0.61 ± 0.18	0.54 ± 0.21	5.9	5.1
Sta. 11	0.53 ± 0.37	0.40 ± 0.21	0.60 ± 0.34	0.63 ± 0.30	5.7	4.0
Average	0.60	0.29	0.76	0.50	4.5	3.4

In winter (December 2012), the beam attenuation coefficient at 370 nm was low at Sta. 1 in the inner bay (0.84 m^{-1}), as well as at all other stations (0.20–0.26 m^{-1}). In contrast, the beam attenuation coefficient at 550 nm was high (1.27 m^{-1}) at Sta. 1, while remaining low at other stations (0.45–0.59 m^{-1}).

In summer, the beam attenuation coefficient at 370 nm (C_{370}) in the surface layer (<5 m) was high at Stas. 1, 2, 4 and 7 (0.71–0.95 m^{-1}) and lower at other stations (0.36–0.58 m^{-1}). The coefficients of the deep layer (>5 m) were all low (0.24–0.47 m^{-1}). The coefficients at 550 nm (C_{550}) in the layer <5 m were high at Stas. 1, 2, 7 and 8 (0.79–1.16 m^{-1}) and lower at the remaining stations (0.55–0.73 m^{-1}). In the layer >5 m, the coefficients were all between 0.39 and 0.63 m^{-1}. These results indicate high turbidity in Kesennuma Bay in the summertime such that light could not be detected from the seabed.

16.3.3 Detection Limit Depth of Oil on Seabed Sediment

Examination of the depth of the detection limit of fuel oil A and C in Kesennuma Bay showed a logarithmic relationship between laser outputand the water depth z:

$$E_Z = E_0 exp(-C_{355} \cdot z), \tag{16.5}$$

where E_0 is the laser intensity just beneath the sea surface, C_{355} is the beam attenuation coefficient at 355 nm and z is the depth (m). It can be thought that the C_{355} value is similar the C_{370}. Here, the beam attenuation coefficient at 370 nm in wavelength (C_{370}) measured is used to estimate the detection limit depth as the C_{355} value.

The strength of the laser output for a given water depth (z) defines the fluorescence intensity (E_F) of fuel oil A arriving at the sea surface as:

$$E_F = 0.76E_Z + 2.6 \tag{16.1$'$}$$

$$E_{0u} = E_F exp(-C_{550} \cdot z), \tag{16.6}$$

where E_{0u} is the upward fluorescence intensity at the sea surface, C_{550} is the beam attenuation coefficient at 550 nm wavelength.

Where fuel oil A occurs on seabed sediments, the upward fluorescence just beneath the sea surface can be expressed by combining Eqs. (16.1′), (16.5) and (16.6) to give:

$$E_{0u} = [0.76(E_0 exp(-C_{355} \cdot z)) + 2.6] exp(-C_{550} \cdot z), \tag{16.7}$$

The detection limit depth of oil on the seabed at each station with an assumed fluorescence output at the sea surface of 0.1 V is shown in Table 16.2.

The wintertime (December) detection limit depth of fuel oil A was shallow (2.9 m) at the inner part of the bay (Sta. 1). At other stations, the detection limit depth ranged from 7.6 to 9.9 m. The summertime (July) detection limit depth was 3.1 m at Sta. 1, and 3.5–7.2 m elsewhere. Thus, seasonal change in the depth of the detection limit was small in the inner bay, but showed modest increases in winter throughout the rest of the bay.

The detection limit depth of fuel oil C on the seabed (Fig. 16.4) was calculated as:

$$E_F = 4.4\ln(E_Z) - 0.51, \tag{16.2'}$$

and the upward fluorescence just beneath the sea surface can be expressed by combining Eqs. (16.2′), (16.5) and (16.6):

$$E_{0u} = [4.4\ln E_0 exp(-C_{355}\cdot z) - 0.51]exp(-C_{550}\cdot z), \tag{16.8}$$

The detection limit depth of fuel oil C in winter (December) in Kesennuma Bay was shallow at Sta. 1 (2.4 m) and 6.8–9.1 m at the other stations (Table 16.2). In summer (July), the detection limit depth was 2.3 m at Sta. 1 and 2.9–5.8 m elsewhere. Overall, the detection limit depth of fuel oil C was shallower (up to 3 m) in summer than in winter.

The water of channel and east part in the Kesennuma Bay is both deeper (>30 m) and more turbid. The detection limit occurred at depths shallower than the seabed, making it difficult to use UV-excited fluorescence to determine the oil distribution on the seabed in Kesennuma Bay. Development of a system that can detect oil on the sediment in seas with high turbidity such as Kesennuma Bay is needed.

In clear seawaters in which the beam attenuation coefficient at 355 nm and 550 nm in wavelength is 0.1 m^{-1}, however, the UV detection system has a depth limit of 33 m and 22 m for fuel oil A and C, respectively. Thus, this method is useful to detect oil on the seabed in areas with clear water.

Marine oil pollution occurs frequently (e.g. Vanem et al. 2008). Rapid assessment of the distribution of oil in the water and on the seabed is imperative after spills to keep or restore the health of the coastal ecosystem, particularly because oil pollution can remain in the sediments for long periods (e.g. Reddy et al. 2002; Peacock et al. 2007).

In the development of this system of oil detection, we examined the relationships between the content of n-hexane extracts and the fluorescence intensity of oil. The n-hexane extracts occur in the sea not only from spilled oil, but also in natural material (Higano 1974). Thus, in the proposed system, one must be careful of fluorescence occurring from substances other than mineral oil. Previously, a detailed distribution of oil on the seabed sediments could not be determined because investigation of oil pollution on the seabed could only be accomplished by sediment collection. Because UV-excited fluorescence can detect oil on the sea bottom continuously over a short time, this system will be useful for measurement of oil pollution in the future.

Acknowledgements Funding was provided by the Tohoku Marine Science of Ministry of Education, Culture, Sports, Science and Technology of Japan. The Miyagi Fisheries Experimental Station, Fishermen's Cooperative Association and Miyagi Karakuwa assisted with the collection of field data. Dr. Sasano gave advice on the laser instrumentation and measurement system. Fuel oils used for this experiment were obtained from Mitsuuroko Co., Ltd.

References

Arakawa H, Nakamura M (2016) Monitoring of oil spill pollution on seabed of the Kesennuma Bay in the wake of Great East Japan Earthquake Disaster. J JIME 51:83–86 (in Japanese)

Blumer MG Souza, Sass J (1970) Hydrocarbon pollution of edible shell fish by an oil spill. Mar Biol 5:195–202

Blumer M (1971) Scientific aspects of the oil spill problem. Boston Coll Environ Aff Law Rev 1:54–73

Brown CE, Fingas MF (2003) Review of the development of laser fluorosensors for oil spill application. Mar Pollut Bull 47:477–484

Fornaca SHH, Agavante BI, Hauss PS, Lee MS, Mussetto JS, Pearlman BH, Stark S, Talmadge, Valles A (1995) Remote oil spill sensing system (ROSSS). In: Proceedings 3rd themat conference remote sensing coast environment, vol 1, pp 473–479

Higano R, Taguchi H, Sue M (1974) Consideration on oil pollution to particle size of bottom sediments. Report Hydrogr Res 10:45–50 (in Japanese)

Hover GL, Plourde JV (1995) A nighttime evaluation of infrared sensors over the Santa Babara oil seeps. In: Proceedings 3rd themat conference remote sensing coast environment, vol 1, pp 7–17

Japan fisheries resource conservation association (2013) Water quality standard for fisheries on 2012, Tokyo

Kim M, Yim UH, Hong SH, Jung JH, Choi HW, An J, Won J, Shim WJ (2010) Hebei spirit oil spill monitored on site by fluorometric detection of residual oil in coastal waters off Taean. Korea Mar Pollut Bull 60:383–389

Ministry of environment of Japan (2012) Analysis methods of sediments. http://www.env.go.jp/water/teishitsu-chousa/ Accessed 30 Jul 2015

Morinaga TH, Arakawa M Shoji, Kiyomiya T (2003) Estimate of the slick thickness for leaked heavy oil from sunken Nakhodka in sea of Japan. La mer 41:114–121

Mori H, Toyama S (1994) Mechanism of spilled oil pollution on the sea. Bull Soc Sea Water Sci Jpn 48:77–84

Ogata M, Fujisawa K (1991) Environmental monitoring and bio monitor ing of petroleum pollution in the sea. Bull Soc Sea Water Sci Jpn 45:259–266

Ogata M, Fujisawa K (1999) Evaluating effect of oil pollution in the sea on the marine organisms and disorder of oil scavengers with its countermeasure. Bull Soc Sea Water Sci Jpn 53:84–94

Peacock EEGR, Hampson RK, Nelson L, Xu GS, Frysinger RB, Gaines JW, Farrington BW Tripp, Reddy CM (2007) The 1974 spill of the Bouchard 65 oil barge: petroleum hydrocarbons persist in Winsor Cove salt marsh sediments. Mar Pollut Bull 54:214–225

Reddy CMTI, Eglinton A, Hounshell HK, White L, Xu RB Gaines, Frysinger GS (2002) The west Falmouth oil spill after thirty years: the persistence of petroleum hydrocarbons in marsh sediments. Environ Sci Technol 36:4754–4760

Sakai K (2013) Marine damage and correspondence in a fisheries experiment station of an eastern Japan great earthquake, Kessen-numa Fisheries Station of Miyagi Prefecture. Nippon Suisan Gakkaishi 79:93–94 (in Japanese with English abstract)

Sasano MK, Hitomi, Ymanouchi H (2008) The fluorescence lidar for monitoring of oil spill. Trans Visual Soc Jpn 28:9–14

Sasano MM, Imasato H Yamano, Oguma H (2016) Development of a regional coral observation method by a fluorescence imaging LIDAR installed in a towable Buoy. Remote Sens 8:48

Shimono IT, Takamura, Nishino M (2004) Fluorescence properties of firing scallop shell. J Ceramic Soc Jpn 112:184–188

Smith RC, Baker KS (1981) Optical properties of the clearest natural waters (200–800 nm). Appl Opt 20:177–184

Vanem EØ, Endresen, Skjong R (2008) Cost-effectiveness criteria for marine oil spill preventive measures. Reliab Eng Syst Saf 93:1354–1368

Yamamoto MK, Yokoyama, Yoshinaga I (2012) Change of heavy metals and oil spilled in Kesennuma Bay and Moune Bay after the disaster 3.11. Aquabiology 34:538–544. (in Japanese with English abstract)

Chapter 17
Retinomotor and Stress Responses of Marbled Sole *Pseudopleuronectes Yokohamae* Under the LEDs

Rena Shibata, Yasuyuki Uto, Kenichi Ishibashi and Takashi Yada

Abstract Retinomotor and stress responses were examined under different LED spectra (blue, green and red), as well as under complete darkness and sunlight (control) for the marbled sole *Pseudopleuronectes yokohamae* juveniles to grasp bases for better culture condition, particularly toward prevention of the biting behaviour in hatchery. No difference was observed between green and control in the both responses, meaning that the retina under green were completely light-adapted. Cortisol concentrations in the peripheral blood plasma suggested the significantly higher stress level under red, while lower under blue and green. The present study demonstrated green was most suitable for the retinal light adaptation, and implied that green and blue were expected to reduce stress of fish.

Keywords *Pseudopleuronectes yokohamae* · Retinomotor response · Stress response · LED

R. Shibata (✉)
National Research Institute of Fisheries Engineering,
Fisheries Research and Education Agency, 7620-7, Hasaki,
Kamisu, Ibaraki 314-0408, Japan
e-mail: renas@fra.affrc.go.jp

Y. Uto
Tokyo Bay Fisheries Laboratory, Chiba Prefectural Fisheries Research Center,
3091, Kokubo, Futtsu, Chiba 293-0042, Japan

K. Ishibashi
Seed Production Research Laboratory, Chiba Prefectural Fisheries Research Center,
2568-38, Kokubo, Futtsu, Chiba 293-0042, Japan

T. Yada
Nikko Station, Freshwater Fisheries Research Division, National Research Institute
of Fisheries Science, Fisheries Research and Education Agency, 2482-3, Chugushi,
Nikko, Tochigi 321-1661, Japan

T. Komatsu et al. (eds.), *Oceanography Challenges to Future Earth*,
https://doi.org/10.1007/978-3-030-00138-4_17

17.1 Introduction

The marbled sole *Pseudopleuronectes yokohamae*, commonly found in the coastal waters around Japan, is a commercially high-valued species, and has been utilized intensely. Thus the larvae production technology for this species toward stock enhancement through the releasing program has been required and established in early 1980' (cf. Hatate 1987). Thereafter the larvae production of the marbled sole reached the second highest next to that of Japanese flounder *Paralichthys olivaceus* in Japan among flatfish species (Seikai 1985). On the other hand, it became a problem that the marbled sole juveniles often exhibited the fin loss, particularly caudal fine, due to biting each other in hatchery, with decrease of the survival due to infections by bacteria (Sugimoto et al. 2007). It would also impair the survival after released because of the inferior swimming performance as reported in tiger puffer (cf. Saito 2013). According to recent data, for instance, the incidences of the caudal fin loss immediately before being released were 100% in individuals losing the fin irrespective of more or less, while 61.3% in those losing half or more area of the fin (Ishibashi 2016). Unfortunately, however, effective methods to prevent biting have not been established yet, and thus become a pressing issue to be solved toward their better survival and healthiness in hatchery. For establishment of the appropriate methods, it would be much helpful to clarify possible environmental condition affecting the biting behaviour. Considering the diurnal rhythm of the marbled sole as shown their feeding activity in daytime (cf. Minami 1981), visual sense probably take primary roles for the biting behaviour. For species relying on the optical detection, colour in the surrounding environment would be a major factor affecting fish survival and growth performance, as reported in several species (Sabri et al. 2012; Yamanome et al. 2009; Blanco-Vives et al. 2010). As a preliminary approach to grasp basis for better culture condition for the marbled sole, we examined retinomotor and stress responses of this species under different colours using LED lamps, of which the responses are expected to offer information on preference of fish to the ambient colour.

17.2 Materials and Methods

Fifty five marbled sole juveniles with an average total length of 55 mm and 159 dph (days post hatching), artificially fertilized and reared at the Seed Production Res. Lab., Futtsu Sea Farming Section, Chiba Prefectural Fisheries Research Center, were used as the experimental animals for the analyses of both retinomotor response through histological observation of retina cells and stress response through determination of plasma cortisol concentration. They were separately exposed under different colour conditions (wave lengths): 470 nm (blue; B), 525 nm (green; G) and 660 nm (red; R) with the illumination intensity of 10 $\mu molm^{-2}s^{-1}$ as the test group, as well as complete darkness (D) and sunlight as the control (C) for the retina cell observation (RCO), while four conditions excluding D for the cortisol analysis (CA). After 3 h exposure under the respective conditions, they were anesthetized with benzocaine just before sampling (N = 3 and 10 per condition for RCO and CA, respectively). Whole body samples for RCO were fixed in Bouin's fluid, processed for embedding in paraffin, 3 μm thick transverse sectioning and the H&E stain of retina. The section were photographed under a light microscope, and the thickness of the visual cell layer (*v*), pigment layer (*p*), and length of the cone myoid (m) were measured according to the method after Ali (1959). Ratios shown as *p/v* and *m/v* were used as the indices of retinomotor response of the retina. For CA, peripheral blood was withdrawn from caudal veins to determine the plasma cortisol concentration by ELISA.

17.3 Results

The retinal index showing the ratio of thickness of pigment layer to that of visual cell layer (*p/v*) under the green light were significantly higher than those under the blue and red lights, while the other index showing the ratio of length of the cone myoid to those of visual cell layer (*m/v*) were significantly lower under the green light, contrarily. No significant differences were detected between the green and the control for both indices (Fig. 17.1).

Cortisol concentrations were 4.0 ± 8.8, 2.5 ± 2.6, 2.6 ± 4.3, and 16.1 ± 18.2 $ngml^{-1}$ in the control, blue, green and red lights, respectively, showing significant differences between the red and the other groups (Fig. 17.2).

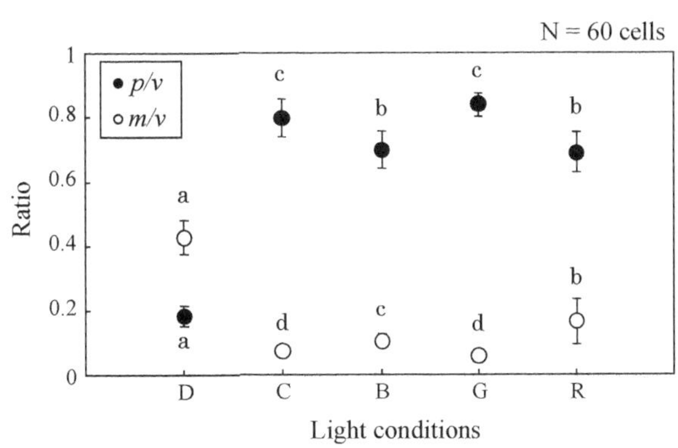

Fig. 17.1 Retinal indices *p/v* (●) and *m/v* (○) of the marbled sole exposed under different light conditions. The mean and standard deviation were shown. Different letters indicate significant difference (One-way ANOVA with Dunnett multiple comparison test, $p < 0.05$)

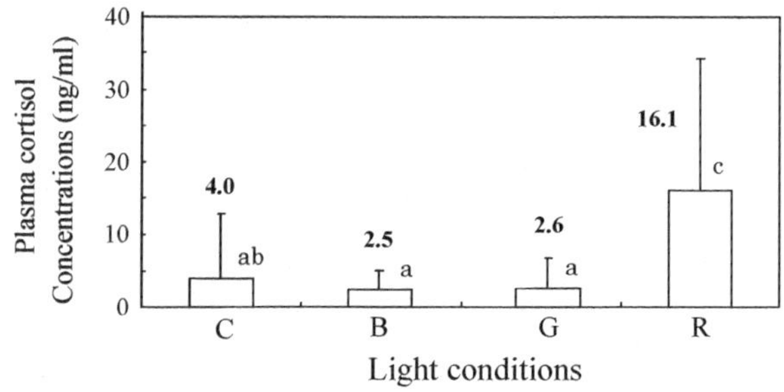

Fig. 17.2 Plasma cortisol concentrations under different light conditions. Vertical bars show the standard deviation (One-way ANOVA with Dunnett multiple comparison test, $p < 0.05$)

17.4 Discussion

We found the LED colour-specific retinomotor and stress responses in the marbled sole. These suggest significance of colour selection in the culture system for this species, as shown in other fish species. For instance, Nile tilapia *Oreochromis niloticus* showed lower mortality under blue light in culture tank (Sabri et al. 2012). Barfin flounder *Verasper moseri* exhibited better growth performance under both blue and green lights (Yamanome et al. 2009). Senegal sole *Solea senegalensis* larvae exposed to red light resulted in a delay in growth: the yolk sac completely absorbed under the blue light was not absorbed under the red light in the identical period (Blanco-Vives et al. 2010). Thus the colour commonly showing better growth and survival over species was proved not red but blue and/or green, and this would be true for the marbled sole as well because importance of green light for the marbled sole culture production were suggested as discussed below.

First, no differences were observed commonly in the retinal indices and in the plasma cortisol concentration between the green light and the control. Judging from such a common ground, retinomoter responses are considered to reflect stress responses of fish to the ambient colours. In addition, the retinal profiles under the green light are considered to be same as those under the light-adapted condition because the control was conditioned by the sunlight with the exposure time enough to be adapted for retina cells. And this probably underlies the basis that the green light allows the marbled sole to acquire full acuity of vision, based on findings of Ali (1959) who demonstrated the dependence of acuity of vision upon the complete light adaptation of cone cells.

Second, the plasma cortisol level in the peripheral blood of the marbled sole was largely decreased under the blue and green lights compared to that under the red light. This result implies that the marbled sole was in an unstressed status under the blue and green lights, being supported by responses in several species as follows. In Nile tilapia *O. niloticus,* the blue light spectrum was found to prevent stress (Volpato and Barreto 2001). Stress reduction was also shown under in both green and blue LED spectra relative to red LED spectra in yellowtail clownfish *Amphiprion clarkii* (Shin et al. 2011).

In conclusion, the present study demonstrated that green light was most suitable for the retinal light adaptation, and implied that green and blue light were expected to reduce stress of fish. This implication would hopefully bring the positive effect to prevent from biting. In our ongoing study, biting behaviours are now examined under the different LED spectra.

Acknowledgements Dr T. Azuma kindly gave us a lot of precious comments and suggestion to complete this manuscript. Dr. K. Sawada generously allowed us to use his LED light system. Dr. Y. Taga helped us for blood samplings. The authors would also like to thank Drs. T. Watanabe, S. Akeda, and Y. Ueno for their support to the present study.

References

Ali MA (1959) The ocular structure, retinomotor and photo- behavioural responses of juvenile pacific salmon. Can J Zool 37:965–996

Blanco-Vives B, Villamizar N, Ramos J, Bayarri MJ, Chereguini O, Sánchez-Vázquez FJ (2010) Effect of daily thermo- and photo- cycles of different light spectrum on the development of Senegal sole (*Solae senegalensis*) larvae. Aquaculture 306:137–145

Hatate K (1987) Business report on technical development for the release program on the marbled sole. In: Oita prefectural farming fisheries center and Oita prefectural coastal fisheries experimental station pp 1–2. (in Japanese)

Ishibashi K (2016) Annual report of the 2014 fiscal year. Chiba Prefectual Fisheries Research Center, pp 19–20. (in Japanese.)

Minami T (1981) The early life of a flounder *Limanda yokohamae*. Bull Jpn Soc Sci Fish 47: 1411–1419 (in Japanese)

Sabri DM, Elnwishy N, Nwonwu F (2012) Effect of environmental color on the behavioral and physiological response of Nile tilapia, *Oreochromis niloticus*. Glob J Front Res 12(4):11–20

Saito K (2013) Larval rearing technology preventing from the caudal fin loss in tiger puffer *Takifugu rubripes*. Report on applicable researches accomplished in the 2012 fiscal year. Division of Agriculture, Forestry and Fisheries, Akita Prefecture, 55–56. (in Japanese)

Seikai T (1985) Effect of population density and tank color on the abnormal coloration of hatchery-reared mud dab, *Limanda yokohamae*. Aquiculture 33:119–128 (in Japanese)

Shin HS, Lee JH, Choi CY (2011) Effects of LED light spectra on oxidative stress and the protective role of melatonin in relation to the daily rhythm of the yellowtail clownfish, *Amphiprion clakii*. Comp Biochem Physiol 160A:221–228

Sugimoto K, Suzuki K, Kumagai A (2007) Influence of rearing density of juvenile marbled sole *Pleuronectes yokohamae* on fin lack caused by biting. Miyagi Pref Rep Fish Sci 7:13–15 (in Japanese)

Volpato GL, Barreto RE (2001) Environmental blue light prevents stress in the fish Nile tilapia. Braz J Med Biol Res 34:1041–1045

Yamanome T, Mizusawa K, Hasegawa E, Takahashi A (2009) Green light stimulates somatic growth in the barfin flounder *Verasper moseri*. J Exp Zool 311A:73–79

Chapter 18
Metabolome Profiling of Growth Hormone Transgenic Coho Salmon by Capillary Electrophoresis Time-of-Flight Mass Spectrometry

Toshiki Nakano, Hitoshi Shirakawa, Giles Yeo, Robert H. Devlin and Tomoyoshi Soga

Abstract Growth in fish is regulated in part by the growth hormone (GH)-insulin-like growth factor (IGF) axis, and salmon transgenic for GH are known to show dramatic increased growth. However, little is known concerning the in vivo global levels of metabolites and the mechanism of enhancement of growth in GH transgenic vertebrates. The present study examined the charged metabolites levels in GH transgenic coho salmon (*Oncorhynchus kisutch*) overexpressing GH by metabolomic analysis. Triplicate groups of size-matched (0 year-old, approx. 60 g) and age-matched (1.5 years-old) GH transgenic (T) and non-transgenic (NT) wild salmon were quantitatively assessed for levels of approximately 200 metabolites in both muscle and liver. The most notable difference found between T and NT fish was that glycolysis metabolite levels were increased in the muscle of transgenic fish. In addition, an increase in some metabolite levels in the transgenic fish muscle was found to be enhanced by ration-restriction. However, these effects observed in muscle were different from that seen in liver. The results suggest that GH

T. Nakano (✉)
Marine Biochemistry Laboratory, Graduate School of Agricultural Science,
Tohoku University, Aoba-ku, Sendai 980-0845, Japan
e-mail: nakanot@tohoku.ac.jp

H. Shirakawa
Nutrition Laboratory, Graduate School of Agricultural Science,
Tohoku University, Aoba-ku, Sendai 980-0845, Japan

G. Yeo
Institute of Metabolic Science, University of Cambridge, Addenbrooke's Site B289,
Addenbrookes Hosp Hills R, Robinson Way, Cambridge CB2 0QQ, UK

R. H. Devlin
Center for Aquaculture and Environment Research, Fisheries and Oceans Canada,
4160 Marine Drive, West Vancouver, BC V7V 1N6, Canada

T. Soga
Institute for Advanced Biosciences, Keio University, Mizukami
Tsuruoka 997-0035, Japan

T. Komatsu et al. (eds.), *Oceanography Challenges to Future Earth*,
https://doi.org/10.1007/978-3-030-00138-4_18

transgenesis can improve the use of carbohydrates as a source of energy associated with rapid growth. These effects are likely to depend on the level of total digestible energy intake and type of tissue in transgenic fish.

Keywords Metabolome · Growth hormone · Transgenic fish · CE-TOFMS

18.1 Introduction

The demand for and production of farmed fish and shellfish has been increasing due to the worldwide decline of ocean fisheries stocks. Enhancement of global fish and shellfish farming could relieve pressure on ocean fisheries; therefore, aquaculture is an important development to assist world food production. In addition, aquaculture has the potential to reduce fishing pressure on threatened stocks and thereby reduce effects on ecosystems and biodiversity (Naylor et al. 2000; Pauly et al. 2002).

More than 220 species of fish and shellfish have been farmed to date. However, the balance between cultured and wild-caught fish, as well as the total supply of fish available for consumption by humans, will depend on future aquaculture practices (Naylor et al. 2000). Carnivorous and piscivorous species used in aquaculture requires large inputs of wild fish for feed in the form of fish meal and fish oil. Feed is the largest production cost for commercial aquaculture, and accordingly, improving feed efficiency (feed conversion rate) and enhancement of growth and fitness of farmed fish will be of benefit (Nakano 2007; Naylor et al. 2000).

The enhancement of growth of cultured fish is one of the primary goals for the development of aquaculture, and is a topic of intense research. In fish, growth is promoted to a major extent by liver-derived insulin-like growth factor (IGF)-1 in response to pituitary-secreted growth hormone (GH) binding to GH receptors (GHR). Thus, this GH-IGF-1 axis plays an important role in the regulation of both growth and development. Secretion of GH is under hypothalamic regulation mediated by many modulators. The growth-promoting actions of IGFs are controlled by GH and IGF binding proteins and a specific IGF receptor on the surface of target cells. Growth is also genetically regulated and is influenced by multiple cellular, endocrinological, and environmental factors (Deane and Woo 2009, 2011; Donaldson et al. 1979; Duan 1998; Moriyama et al. 2000; Nakano 2011, 2016; Nakano et al. 2013, 2015, 2011; Reineck 2010). Selective breeding has increased growth rates of fish over many generations and its use is well established. Fish growth can also be significantly stimulated by treatment with exogenous GH (Donaldson et al. 1979). Recently, transgenic technologies to stimulate growth have also been researched the transfer of GH transgenes to fish can result in strong stimulation of growth (Devlin et al. 2001; Devlin et al. 1994; Du et al. 1992; Leggatt et al. 2014; Nakano 2016; Nakano et al. 2011). More than 35 economically important fish species, such as coho salmon (*Oncorhynchus kisutch*), Atlantic salmon (*Salmo salar*), Nile tilapia (*Oreochromis niloticus*), channel catfish (*Ictalurus punctatus*) and common carp (*Cyprinus carpio*), have been using for transgenic research (Devlin et al. 2015, 1994; Reichardt 2000; Stokstad 2002). Atlantic salmon containing a growth hormone gene from the Pacific Chinook salmon (*Oncorhynchus tshawytscha*) driven by a powerful promoter sequence can accelerate the growth rate of this species dramatically, genetically modified fish to reach market size more rapidly (Du et al. 1992). This genetically modified Atlantic salmon had now been approved for production and human consumption by US Food and Drug Administration (FDA), and by Health Canada and Environment Canada, being the first transgenic animal as available as food for human consumption (Ledford 2015).

In GH transgenic salmon, the levels of GH and IGF-1 in plasma, and mRNA expression of *igf1* and *ghr* in tissues, are increased (Nakano et al. 2011; Raven et al. 2008). Growth abnormalities, altered muscle and pituitary structures, and high feeding motivation have been observed in GH transgenic salmon relative to non-transgenic wild type fish (Devlin et al. 2004; Devlin et al. 2015; Devlin et al. 1995; Mori and Devlin 1999; Rise et al. 2006). This expression of transgenes can affect the many aspects of both physiology (metabolism) and behavior in the GH transgenic fish. GH transgenic fish might differently respond and adapt to shift in environmental conditions, compared with non-transgenic wild fish (Devlin et al. 2015; Lohmus et al. 2010). The potential application of GH transgenic fish technology has both scientific and public interests concerning the prospective economic benefits as well as the potential risks in the fields of food safety and environmental consequences should the transgenic fish escape into the wild or enter the human food supply (Devlin et al. 2015; Ledford 2015). However, the exact mechanism of enhancement of growth in GH transgenic coho salmon is unknown and there are no details concerning the status of various metabolites which may provide clues to the growth enhancement mechanisms involved. In the present study, we examined the charged metabolite profiles of NT and GH transgenic coho salmon in which GH is overexpressed, using capillary electrophoresis mass spectrometry (CE-MS)-based metabolomic analysis.

18.2 Materials and Methods

18.2.1 Fish and Sample Collection

Three groups of coho salmon (approx. 60 g) were utilized: (1) non-transgenic wild type (NT, 21 months old), (2) GH transgenic salmon growing at their full rate (T, 9 months old) and (3) ration-restricted GH transgenic salmon (TR, 21 months old) that were pair fed a NT ration level to have a growth rate equivalent to that of non-transgenic salmon. All fish possessed a wild type genetic background from the Chehalis River, BC, Canada strain. Transgenic salmon possessed a single copy of the OnMTGH1 gene construct in strain M77 (Devlin et al. 2004; Devlin et al. 1994). All fish were reared in the Center for Aquaculture and Environment

Research (CAER) aquarium facility in tanks supplied with aerated running well water (10–11 °C) under natural photoperiod. Fish (transgenic and non-transgenic) were fed by hand with commercially available diets (Skretting Canada, Canada) and their size matched according to the method reported previously (Nakano et al. 2011; Raven et al. 2008; Rise et al. 2006). Tissues from transgenic fish and non-transgenic fish were sampled 4 h after feeding (feeding was at 9:00 AM). Fish were anaesthetized with 100 mg/L tricane methane sulphonate buffered with an equal weight of $NaHCO_3$, and rapidly team-sampled for tissues. The excised tissues were cut into small pieces, immediately frozen in liquid nitrogen, and then stored at −80 °C until analysis. Fish rearing and handling was Canadian Council of Animal Care Guidelines.

18.2.2 Determination of Metabolites

Deep-frozen samples were weighed and homogenized by a cell disrupter, after adding methanol containing internal standards (methionine sulfone, 2-(N-morpholino)-ethanesulfonic acid, and D-camphol-10-sulfonic acid). The homogenate was then mixed with chloroform and Milli-Q water and centrifuged at 4,600 × g for 15 min at 4 °C. The resultant aqueous phase was centrifugally filtered through a 5 kDa cutoff filter (Millipore, MA, USA). The filtrate was dried and dissolved in Milli-Q water containing 3-aminopyrrolidine and trimesate as a reference compound before capillary electrophoresis time-of-flight mass spectrometry (CE-TOFMS) analysis.

Metabolites in the samples obtained were determine by CE-TOFMS analysis with an Agilent CE Capillary Electrophoresis System (Agilent Technologies, CA, USA) according to Soga et al. (2003) and Hirayama et al. (2009). For the control and data acquisition in this system, an Agilent 3D CE-MSD ChemStation software was used.

18.2.3 Statistical Analysis

For each sample, measured metabolite concentrations were normalized using tissue weight to obtain the amount of metabolite contained per gram of sample. Results are reported as mean ± SEM (nmol/g tissue, n = 3–4). All data were subjected to one-way analysis of variance (ANOVA). Multiple comparisons between groups were made by Fisher's least-square difference (LSD) test and results were determined statistically significant at $p < 0.05$ (Nakano et al. 2013; Yanai 1998).

18.3 Results and Discussion

The CE-TOFMS systems in two different modes for cation and anion analyses quantified 170 and 191 metabolites involved in glycolysis, pentose phosphate pathway, tricarboxylic acid (TCA) cycle, amino acid, etc. in muscle and liver respectively. Levels of metabolites involved in central carbon metabolism, such as glycolysis and TCA cycle, are shown in Figs. 18.1, 18.2, 18.3 and 18.4.

Several metabolites such as glyceraldehyde-3-phosphate (G3P) and 1,3-diphosphoglycerate (1,3DPG; 1,3-bisphosphoglycerate) of the glycolysis, and oxaloacetate, 2-oxoglutarate, etc. of the TCA cycle were not detected. The most notable difference found between GH transgenic and NT fish was that glycolysis metabolite levels were higher in the muscle of GH transgenic fish (Fig. 18.1).

These levels were affected by ration-restriction in GH transgenic fish (GH transgenic fish that had their growth rate limited to that of NT). Surprisingly, the muscle metabolites levels of the initial part of the glycolytic pathway, including glucose 1-phosphate (G1P), glucose 6-phosphate (G6P), and fructose 6-phosphate (F6P), were observed to be markedly higher in ration-restricted GH transgenic fish than in both fully fed GH transgenic and NT fish. This observation suggests that, when supply of dietary energy was restricted, the GH transgenic fish had increased emphasis on carbohydrate metabolism for energy production, compared with NT fish. Accordingly, the preferential use of carbohydrates for energy production in GH transgenic fish appears to strongly depend on ration level (food intake and digestion; total digestible energy intake) as well as the dietary level of carbohydrate (Higgs et al. 2009; Kim et al. 2015; Leggatt et al. 2009).

Several muscle metabolites (organic acids) of the TCA cycle such as succinate, fumarate, and malate were decreased in fed GH transgenic fish compared to NT fish (Fig. 18.2).

Furthermore, the muscle level of pyruvate was significantly higher in GH transgenic fish than that in NT fish muscle (Fig. 18.1). These data suggest a high dependence of GH transgenic fish muscle on aerobic breakdown of pyruvate. Interestingly, these effects on muscle metabolite for glycolysis and the TCA cycle were different from effects seen in liver (Figs. 18.3 and 18.4). For example, the glycolysis metabolite levels

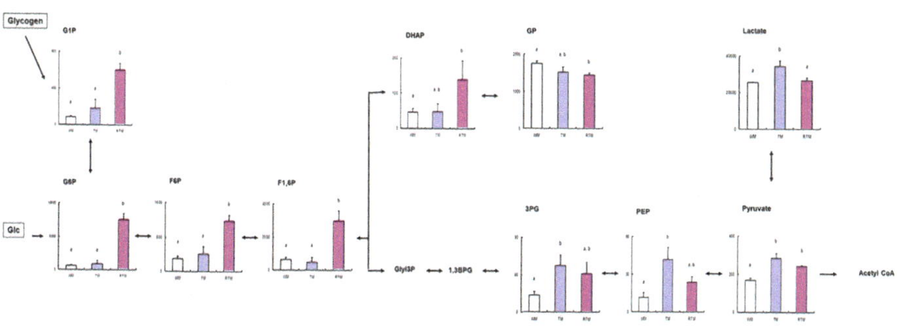

Fig. 18.1 Levels of glycolysis metabolites in the muscle of coho salmon *O. kisutch* Data represent means ± SEM ($n = 3$). Statistical relationships among groups are indicated by letters where significant differences were detected ($p < 0.05$)

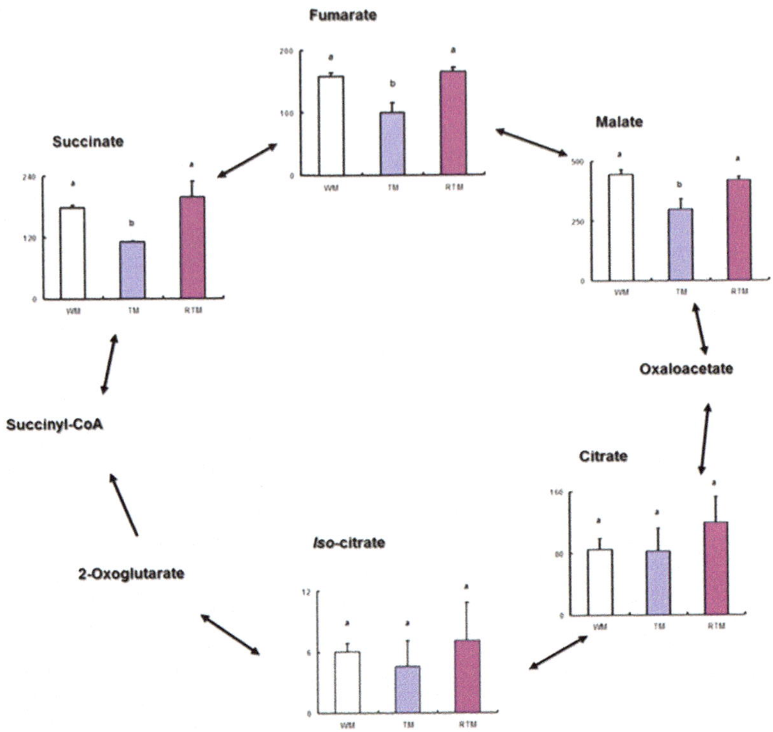

Fig. 18.2 Levels of TCA cycle metabolites in the muscle of coho salmon *O. kisutch* Data represent means ± SEM (n = 3). Statistical relationships among groups are indicated by letters where significant differences were detected ($p < 0.05$)

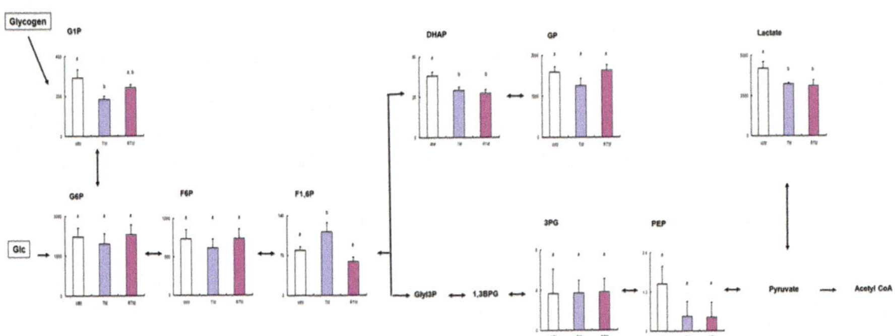

Fig. 18.3 Levels of glycolysis metabolites in the liver of coho salmon *O. kisutch* Data represent means ± SEM (n = 4). Statistical relationships among groups are indicated by letters where significant differences were detected ($p < 0.05$)

in the liver of GH transgenic fish were found to be similar to that of NT fish (Fig. 18.3). Accordingly, these results suggest slightly different effects of GH transgenesis on metabolic pathways such as glycolysis and TCA cycle in the liver. The levels of several

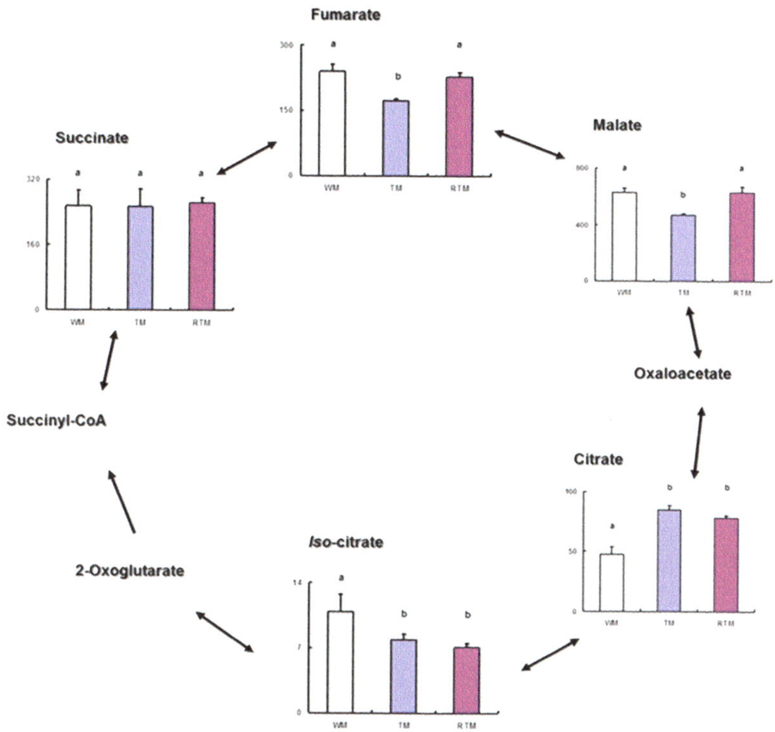

Fig. 18.4 Levels of TCA cycle metabolites in the liver of coho salmon *O. kisutch* Data represent means ± SEM ($n = 4$). Statistical relationships among groups are indicated by letters where significant differences were detected ($p < 0.05$)

amino acids, nucleotides and peptides in both muscle and liver also seemed to be affected by GH transgenesis (data not shown).

Expression of the GH transgene has been detected in many tissues, including muscle liver, kidney, stomach, intestine, etc. in GH transgenic coho salmon (Mori and Devlin 1999; Raven et al. 2008). Accordingly, it is not unexpected that tissue-specific differences concerning metabolites involved in central carbon metabolism in GH transgenic fish might exist. In both GH transgenic fish and in fish treated with exogenous GH, high levels of GH could promote gluconeogenic and lipolytic metabolism (Björnsson 1997; Björnsson et al. 2002; Devlin et al. 2009; Higgs et al. 2009; Leung and Woo 2010; Overturf et al. 2013; Panserat et al. 2014). Higher glycolysis (carbohydrate metabolism) in muscle has also observed in growth-enhanced transgenic fish and fast-growing domesticated fish strains (Devlin et al. 2009; Hill et al. 2000; Krasnov et al. 1999). The level of food intakes, food-conversion efficiency, carbohydrate degradation, utilization of lipids and proteins (rate of protein synthesis and lipid mobilization) of GH transgenic fish are greater than those of non-transgenic fish (Higgs et al. 2009; Huang et al. 2004; Leggatt et al. 2009; Oakes et al. 2007; Raven et al. 2006). Increased activities of glycolytic enzymes and decreased activities of lipolytic enzymes in GH transgenic fish have also been confirmed (Devlin et al. 2009; Hill et al. 2000; Leggatt et al. 2009; Panserat et al. 2014). However, not until the present study have the actual levels of metabolites arising from these changes in gene expression and enzyme kinetics been measured. Together, these studies and present results suggest that GH transgenesis might differentially regulate energy metabolism in fish. In fact, GH transgenic coho salmon were reported to maintain growth rate even when fed high in carbohydrates or low digestible energy (Higgs et al. 2009; Raven et al. 2006). GH transgenic coho salmon also have an ability for different usage of exogenously supplied glucose (Panserat et al. 2014), which our data are consistent with. In addition, several of the biochemical features (altered glycolysis, pentose phosphate pathway, and TCA cycle) detected in GH transgenic fish suggest they may be in a partially starved state (Sugiyama et al. 2012). Accordingly, GH transgenic fish are characterized by altered metabolism and energy utilization.

Energy production from dietary carbohydrates is known to be inefficient in carnivorous fish such as salmonids (Krogdahl et al. 2005; Leggatt et al. 2009; Takeuchi 2009). However, the present results suggest that GH transgenesis can improve the use of carbohydrates for energy production which could be advantageous for improving production efficiency and reducing costs in aquaculture. Further study is required to assess these findings under a range of nutritional and environmental conditions, and at different developmental stages.

18.4 Conclusion

It is important to assess perturbations, recovery, and resilience processes in marine ecosystems, and to manage socio-ecological systems arising from fishing and aquaculture to sustainably deliver environmental benefits and food production linked to human well-being (Hadjimichael et al. 2013; Naylor et al. 2000; Pauly et al. 2002). Feed is one of the largest production costs for commercial aquaculture. Fishmeal has historically been the major protein source using in many aquaculture diets (aquafeeds), especially for carnivorous fish species. However, costs of fish meal and fish oil have been rising, such that a major challenge of current aquaculture is the availability of suitable sources of protein in aquafeeds. Alternative major ingredient sources such as plant protein are now being used widely to replace fishmeal in aquafeeds (Gatlin III et al. 2007; Naylor et al. 2000; Naylor et al. 2009; Overturf et al. 2013; Tacon and Metian 2008; Yamamoto et al. 2015). However, plant-based fishmeal replacement diets can result in reduced growth performance of fish due to the presence of anti-nutritional factors in plant feedstuffs and insufficient nutrition level compared to fishmeal-based diets (Gatlin III et al. 2007; Overturf et al. 2013; Yamamoto et al. 2015). Therefore, an enhancement of nutritional utilization of alternative ingredient such as plant protein and carbohydrate in aquafeeds for farmed fish species may be of interest and benefit.

The results of this study on GH transgenic fish can provide information that is useful for improving fish fitness and production of fishing and aquaculture. Further research regarding metabolic enzymes, energy production, and growth-related factors is needed to elucidate the exact mechanism of accelerated growth in GH transgenic fish.

These data also have implications for environmental risk assessments (Devlin et al. 2015) since it appears that GH transgenic salmon may be able to differentially access and utilize different prey items in their ecosystem compared to those used by non-transgenic animals. Whether this would provide them with a disadvantage could depend in part on whether they have a higher requirement for carbohydrate which may not be available in their ecosystem, or cause them to have for example an altered feeding behavior and related predator avoidance responses that reduces their fitness. Alternatively, GH transgenic salmon may have a greater scope for using carbohydrate energy source when available, providing them with greater options for acquiring sufficient energy through development to meet growth, survival and reproductive demands which may enhance their fitness and influence their consequences in ecosystems.

Acknowledgements The authors thank staff at Center for Aquaculture and Environment Research, Fisheries and Oceans Canada, Institute for Advanced Biosciences, Keio University, Japan, and Dr. T. Fujimoto at Hokkaido University, Japan for assistance with lab work. TN acknowledges Dr. J. G. Richards, Mr. D. Chung and Miss. H. Bryant at Department of Zoology, The University of British Columbia, Canada, and Drs. M. Sato, T. Yamaguchi, Y. Ochiai, and I. Gleadall at Tohoku University, Japan for valuable discussion. This study was supported in part by KAKENHI grant (#23580277) and a fund to Core-to-Core Program (A. Advanced Research Networks) entitled "Establishment of international agricultural immunology research-core for a quantum improvement in food safety" from JSPS to TN and by the Canadian Regulatory System for Biotechnology to RHD.

References

Björnsson BT (1997) The biology of salmon growth hormone: from daylight to dominance. Fish Physiol Biochem 17:9–24

Björnsson BT, Johansson V, Benedet S, Einarsdottir E, Hildahl J, Agustsson T, Jonsson E (2002) Growth hormone endocrinology of salmonids: regulatory mechanisms and mode of action. Fish Physiol Biochem 27:227–242

Deane EE, Woo NYS (2009) Modulation of fish growth hormone levels by salinity, temperature, pollutants and aquaculture related stress: a review. Rev Fish Biol Fish 19:97–120

Deane EE, Woo NYS (2011) Advances and perspectives on the regulation and expression of piscine heat shock proteins. Rev Fish Biol Fish 21:153–185

Devlin RH, Yesaki TY, Biagi C, Donaldson EM, Swanson P, Chan W-K (1994) Extraordinary salmon growth. Nature 371:209–210

Devlin RH, Yesaki TY, Donaldson EM, Hew CL (1995) Transmission and phenotypic effects of an antifreeze/GH gene construct in coho salmon (*Oncorhynchus kisutch*). Aquaculture 137:161–169

Devlin RH, Biagi C, Yesaki TY, Smailus DE, Byatt JC (2001) Growth of domesticated transgenic fish. Nature 409:781–782

Devlin RH, Biagi C, Yesaki TY (2004) Growth, viability and genetic characteristics of GH transgenic coho salmon strains. Aquaculture 236:607–632

Devlin RH, Sakhrani D, Tymchuk WE, Rise ML, Goh B (2009) Domestication and growth hormone transgenesis cause similar changes in gene expression in coho salmon (*Oncorhynchus kisutch*). Proc Nat Acad Sci USA 106:3047–3052

Devlin RH, Sundstrom LF, Rosalind AL (2015) Assessing ecological and evolutionary consequences of growth-accelerated genetically engineered fishes. Bioscience 65:685–700

Donaldson EM, Fagerlund UHM, Higgs DA, McBride JR (1979) Hormonal enhancement of growth. In: Hoar WS, Randall DJ, Brett JR (eds) Bioenergetics and Growth, Fish Physiology. Academic Press, New York, pp 455–597

Du SJ, Gong Z, Fletcher GI, Shears MA, King MJ, Idler DR, Hew CL (1992) Growth enhancement in transgenic Atlantic salmon by the use of an "all fish" chimeric growth hormone gene construct. Nat Biotechnol 10:176–181

Duan C (1998) Nutritional and developmental regulation of insulin-like growth factors in fish. J Nutr 128:306S–314S

Gatlin DM III, Barrows FT, Brown P, Dabrowski K, Gaylord TG, Hardy RW, Herman E, Hu G, Krogdahl A, Nelson R, Overturf K, Rust M, Sealey W, Skonberg D, Souza EJ, Stone D, Wilson R, Wurtele E (2007) Expanding the utilization of sustainable plant products in aquafeeds: a review. Aquacult Res 38:551–579

Hadjimichael M, Delaney A, Kaiser MJ, Edwards-Jones G (2013) How resilient are Europe's inshore fishing communities to change? differences between the north and the south. Ambio 42:1037–1046

Higgs DA, Sutton JN, Kim H, Oakes JD, Smith J, Biagi C, Rowshandeli M, Devlin RH (2009) Influence of dietary concentrations of protein, lipid and carbohydrate on growth, protein and energy utilization, body composition, and plasma titres of growth hormone and insulin-like growth factor-1 in non -transgenic and growth hormone transgenic coho salmon, *Oncorhynchus kisutch* (Walbaum). Aquaculture 286:127–137

Hill AJ, Kiessling A, Devlin RH (2000) Coho salmon (*Oncorhynchus kisutch*) transgenic for a growth hormone gene construct exhibit increased rates of muscle hyperplasia and detectable levels of differential gene expression. Can J Fish Aqua Sci 57:939–950

Hirayama A, Kami K, Sugimoto M, Sugawara M, Toki N, Onozuka H, Kinoshita T, Saito N, Ochiai A, Tomita M, Esumi H, Soga T (2009) Quantitative metabolome profiling of colon and stomach cancer microenvironment by capillary electrophoresis time-of-flight mass spectrometry. Cancer Res 69:4918–4925

Huang C-H, Higgs DA, Balfry SK, Devlin RH (2004) Effect of dietary vitamin E level on growth, tissue lipid peroxidation, and erythrocyte fragility of transgenic coho salmon, Oncorhynchus kisutch. Comp Biochem Physiol 139A:199–204

Kim J-H, Leggatt RA, Chan M, Vokoff H, Devlin RH (2015) Effects of chronic growth hormone overexpression on appetite-regulating brain gene expression in coho salmon. Mol Cell Endocrinol 413:178–188

Krasnov A, Agren JJ, Pitkanen TI, Molsa H (1999) Transfer of growth hormone (GH) transgenes into Arctic charr (*Salvelinus alpinus* L.) II. Nutrient partitioning in rapidly growth fish. Gen Analysis 15:99–105

Krogdahl A, Hemre G-I, Mommsen TP (2005) Carbohydrates in fish nutrition: digestion and absorption inpostlarval stages. Aquac Nutr 11:103–122

Ledford H (2015) Transgenic salmon leaps to the dinner table. Nature 527:417–418

Leggatt RA, Raven P, Mommsen TP, Sakhrani D, Higgs DA, Devlin RH (2009) Growth hormone transgenic influences carbohydrate, lipid and protein metabolism capacity for energy production in coho salmon (*Oncorhynchus kisutch*). Comp Biochem Physiol 154B:121–133

Leggatt RA, Hollo T, Vandersteen WE, McFarlane K, Goh B, Prevost J, Devlin RH (2014) Rearing in seawater mesocosms improves the spawning performance of growth hormone transgenic and wild type coho salmon. PLoS ONE 9:e105377

Leung LY, Woo NYS (2010) Effects of growth hormone, insulin-like growth factor I, triiodothyronine, thyroxine, and cortisol on gene expression of carbohydrate metabolic enzymes in sea bream hepatocytes. Comp Biochem Physiol 157A:272–282

Lohmus M, Sundstrom LF, Bjorklund M, Devlin RH (2010) Genotype-temperature interaction in the regulation of development, growth, and morphometrics in wild-type, and growth-hormone transgenic coho salmon. PLoS ONE 5:e9980
Mori T, Devlin RH (1999) Transgene and host GH gene expression in pituitary and nonpituitary tissues of normal and GH transgenic salmon. Mol Cell Endocrinol 149:129–139
Moriyama S, Ayson FG, Kawauchi H (2000) Growth regulation by insulin-like growth factor-I in fish. Biosci Biotech Biochem 64:1553–1562
Nakano T (2007) Microorganisms. In: Nakagawa H, Sato M, Gatlin DM III (eds) Dietary supplements for the health and quality of cultured fish. CAB International, Oxfordshire, UK, pp 86–108
Nakano T (2011) Stress in fish. Yoshoku (Aquaculture Magazine) 48:64–67
Nakano T, Shoji Y, Shirakawa H, Suda Y, Yamaguchi T, Sato M, Devlin RH (2011) Daily expression patterns of growth-related genes in growth hormone transgenic coho salmon, *Oncorhynchus kisutch*. La mer 49:111–117
Nakano T, Afonso LO, Beckman BR, Iwama GK, Devlin RH (2013) Acute physiological stress down-regulates mRNA expressions of growth-related genes in coho salmon. PLoS ONE 8: e71421
Nakano T, Kameda M, Yamaguchi T, Sato M, Afonso LOB, Beckman BR, Iwama GK, Devlin RH (2015) Effect of thermal stressors on growth-related gene expressions in cultured fish. In: Ceccaldi H-J, Hénocque Y, Koike Y, Komatsu T, Stora G, Tusseau-Vuillemin M-H (eds) Marine productivity: perturbations and resilience of socio-ecosystems. Springer International Publishing AG, Cham, Switzerland, pp 147–157
Nakano T (2016) Studies on stress and stress tolerance mechanisms in fish. Nippon Suisan Gakkaishi 82:278–281
Naylor RL, Goldburg RJ, Primavera JH, Kautsky N, Beveridge MCM, Clay J, Folke C, Lubchenco J, Mooney H, Troell M (2000) Effect of aquaculture on world fish supplies. Nature 405:1017–1024
Naylor RL, Hardy RW, Bureau DP, Chiu A, Elliott M, Farrell AP, Forster I, Gatlin DM, Goldburg RJ, Hua K, Nichols PD (2009) Feeding aquaculture in an era of finite resources. Proc Nat Acad Sci USA 106:15103–15110
Oakes JD, Higgs DA, Eales JG, Devlin RH (2007) Influence of ration level on the growth performance and body composition of non-transgenic and growth-hormone-transgenic coho salmon (*Oncorhynchus kisutch*). Aquaculture 265:309–324
Overturf K, Barrows FT, Hardy RW (2013) Effect and interaction of rainbow trout strain (Oncorhynchus mykiss) and diet type on growth and nutrient retention. Aquacult Res 44:604–611
Panserat S, Kamalam BS, Fournier J, Plagnes-Juan E, Woodward K, Devlin RH (2014) Glucose metabolic gene expression in growth hormone transgenic coho salmon. Comp Biochem Physiol 170A:38–45
Pauly D, Christensen V, Guenette S, Pitcher TJ, Sumaila UR, Walters CJ, Watson R, Zeller D (2002) Towards sustainability in world fisheries. Nature 418:689–695
Raven P, Devlin RH, Higgs DA (2006) Influence of dietary digestible energy content on growth, protein and energy utilization and body composition of growth hormone transgenic and non-transgenic coho salmon (*Oncorhynchus kisutch*). Aquaculure 254:730–747
Raven P, Uh M, Sakhrani D, Beckman BR, Cooper K, Pinter J, Leder EH, Silverstein J, Devlin RH (2008) Endocrine effects of growth hormone overexpression in transgenic coho salmon. Gen Comp Endocrinol 159:26–37
Reichardt T (2000) Will souped up salmon sink or swim? Nature 406:10–12
Reineck M (2010) Influences of the environment on the endocrine and paracrine fish growth hormone-insulin-like growth factor-I system. J Fish Biol 76:1233–1254
Rise ML, Douglas SE, Sakhrani D, Williams J, Ewart KV, Rise M, Davidson WS, Koop BF, Devlin RH (2006) Multiple microarray platforms utilized for hepatic gene expression profiling of GH transgenic coho salmon with and without ration restriction. J Mol Endocrinol 37:259–282

Soga T, Ohashi Y, Ueno Y, Naraoka H, Tomita M, Nishioka T (2003) Quantitative metabolome analysis using capillary electrophoresis mass spectrometry. J Proteome Res 2:488–494

Stokstad E (2002) Engineered fish: Friend or foe of the environment? Science 297:1797–1799

Sugiyama M, Takenaga F, Kitani Y, Yamamoto G, Okamoto H, Masaoka T, Araki K, Nagoya H, Mori T (2012) Homozygous and heterozygous GH transgenesis alters fatty acid composition and content in the liver of Amago salmon (*Oncorhynchus masou ishikawae*). Biology Open 1:1035–1042

Tacon AGJ, Metian M (2008) Global overview on the use of fish meal and fish oil in industrially compounded aquafeeds: Trends and future prospects. Aquaculture 285:146–158

Takeuchi T (2009) Carbohydrate. In: Watanabe T (ed) Nutrition and Feeding in Fish and Crustaceans. Kouseisha Kouseikaku, Tokyo, pp 106–114

Yamamoto T, Murashita K, Matsunari H, Oku H, Furuita H, Okamoto H, Amano S, Suzuki N (2015) Selectively bred juvenile F2 amago salmon *Oncorhynchus masou ishikawae* fed a low fishmeal diet exhibit growth comparable to unselected juveniles fed a fishmeal-based diet. Fish Sci 81:83–93

Yanai H (1998) Excell Statistics. OMS Publishing, Tokorozawa, Saitama, Japan

Chapter 19
Estimating the Diets of Fish Using Stomach Contents Analysis and a Bayesian Stable Isotope Mixing Models in Sendai Bay

Hiroyuki Togashi, Yukinori Nakane, Yosuke Amano and Yutaka Kurita

Abstract Food web structures are well-studied, central theme of fisheries science that depicts the pathways of energy flow. Researchers have long been studied on the importance of quantitative estimation of the relative contribution of prey animals in fish by stomach contents analysis. However, this analysis requires experienced technique, and it only reflects recent diets. On the other hand, carbon and nitrogen stable isotope ratios representing average feeding habits and trophic position have become a powerful tool in food web analyses. In this study, to clarify the main resources (feeding habits) and contribution rate of prey animals using these two analyses in Sendai Bay. As a result, the feeding habits by stomach content analysis of dominant ten fish species were classified into four groups; fish feeders, crustacean feeders, benthic feeders and plankton feeders. A dual isotope plot of $\delta^{13}C$ and $\delta^{15}N$ values supported the classification by stomach content analysis. Most of fish feeders depended on bait fishes (mainly feeding on plankton feeders), and crustacean and benthic feeders depended on crustacean and/or benthos. The contribution rate of prey animals by mixing models for $\delta^{13}C$ and $\delta^{15}N$ within Bayesian framework, *Paralichthys olivaceus* in June 2012 was 67% *Engraulis japonicus*, and 32% *Ammodytes personatus*. The results of stable isotope analysis were well consistent with the stomach contents analysis.

Keywords Bayesian mixing models · C and N stable isotopes
Food webs · Stomach contents analysis

H. Togashi (✉) · Y. Amano · Y. Kurita
Tohoku National Fisheries Research Institute, Japan Fisheries Research and Education Agency, 3-27-5, Shinhama-cho, Shiogama, Miyagi 985-0001, Japan
e-mail: htogashi@affrc.go.jp

Y. Nakane
Central Research Institute of Electric Power Industry,
1646, Abiko, Chiba 270-1194, Japan

T. Komatsu et al. (eds.), *Oceanography Challenges to Future Earth*,
https://doi.org/10.1007/978-3-030-00138-4_19

19.1 Introduction

The major two analysis techniques can be used to elucidate the structure of food webs—one is observation of stomach contents and the other stable isotope ratios. Observation of stomach contents provides direct evidence what the fish ate. This analysis makes it possible to understand the predator—prey relationships in the food webs. However, observation of the stomach contents only allows a snapshot data when the fish ate feeds. Additonally, it requires proficiency and experience to identify the prey animals found inside a fish's stomach and, as digestion progresses, it becomes more difficult to identify the prey animals. On the other hand, by measuring carbon and nitrogen stable isotope ratios (δ^{13}C, δ^{15}N) in the organisms, it can be estimated trophic position and primary production origins of the food webs. It is known that stable isotope ratios are concentrated with a δ^{13}C of 0–1.5‰ (Deniro and Epstein 1978; Rau et al. 1983; Fry and Sherr 1989) and a δ^{15}N of 3–4‰ (Deniro and Epstein 1981; Minagawa and Wada 1984; Post 2002) per increasing trophic position. Thus, by measuring the δ^{13}C and δ^{15}N in the organisms, it is possible to reveal food web structure in terms of resources importance and trophic pathways. Moreover, stable isotope ratios include the information after assimilation, in contrast with stomach contents analysis. Thus, the effects of temporary or incidental prey animals are minimized and typical feeding habits are averaged. In addition, it requires only a small amount (about 0.5–1 mg dry weight) for the measurement, making this a widely available and powerful tool for explaining food web structures. However, even with stable isotope analysis, there are some issues, such as large spatiotemporal fluctuations in the stable isotope ratios of phytoplankton (primary producers) and the fact that concentration factor per trophic position vary by animals (McCutchan et al. 2003; Caut et al. 2009). In the present study, observation of the stomach contents was used as a qualitative data for clarifying the main resources of fish, and stable isotopeanalysis used a quantitative data for estimating contribution rate of prey animals in the food webs.

Using of stable isotope ratios in the fisheries science is applied not only the clarification of food web structures, but also to various research, including estimates of seasonal fish migration (Hansson et al. 1997; Suzuki et al. 2008), mass transport to terrestrial ecosystems by runs of anadromous salmonids (Kline et al. 1990; Kaeriyama et al. 2004), and improvements in aquaculture technology by means of clarifications of food resources during the juvenile fish stage (Miller et al. 2013; Tanaka et al. 2014). Against these backgrounds, it is felt that, due to advances in measurement technology and the spread of mass spectrometers made measurement of stable isotope ratios easier. Moreover, even without observing stomach content, we can estimate contribution rate of prey animals from stable isotope ratios. However, two element isotopes such as δ^{13}C and δ^{15}N can only be estimated up three resources. To overcome this problem, we thus had to increase the number of element isotopes. In recent years, the method for estimating contribution rates have been developed in accordance with probability distribution in Bayesian framework that does not rely on the number of element isotopes. The aims of this study to clarify the main resources of fish (feeding habits) by stomach contents analysis and contribution rate of prey animals within Bayesian framework using stable isotope ratios in a coastal ecosystem in Sendai Bay.

19.2 Materials and Methods

19.2.1 Study Site and Survey

The survey was carried out at six locations in Sendai Bay at a water depth of from 30 to 80 m during September 2011 to June 2012, with one survey carried out per season (spring; June-July: summer; August-September: autumn; November-December and winter; January-February) (Fig. 19.1).

The fish samples were collected by bottom trawling (mesh size: 5 mm) using the Wakataka Maru, a 692 ton research vessel belonging to the Tohoku National Fisheries Research Institute, and smaller trawling boats (9.7 ton). Prey animals were collected by the Wakataka Maru using a dredge (mesh size: 5 mm) and sledge net (mesh size: 0.45 mm) system. A ship speed of 2.5–3.0 knots was maintained for both methods; bottom trawling was carried out for 30 min, dredging for 3 min and sledge netting for 8 min. Once brought on board, fish were stored in the refrigerator or freezer until they were delivered to the laboratory. Benthos collected by dredgenet system was putin a 1 mm mesh screen, separated the animals from the sediment. Benthos was identified even low-ranking taxon and then frozen. Zooplankton collected by sledge net system was frozen on board the vessels. Phytoplankton was filtered on board using 100 μm and 10 μm mesh nets, with these filtered species used as specimens.

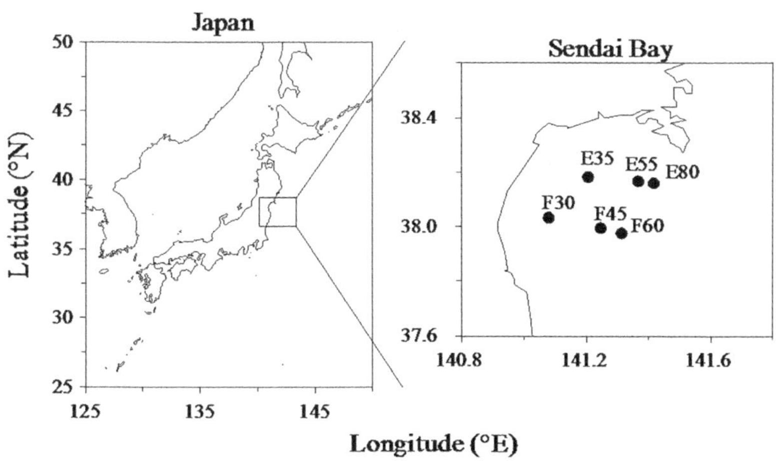

Fig. 19.1 Maps showing the location of study sites in Sendai Bay. The site number represents the water depth in Sendai Bay

19.2.2 Analyzing Feeding Habits by Observing Stomach Contents

Ten major coastal fish species (*Paralichthys olivaceus, Kareius bicoloratus, Lateolabrax japonicus, Gadus macrocephalus, Hexagrammos otakii, Lepidotrigla microptera, Pleuronectes herzensteini, Pleuronectes yokohamae, Ammodytes personatus* and *Engraulis japonicus*) were analyzed. After thawing out the fish, samples were measured the total length (mm) and wet weight (g), and dissected them to check their stomach contents. We tested 30 or more specimens of fish each collection day. The observation of stomach contents was identified by the naked eye and, when necessary, by using a stereoscopic microscope to identify even lower ranking taxon, measuring up to 0.1 g of wet weight for each taxon. As an indicator of feeding habits, we calculated species *i* prey animals as a ratio of the weight of all stomach contents (W_i: %) using Eq. 19.1:

$$W_i(\%) = \left(\Sigma_j \mathrm{SCW}_{i,j} / \Sigma_j \mathrm{SCW}_{all,j}\right) \times 10^2 \quad (19.1)$$

In this equation, $\mathrm{SCW}_{i,\ j}$ indicates the wet weight (g) of species *i* prey animal found amongst the stomach contents of specimen *j* and $\mathrm{SCW}_{all,\ j}$ indicates the total wet weight of the stomach contents of specimen *j*. We defined the most abundant feeding four categories (fishes, crustaceans, non-crustacean benthos (hereinafter, "benthos") and plankton) as particular feeding habits.

19.2.3 Analyzing the Food Web Using Stable Isotope Ratios

In preparation of analysis, after fish samples (ca. 30 mg) and prey animal samples (ca. 10 mg) were freeze-dried and powdered, were defatted using a 2:1 mixture of chloroform and methanol for 24 h to minimize the difference in δ^{13}C values caused by variable lipid content. Weighted all samples (ca. 0.5–1.0 mg) were placed into tin capsules and crimped for combustion. Stable isotope analysis was carried out using either a mass spectrometer (Delta V Advantage, Thermo Fisher Scientific, US) connected to an elementary analyzer (Flash EA 1112, Thermo Fisher Scientific, US) via an interface (Conflo III, Thermo Fisher Scientific, US), or else a mass spectrometer (ANCA-GSL, Sercon, UK) directly connected to an elementary analyzer (Hydra 20–22, Sercon, UK). Stable isotope ratios are expressed in δ notation in per mil units (‰), according to the following Eq. 19.2:

$$\delta^{13}\mathrm{C}\ \mathrm{or} \delta^{15}\mathrm{N}\ (\%) = \left(\mathrm{R}_{\mathrm{sample}} / \mathrm{R}_{\mathrm{standard}} - 1\right) \times 10^3 \quad (19.2)$$

where R indicates the ratios of ^{13}C/^{12}C or ^{15}N/^{14}N, with $\mathrm{R}_{\mathrm{sample}}$ indicating measured samples and $\mathrm{R}_{\mathrm{standard}}$ values indicating based on the Vienna Pee Dee Belemnite (VPDB) for δ^{13}C and nitrogen in air for δ^{15}N. The measured data were corrected using the international standards (CERKU-01, CERKU-02 and CERKU-05) indicated by Tayasu et al. (2011). The standard deviations of the δ^{13}C and δ^{15}N measurements were typically within 0.2‰.

19.2.4 Contribution Rate Estimates for Prey Animals Using Bayesian Framework

The relative contributions of prey animals consumed by fish were estimated using their δ^{13}C and δ^{15}N values within Bayesian framework. Bayesian estimate type of mixing model was utilized SIAR 4.1 (Parnell et al. 2010), an open source R 3.2.1 package (R Development Core Team 2015). The present analysis involved calculating prey animal contribution based on Markov Chain Monte Carlo (MCMC) methods. To be more precise, based on random number simulation, first, the simulation was made a virtual calculation of the stable isotope ratios of predators using the stable isotope ratios of each prey animals. Next, to better approximate estimated predator stable isotope ratios, was varied the absorption rate of each prey animals. This process using hundreds of thousands of orders were repeated, finally coming up with a probability of distribution that close to reality. This analysis had many advantages over traditional mixed models, such as the ability to estimate contribution rate for four or more prey animals from two element isotopes (for example, δ^{13}C and δ^{15}N). The present study incorporated the four prey animals that made up the greatest weight ratio of the stomach contents as prey animal parameters (Table 19.1). Because *P. olivaceus* had preyed upona total of *A. personatus* and *E. japonicus* more than 90%, *Loliolus japonica* was added as three prey animal parameters. In Bayesian stable isotope mixing models within the 500,000 iterations

Table 19.1 Incorporated prey animal parameters into the mixing model using Bayesian framework to estimate the contribution rates

Species	Prey animal parameters			
	1	2	3	4
Paralichthys olivaceus	*Engraulis japonicus*	*Ammodytes personatus*	*Loliolus japonica*	–
Lateolabrax japonicus	*Engraulis japonicus*	*Ammodytes personatus*	*Metapenaeopsis dalei*	Brachyura
Hexagrammos otakii	*Metapenaeopsis dalei*	Brachyura	*Crangon* sp.	*Ammodytes personatus*
Pleuronectes herzensteini	Polychaeta	Philinidae	*Metapenaeopsis dalei*	*Ammodytes personatus*
Ammodytes personatus	Zooplankton	Phytoplankton	–	–
Engraulis japonicus	Zooplankton	Phytoplankton	–	–

were performed, 1–50,000 repetitions were removed from analysis since before getting the steady state, with samples of the remainder taken at a rate of once every fifteen repetitions (number of samples: (500,000–50,000)/15 = 30,000). It was evaluated by the simulated median values, adopting concentration factors of 3.4 ± 0.98‰ for $\delta^{15}N$ and 0.39 ± 1.3‰ for $\delta^{13}C$ in accordance with Post (2002).

19.3 Results and Discussion

In the stomach content analysis, we defined the most abundant feeding four categories (fish, crustacean, benthos and plankton) as fish feeding habits. As a result, *P. olivaceus* preyed upon a total of *A. personatus* and *E. japonicus* more than 90%. Dominant prey animals of *H. otakii* and *P. herzensteini* were *Metapenaeopsis dalei* and Polychaeta, respectively. Therefore, *P. olivaceus*, *K. bicoloratus*, *L. japonicus* and *G. macrocephalus* were classified as fish feeders, *H. otakii* and *L. microptera* were classified as crustacean feeders, *P. herzensteini* and *P. yokohamae* as benthos feeders, and *A. personatus* and *E. japonicus* as plankton feeders (Table 19.2).

The $\delta^{13}C$ and $\delta^{15}N$ of fish and prey animals that we collected from Sendai Bay in June 2012 are showed in Fig. 19.2.

The results of feeding habits as ascertained by observations of stomach contents and stable isotope ratios were very consistent. The $\delta^{13}C$ (± SD) for *P. olivaceus* defined as fish feeders was −17.3 ± 0.44‰ while the $\delta^{15}N$ (± SD) was 12.9 ± 0.75‰. The values of $\delta^{13}C$ and $\delta^{15}N$ were greater than those for the *E. japonicus* ($\delta^{13}C$: −19.2 ± 0.33‰, $\delta^{15}N$: 9.01 ± 0.60‰) as major prey animals, at 1.89‰ and 3.90‰. *H. otakii* ($\delta^{13}C$: −16.4 ± 0.59‰, $\delta^{15}N$: 13.6 ± 0.53‰), a crustacean feeder, had higher $\delta^{13}C$ and $\delta^{15}N$ than *M. dalei*, at 0.46‰ and 2.48‰ while the benthos-feeding *P. herzensteini* ($\delta^{13}C$: −15.8 ± 0.28‰, $\delta^{15}N$: 12.2 ± 0.59‰) had higher $\delta^{13}C$ (1.47‰) and $\delta^{15}N$ (2.68‰) than Polychaeta *A. personatus* and *E. japonicus*, both plankton feeders, were verified as feeding on zooplankton, with both coming in higher than zooplankton for $\delta^{15}N$ at 3.12‰ and 1.47‰ respectively. Stable isotope ratios supported observations of the stomach contents, which detected that bait fishes (*A. personatus* and *E. japonicus*), crustaceans and benthos were major prey animals of each fish. The food webs in Sendai Bay, using phytoplankton as a starting food chain, were able to confirm the existence of zooplankton and plankton feeding fish (*A. personatus* and *E. japonicus*), and fish feeders (*P. olivaceus*, *K. bicoloratus*, *L. japonicus* and *G. macrocephalus*) related based on plankton groups, as well as the crustaceans and benthos group that mainly feeds on *M. dalei* and Polychaeta.

The contribution rate using Bayesian framework for *P. olivaceus*, *L. japonicus*, *H. otakii* and *P. herzensteini* in June 2012 were shown in Fig. 19.3.

In general rules, these conformed to the results gained from observations of stomach contents, with the fish-feeding *P. olivaceus* estimating at 67% for *E. japonicus* and 32% for *A. personatus*. The contribution rate of the prey animals consumed by *H. otakii*, a crustacean feeder, estimated 31%, 20% and 13%

Table 19.2 Prey animals from the stomach content analysis of dominant ten species measured during September 2011–June 2012 in Sendai Bay. The symbols (+) indicate including stomach content of species. Abbreviation of prey animals more than 20% were shown EJ: *Engraulis japonicas*; AP: *Ammodytes personatus*; MD: *Metapenaeopsis dalei*; CR: *Crangon* sp.; PL: Polychaeta; EU: Euphausiidae and CO: Copepoda—Feeding habits with FF: fish feeders; CF: crustacean feeders; BF: benthic feeders and PF: plankton feeders

Species	Prey animals										
	Fish						Crustacean				
	Ammodytes personatus (AP)	*Engraulis japonicas* (EJ)	*Pennahia argentata*	Pleuronectidae	Callionymidae	Others	Brachyura	*Crangon* sp. (CR)	*Metapenaeopsis dalei* (MD)	*Trachysalambria curvirostris*	Others
Paralichthys olivaceus	+	+	+	+	+	+		+	+		
Kareius bicoloratus	+	+					+	+	+		+
Lateolabrax Japonicus	+	+	+			+	+	+	+	+	+
Gadus macrocephalus	+	+	+	+	+	+	+	+	+		+
Hexagrammos otakii	+	+	+		+	+	+	+	+	+	+
Lepidotrigla microptera	+	+				+	+	+	+	+	+
Pleuronectes herzensteini	+	+					+	+	+		+
Pleuronectes yokohamae	+	+					+	+	+		+
Ammodytes personatus											
Engraulis japonicus											

Species	Prey animals									Prey animals more than 20%		Feeding habits
Benthos							Plankton					
Polychaeta (PL)	Oligochaeta	Philinidae	*Ophioplocus japonicus*	Actiniaria	Bivalve	Others	Euphausiidae (EU)	Copepoda (CO)	Others			
									+	EJ	AP	FF
+					+				+	AP		FF
					+				+	AP	EJ	FF

(continued)

Table 19.2 (continued)

Species	Prey animals									Prey animals more than 20%		Feeding habits
Benthos							Plankton					
Polychaeta (PL)	Oligochaeta	Philinidae	*Ophioplocus japonicus*	Actiniaria	Bivalve	Others	Euphausiidae (EU)	Copepoda (CO)	Others			
					+				+	AP		FF
					+		+		+	MD		CF
+		+			+		+		+	CR	MD	CF
+	+	+	+	+	+	+	+		+	PL		BF
+	+	+	+	+	+	+	+		+	PL		BF
							+	+		EU	CO	PF
							+	+		EU	CO	PF

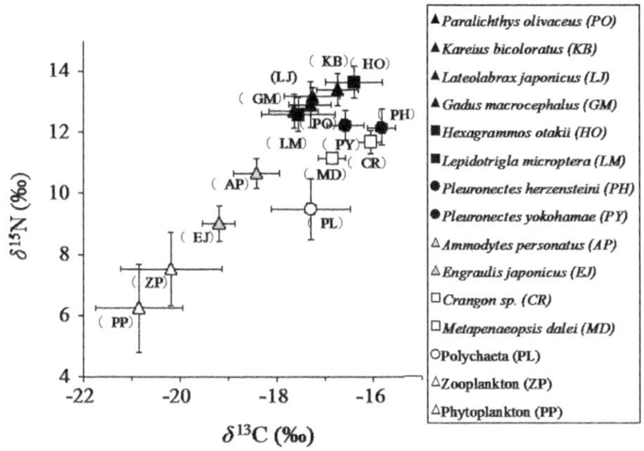

Fig. 19.2 Stable isotope ratios of carbon (δ^{13}C) versus nitrogen (δ^{15}N) for the fish and prey animals in Sendai Bay in June 2012. Open symbols indicate prey animals (plankton, crustacean and benthos). Solid triangles; fish feeders: solid squares; crustacean feeders: solid circles; benthos feeders and gray triangles; plankton feeders are shown, respectively

respectively for *M. dalei*, Brachyura and *Crangon* sp. The benthos-feeding *P. herzensteini* assimilated 30% Polychaeta and 43% Philinidae; our preliminary calculation showed that over 70% of this species' diet was made up ofbenthos.

Targeting the coastal ecosystem in Sendai Bay, in the present study we reported on food web analysis carried out by observing stomach contents and carbon/nitrogen stable isotope ratios. Moreover, putting mixed models that utilize Bayesian

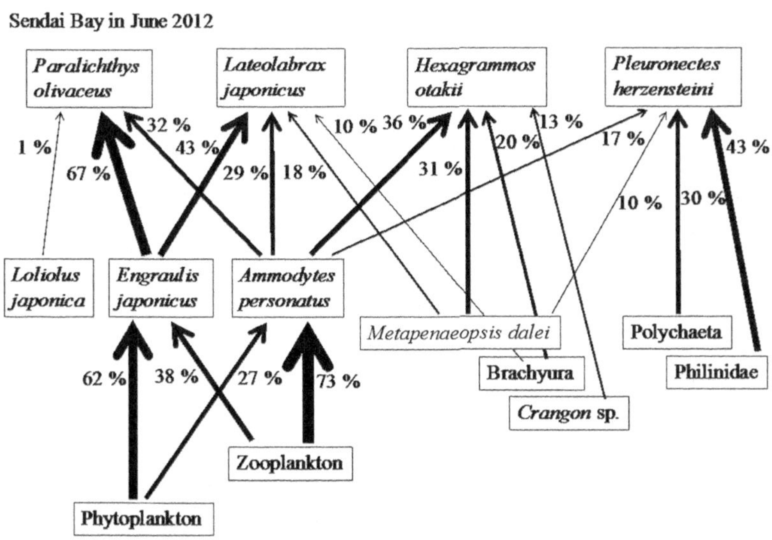

Fig. 19.3 Contribution rate for prey animals by mixing models for δ^{13}C and δ^{15}N within a Bayesian framework in Sendai Bay in June 2012

estimation to practical use, we also estimated the contribution rate of prey animals in fish species. The results of stomach content observations and stable isotope ratios were similar. However, as stated at the beginning, these analysis techniques both involve advantages and disadvantages. We probably need to decide whether to adopt one of the methods in accordance with the targeted species and objectives.

Although the present study only showed the June 2012 results, identical analysis was evaluated for other seasons (Togashi et al. 2015). For example, the contribution rate of *A. personatus* consumed by fish feeders was reduced from summer to autumn. According to Hashimoto (1991), who investigated the development stages and annual life cycle of *A. personatus* in Sendai Bay, this species takes almost one year to mature after birth. *A. personatus* spawns from December to January, vigorously feeding and growing from February to July. This species is thought to originate in cold waters; it is now known that, with rising water temperatures from August to November, they bury themselves in sand and become dormant. Therefore, the declining contribution rate for *A. personatus* by fish fedders may be fewer opportunities of predation.

In recent years, the isotope analysis carried out at the molecular level in the food web studies. For example, with nitrogen stable isotope ratios of amino acids, (absolute) trophic position can be assumed from differences in amino acid metabolism. Glutamic acid increased 8.0‰ per trophic position. In contrast, phenylalanine concentrated only 0.4‰ (Chikaraishi et al. 2009). It is because organisms can not synthesize phenylalanine themselves. The measurement of nitrogen stable isotope ratios of amino acids has advantage not to require stable isotope ratios of primary producer for assumptions at the trophicposition. Moreover, food web analysis that utilizes radiocarbon 14 included in muscle (Ishikawa et al. 2016), trace elements found in otolith and bone (e.g., strontium), and heavy element isotopes (e.g., neodymium) will continue to be applied in natural ecosystems. The development and application of these analytical techniques will provide usthe understanding of biodiversity and ecosystem functions.

Acknowledgements The crews of the research vessel Wakataka Maru, and the Daiei Maru and Seiko Maru -both small trawling boats -provided invaluable assistance in collecting fish and prey animals. Mr. Takahiro Yamaguchi, Ms. Shizuka Sasaki and Ms. Junko Suzuki of the Tohoku National Fisheries Research Institute provided invaluable assistance in the pre-processing and measurement of samples. The present study was supported by the Fisheries Agency, Ministry of Agriculture, Forestry and Fisheries, Japan, and CREST, Japan Science and Technology Agency.

References

Caut S, Angulo E, Courchamp F (2009) Variation in discrimination factors (Δ^{15}N and Δ^{13}C): the effect of diet isotopic values and applications for diet reconstruction. J Appl Ecol 46:443–453

Chikaraishi Y, Ogawa NO, Kashiyama Y, Takano Y, Suga H, Tomitani A, Miyashita H, Kitazato H, Ohkouchi N (2009) Determination of aquatic food-web structure based on compound-specific nitrogen isotopic composition of amino acids. Limnol Oceanogr Methods 7:740–750

DeNiro MJ, Epstein S (1978) Influence of diet on the distribution of carbon isotopes in animals. Geochim Cosmochim Acta 42:495–506

DeNiro MJ, Epstein S (1981) Influence of diet on the distribution of nitrogen isotopes in animals. Geochim Cosmochim Acta 45:341–351

Fry B, Sherr EB (1989) δ^{13}C measurements as indicators of carbon flow in marine and freshwater ecosystems. In: Rundel PW, Ehleringer JR, Nagy KA (eds) Stable isotopes in ecological research. Springer, New York, pp 196–229

Hansson S, Hobbie JE, Elmgren R, Larsson U, Fry B, Johansson S (1997) The stable nitrogen isotope ratio as a marker of food-web interactions and fish migration. Ecology 78:2249–2257

Hashimoto H (1991) Population Ecology of Japanses Sandeel. J Fac Appl Biol Sci Hiroshima Univ 30:135–192 (in Japanese)

Ishikawa NF, Togashi H, Kato Y, Yoshimura M, Kohmatsu Y, Yoshimizu C, Ogawa NO, Ohte N, Tokuchi N, Ohkouchi N, Tayasu I (2016) Terrestrial–aquatic linkage in stream food webs along a forest chronosequence: multi-isotopic evidence. Ecology 97:1146–1158 (first two authors contributed equally)

Kaeriyama M, Nakamura M, Edpalina R, Bower J, Yamaguchi H, Walker R, Myers K (2004) Change in feeding ecology and trophic dynamics of Pacific salmon (*Oncorhynchus* spp.) in the central Gulf of Alaska in relation to climate events. Fish Oceanogr 13:197–207

Kline TC Jr, Goering JJ, Mathisen OA, Poe PH, Parker PL (1990) Recycling of elements transported upstream by runs of Pacific salmon: I, δ^{15}N and δ^{13}C evidence in Sashin Creek, Southeastern Alaska. Can J Fish Aquat Sci 47:136–144

McCutchan JH, Lewis WM, Kendall C, McGrath CC (2003) Variation in trophic shift for stable isotope ratios of carbon, nitrogen, and sulfur. Oikos 102:378–390

Miller MJ, Chikaraishi Y, Ogawa NO, Yamada Y, Tsukamoto K, Ohkouchi N (2013) A low trophic position of Japanese eel larvae indicates feeding on marine snow. Biol Lett 9:20120826

Minagawa M, Wada E (1984) Stepwise enrichment of ^{15}N along food chains: further evidence and the relation between δ^{15}N and animal age. Geochim Cosmochim Acta 48:1135–1140

Parnell AC, Inger R, Bearhop S, Jackson AL (2010) Source partitioning using stable isotopes: coping with too much variation. PLoS ONE 5:e9672

Post DM (2002) Using stable isotopes to estimate trophic position: models, methods, and assumptions. Ecology 83:703–718

Rau G, Mearns A, Young D, Olson R, Schafer H, Kaplan I (1983) Animal ^{13}C/^{12}C correlates with trophic level in pelagic food webs. Ecology 64:1314–1318

R Development Core Team (2015) R: A language and environment for statistical computing. R Foundation for Statistical Computing, Vienna, Austria. http://www.R-project.org

Suzuki KW, Kasai A, Ohta T, Nakayama K, Tanaka M (2008) Migration of Japanese temperate bass (*Lateolabrax japonicus*) juveniles within the Chikugo River estuary revealed by δ^{13}C analysis. Mar Ecol Prog Ser 358:245–256

Tanaka Y, Minami H, Ishihi Y, Kumon K, Higuchi K, Eba T, Nishi A, Nikaido H, Shiozawa S (2014) Relationship between prey utilization and growth variation in hatchery-reared Pacific bluefin tuna, Thunnus orientalis (Temminck et Schlegel), larvae estimated using nitrogen stable isotope analysis. Aquac Res 45:537–545

Tayasu I, Hirasawa R, Ogawa NO, Ohkouchi N, Yamada K (2011) New organic reference materials for carbon-and nitrogen-stable isotope ratio measurements provided by Center for Ecological Research, Kyoto University, and Institute of Biogeosciences, Japan Agency for Marine-Earth Science and Technology. Limnology 12:261–266

Togashi H, Kurita Y, Yamaguchi T, Tamate T, Shigenobu Y (2015) Transfer of cesium-137 in the ecosystem in Sendai Bay and northern coast off Fukushima. The scientific reports of nuclear accident investigations by the Fisheries Research Agency, pp 43–47 (in Japanese)

Part V
Coastal Ecosystem and Management

Chapter 20
Ecological Status of Atlantic Salmon (*Salmo Salar* L.) in France: Need for an Ecosystemic Approach

Patrick Prouzet and Nicolas Michelet

Abstract Atlantic salmon (*Salmo salar* L.) is, in France, a totemic species and a symbol of the quality of aquatic environment. Formerly, very abundant in numerous river basins entering the Channel or the Atlantic coast, this species, unfortunately, is declining since the beginning of the 20th century and particularly after the end of the 2nd World War. Presently, this anadromous fish migrates on 40 river systems located between the Belgium border and the Spanish one. Salmon is mainly exploited by rod anglers but there still exists a professional fishery in the Adour-Gaves basin (located in the South-West part of France) where salmon catches are allowed with drift net. In all other estuaries and coastal waters where they are not prohibited, salmon catches by (professional) fishermen remain incidental and very few. Commercial salmon fishing, whether targeted or accidental, is strictly regulated in France, including a fishing licensing regime specific to diadromous fish species." The total catch is currently estimated to 10 metric tons around 3000 fish. The decline of Atlantic salmon in France is due to multiple anthropogenic pressures that have affected the size and the quality of its freshwater habitat during the 20th century such as: edification of dams on a great number of rivers for energy production or for agricultural and domestic purposes; increase of the water pollution with a degradation of the quality of the spawning areas; impediments to free migration due to an insufficient number of fish ladders or non effective fishways that prevent salmon to reach safely and rapidly their spawning areas. The final result is a large decrease of productivity of salmon stocks in most part of French salmon rivers and some difficulties to maintain a professional or leisure exploitation. Despite the ban of salmon catch in large river systems such as the Loire, Gironde, Seine, Garonne, Dordogne basins and the drastic decrease of the number of salmon fishers (presently less than 3000 for anglers and less than 30 for professional fishers), managers have some great difficulties to rebuild the population. Most of restoration programs failed due to a too fragmented approach without a global view of salmon

P. Prouzet (✉)
SFJO France, Paris, France
e-mail: patrick.prouzet@orange.fr

N. Michelet
CNPMEM/CONAPPED, Paris, France

T. Komatsu et al. (eds.), *Oceanography Challenges to Future Earth*,
https://doi.org/10.1007/978-3-030-00138-4_20

needs; fishing regulations without improvement of salmon environment have limited effects on the stock restoration if the main cause of decline is the decrease of the quality of salmon habitat. So, the future of Atlantic salmon in France lies in the implementation of an ecosystemic approach in order to have perennial and significant effects on the abundance of this population. The examples of restoration programs undertaken in the Adour-Gaves basin or in small river systems such as the Elorn river in Brittany confirm the need for management plans including not only fishing regulations but also improvement of salmon environment. On the Adour-Gaves river system for example, a restoration program including regulation of fishery, improvement of the free migration of individuals for a better access to the best spawning areas located in the upper course of the salmon tributaries, has given significant results on the stock abundance and on the percentage of large spawners in the population.

Keywords Atlantic salmon · France · Biology · Exploitation and historical evolution of the population · Ecosystemic approach

20.1 Introduction

Atlantic salmon (*Salmo salar*) is an amphihaline species with a reproduction in freshwater and a feeding phase at sea. It is considered in France as in the whole of its colonization area as a totemic species, a symbol of the quality of freshwater ecosystem.

This species is largely distributed in the North Atlantic Ocean both in the western and eastern parts and colonizes a lot of rivers located in Canada, Iceland, Ireland, British Islands, Norway, Sweden, Finland, Poland and in the southern part of its distribution area: France and Northern Spain.

In France, if we except the Mediterranean sea, many rivers constituted suitable habitat for the spawning of Atlantic salmon and the production of juveniles. Large river systems as the Rhine, the Loire-Allier, the Gironde-Garonne-Dordogne and the Adour-Gaves produced a large number of salmon at least until the end of the 19th century.

Since the beginning of the 20th century and particularly since the second half of the 20th after the Second World War, a great number of salmon stocks were in danger. The abundance of spawners decreased on a great number of European rivers mainly due to the degradation of the quality of spawning areas and the edification of dams not equipped of fish ladders. Despite a strong regulation of professional and sport fisheries, salmon restoration in most part of salmon rivers failed. It is the case in France and the principal reason is the implementation of a narrowly sector-based approach in salmon management. We have confined ourselves to constraint the salmon exploitation but without in the same period improved the productivity of the salmon population. So, during the last century, the productivity of the population have decreased drastically in many river systems and the consequence is a sharp decrease of the maximum sustainable yield of the different stocks, the level of which is strictly link with the quality of the salmon environment and the connectivity among critical salmon habitats (spawning, nursery areas and feeding) (Dumas and Prouzet 2003; Prouzet 2010). When a true ecosystemic approach is undertaken, combining regulation of fisheries and improvement of salmon habitat, we observe some significant progress on a long-term period. This is the case either on small rivers such as the Elorn river in Brittany (Kermarrec and Le Maout 2013) or in the Adour–Gave system where we have kept a professional salmon fishery in the Adour estuary and observe a significant increase of the production of juveniles since the beginning of the 21st century (Prouzet et al. 2010). This also implies the adhesion of all the partners implicated in the salmon management in order to develop a participatory research project to share traditional and academic knowledge and know-how.

20.2 Brief Review on the Atlantic Salmon Biology

20.2.1 Location of Its Colonization Area

According to Saunders and Bailey (1980), the population of Atlantic salmon is constituted by 2000 biological units, each considered as a biological stock. There may be at least one stock by salmon river, but in some large salmon river systems we can observe several stocks, each of them representing an independent biological unit with its own population dynamic. In fisheries science, the concept of stock is fundamentally different and is made up by a group of fish issued from different biological stocks and exploited as a whole by different fishing fleets. So, during its feeding phase at sea, different Atlantic salmon stocks provide, on the feeding grounds, some large schools of fish constituting a fish stock as for example the stock of salmon of the Labrador sea.

Figure 20.1 shows the approximate limits of the natural Atlantic salmon range. According to the observations made on the genetic characteristics of the North-Atlantic salmon population, there exits two sub-populations genetically well differentiated: one in North America and the other one in Europe (Guyomard 1994; King 2001). These genetic differences are expressed at the phenotypic level as observed for example, in scale shape differences between the two sub-populations (de Pontual and Prouzet 1987; Reddin et al. 1992).

20.2.2 Geographical Distribution of Atlantic Salmon in France

The Fig. 20.2 shows the locations of the main salmon rivers in France. There exist 3 large salmon river systems: The Loire-Allier axis (1), the Gironde-Garonne-Dordogne river system (2) and the Adour- Gaves basin (3). In the river Rhine (4), salmon is back

Fig. 20.1 Colonization area of Atlantic salmon (after migrateur-Loire.fr). The dark area represents schematically the locations of the estuary of the salmon rivers and the migration ways in the North Atlantic Ocean

and its return is linked to the restoration of the ecosystem and the reintroduction of the species, but the Atlantic salmon population remains very scarce (just few tens of individuals in 2013—statistics MEDDE). The others rivers, generally small with a steep gradient are located in Brittany (5) and Lower and Upper Normandy (6).

In Brittany, we count 23 rivers where salmon is caught with rod and line.[1] The rod catches are estimated in average to 1000 salmon a year.[2] In Lower and Upper Normandy, salmon is present on 10 rivers and the number of salmon caught by anglers is estimated in average to 550 individuals annually.

On the Adour-Gaves basin,[3] Atlantic salmon is observed on 4 rivers: the Nivelle, the Nive, the gave d'Oloron and the gave de Pau. In average, 240 salmon catches have been reported annually from the sport fishery since 2003 and 1370 catches from the profes-

[1]There exists 25 salmon rivers 22 of which are a drainage area of less than 1000 km² (Valadou 2014)

[2]Ten year average since 2003

[3]Including a neighborhood river: The Nivelle river

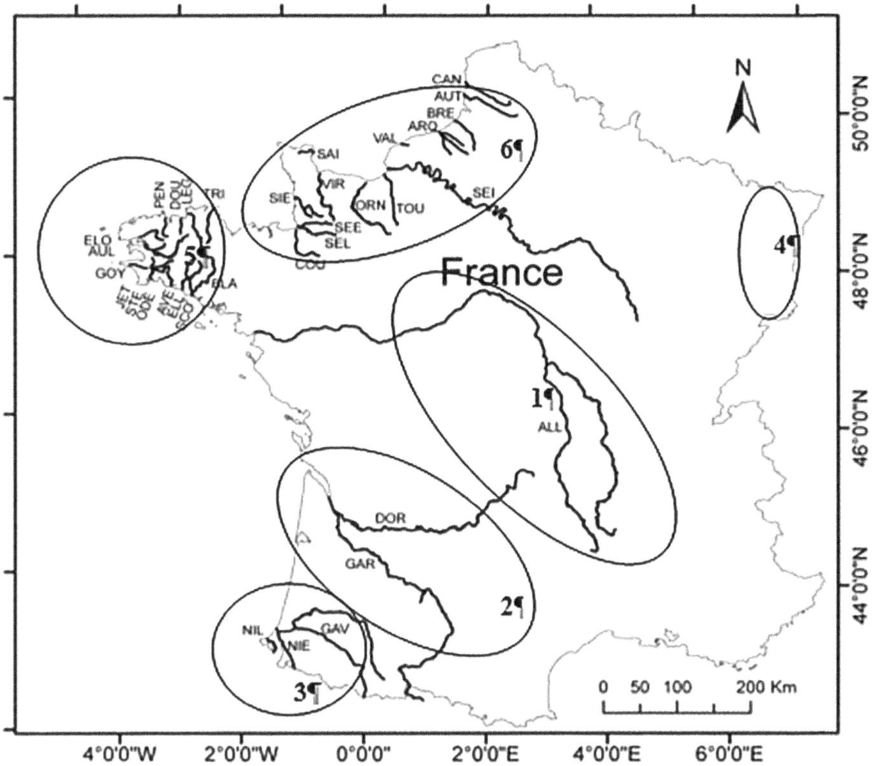

Fig. 20.2 Geographical location of the main Salmon River systems in France—(ELO: Elorn river, AUL: Aulne river; ALL: Allier river; DOR: Dordogne river; GAR: Garonne river; GAV: Gaves river; NIE: Nive river; NIL: Nivelle river)

sional fishery located in the Adour estuary.[4] Salmon are caught off the coast line in the vicinity of Adour estuary with fixed nets but their numbers are difficult to estimate.

20.2.3 Main Biological Features

As all the amphihaline or amphibiotic species, the salmon is characterized by two different periods during its life cycle: the first one spent in freshwater and the second one spent at sea. These two biological phases are separated by a specific physiological feature called smoltification process. Before this smoltification period, the juvenile of salmon looks like a trout and have a similar behavior. It is called "parr" and displays a territorial behavior. It is during this period that the size of the future salmon population is mainly determined and depends on the number of territories the size of which for a given nursery area is link to the productivity of the

[4]The Adour estuary is the only estuary where a professional fishery for salmon is allowed.

river[5] (Gueguen and Prouzet 1994). In spring, the "parr" is doing a true metamorphosis and gives the "smolt" after the smoltification process. The behavior becomes gregarious, the fish becomes silvery and deep physiological changes occur allowing the "smolt" to adapt to a more salty environment (Boeuf 1994). As all migratory salmonids or shads, Atlantic salmon is characterized by its homing behavior that allows the salmon to return in its home river in using chemical and physical cues (Hasler 1960; Quinn 1982; Davaine and Prouzet 1994; Putman et al. 2013).

20.2.3.1 Age

In the French ecological context, the salmon spent generally 1 or 2 years in freshwater before the smoltification process and the migration at sea.[6] The mean age of the juvenile population is ranged between 1.1 years old in the Adour basin to 1.5 years old in Northern Brittany (Cuende and Prouzet 1994). Some male "parr" mature prior to becoming "smolt" and stay in river to participate, during the winter period, to the breeding phase (Beall 1994; Baglinière 2013). At sea, the salmon has a rapid growth and spends one to 4 years (Davaine and Prouzet 1994) before to return to its home river. The sea age distribution of the adult population varies according the different rivers system (Prouzet 1990). This variation is mainly link to the length of the river system and the distance from the sea of the main spawning areas (Prouzet 1990; Baglinière and Porcher 1994) (Table 20.1). Thus, according to the hydrological characteristics of the river system, different categories of salmon migrate up the river at various period of the year: (i) very large and large winter and summer salmon that spend 3 or 4 years at sea; (ii) spring and summer salmon with 2 years at sea and (iii) *grilse* that have been at sea for one winter only and that return to their home river either in summer (before the season of low flow levels) or in autumn (called late *grilse*) (Fig. 20.3).

This age distribution can also change over time for a given river system. For example, the comparison of salmon stocks exploited by commercial fisheries in the Adour basin at the end of the 19th and during the 20th century (Cuende and Prouzet 1992; Prouzet 2013) shows significant differences among the proportions of the different components of the salmon population at the end of the two last centuries. In the 19th century, the proportion of grilse was lower in the commercial salmon landings between March and July, but the proportion of large multi-sea winter fish (3 and 4 years at sea) was much higher (Fig. 20.4).

This decrease of the mean age of the salmon stock between March and July[7]: 2.17 at the end of the 19th *versus* 1.35 at the end of the 20th can be explained by the

---

[5]In a succinct way, more the productivity level is high more the size of the territory is small and more the density of individuals is high.

[6]In the highest latitudes such as Scandinavian rivers or rivers flowing in Newfoundland for example, the freshwater phase can be longer: 4 to 7 years (Lee and Power 1976; Chadwick 1985)

[7]We have taken this period because in the 19th century the salmon fishing was allowed from January to November and at the end of the 20th century, the drif net fishery was allowed from mid-March until the end of July.

Table 20.1 Some examples of the importance (in %) of sea age categories according to the length of the river system

	1 Sea Winter fish	2 Sea Winter fish	3 Sea Winter fish	4 Sea Winter fish	River length in km	References
Elorn river[a]	78	21	1		58	Dartiguelongue (2013) Counting with a salmon trap—(average 2007–2012)
Scorff river	86,6	13,4			79	Caudal and Prévost (2014) Counting with a salmon trap—(average 1995–2013)
Adour-Gaves basin[b]	57	41	2		309	Prouzet et al. (1997) Average of the age distribution of catches in estuary (1985–1996)
Nivelle river	66.3	33.7			39	Lange et al. (2015) Counting with a salmon trap in 2013
Allier river	1	63	36		960	Logrami (2013) Counting with a salmon trap.

[a]the discrimination between the different categories of salmon is made from the size estimated by a video camera: lower than 70 cm (1SWF); between 70 and 90 cm (2 SWF); higher than 90 cm (3 SWF), thus the percentage of 3 SWF could be underestimated (see Table 20.3)

[b]as the fishing season begins in mid March, the proportion of 3SWF in the population is probably underestimated

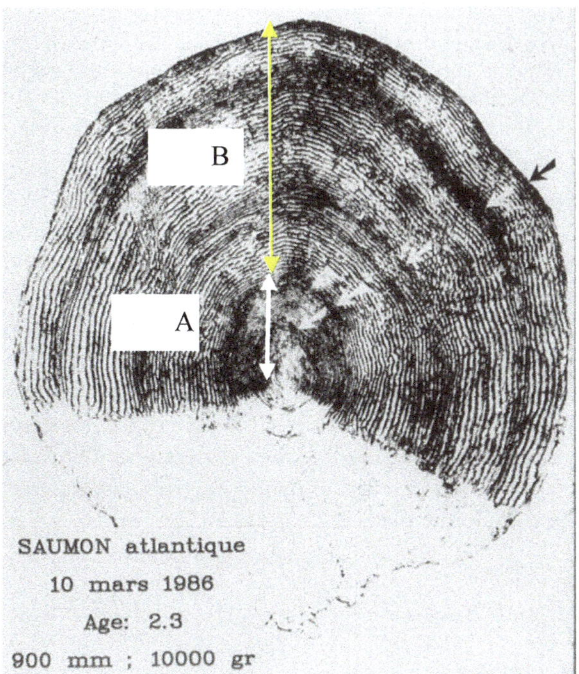

Fig. 20.3 Scale of a three sea winter fish caught on the Adour estuary the total age of which was 5 years old; 2 years in freshwater and 3 years at sea; (fork length: 90 cm and fresh weight: 10 kgs) —The white and black arrows show the positions of winter rings (annuli), the first two rings corresponding to the freshwater period. **a** Fresh water period; **b** sea water period

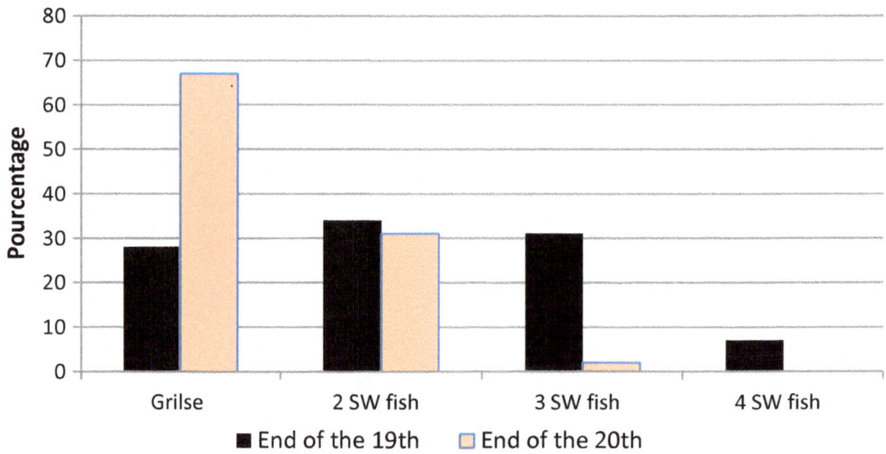

Fig. 20.4 Comparison of sea age distribution of salmon catches between March and July at the end of the 19th and the 20th centuries (The results take into account the possible difference of selectivity between the "baro" fisheries and the drift net fisheries. The "baro" was a fixed gear, equipped with a kind of paddle wheel with one of two blades carrying a net bag, with a stretched mesh size of 100 to 120 mm. The drift gill net is an upright trammel net having a stretched mesh of 110 mm and since the eighties is the only method authorized nowadays for professional salmon fishery in the Adour estuary (Cuende and Prouzet 1992; Prouzet 2013)). (after Prouzet 2013)

edification of dams throughout the 20th century that prevent the salmon to reach the spawning areas located in the upper course of the river system (Cuende and Prouzet 1992). So, after the Second World War, the most part of the spawning activity took place on the medium courses of the salmon tributaries at a lower distance from the sea (Prouzet 2013). As demonstrated by Schaffer and Elson (1975), there exists a correlation between the sea salmon age and the distance of the spawning areas from the sea: more the distance is high and more the proportion of multi-sea winter fish in the population is high. Thus, the increasingly challenge for the salmon to reach the upper spawning grounds on the Adour-Gaves system after the Second World War has led to a decrease of the mean age of that stock with a negative consequence on its reproductive potential (see Sect. 20.2.3.1.).

20.2.3.2 Length and Weight

The length and weight of salmon vary according to the salmon categories corresponding to various duration of its marine life (Fig. 20.5).

In Table 20.2, the observations show the size differences of salmon migrating up the Adour-Gave basin according to their number of winters spent at sea.

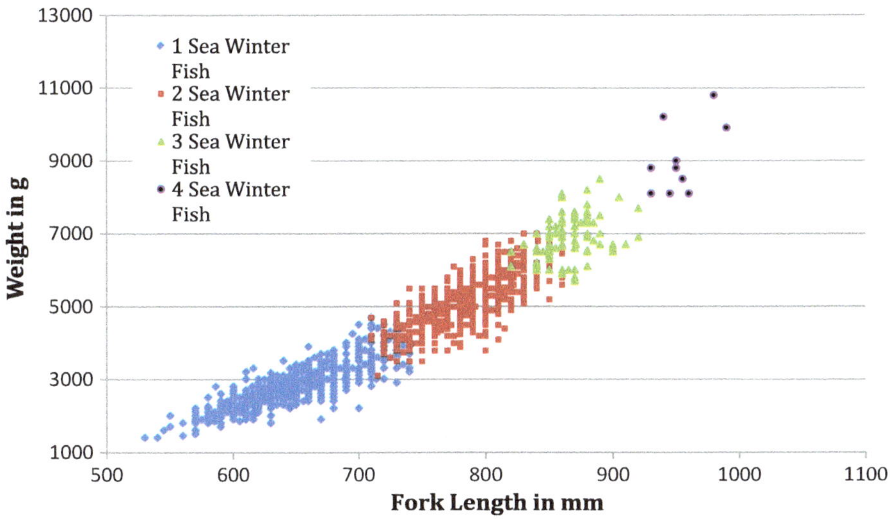

Fig. 20.5 Weight and Length scatter plot for a sample Atlantic salmon caught off the Adour Estuary during the period (1992–2004) (*Source* Ifremer data)

Table 20.2 Length and weight characteristics of Atlantic salmon caught off the Adour estuary during the period (1992–2004)

Salmon types	Number sampled	Mean Fork Length (standard deviation)	Mean Weight (standard deviation)	Minimum observed (Length—Weight)	Maximum observed (Length—weight)
1 SW Fish (Grilse)	957	647 mm (37)	2801 g (529)	530 mm–1400 g	740 mm–4700 g
2 SW Fish (Spring and summer salmon)	747	784 mm (31)	5167 g (685)	710 mm–3100 g	860 mm–7000 g
3 SW Fish (Large spring and summer salmon)	90	865 mm (21)	6168 g (597)	820 mm–5700 g	920 mm–8500 g
4 SW Fish (Very large salmon)	10	953 mm (20)	9030 g (957)	930 mm–8100 g	990 mm–10800 g

Some variations of the mean size for a given category of salmon are also observed according to the size of the river basin. For a given sea age, the salmon size is greater on large and mighty river systems than on small rivers as shown in Table 20.3.

20.2.3.3 Sex Ratio

In Atlantic salmon stock, sex ratio fluctuates according to the number of winters spent at sea and the period of return to the home river. The number of male is greater in the grilse (1 SW) stock than in the Multi Sea Winter stock. Furthermore, a higher proportion of female is found in the late run of grilse (migrating from September to December) than in the early run of grilse (Table 20.4).

If we except the early run of grilse, the proportion of female is higher than the proportion of male in the different salmon types. This distortion in favor of female is offset on the spawning ground by the participation of mature male *parr* in the reproduction process.[8]

20.2.3.4 Fecundity

Different observations made in hatchery from the artificial spawning of wild salmon show that the number of eggs per kg is ranged between 1457 ova (Elorn river, Prouzet and Gaignon 1985) to 2189 ova (Nive river, Prouzet and Martinet 1989). The absolute fecundity for an average spawner in the migrating whole stock will depend of 3 parameters: (i) the mean weight of females for each sea age group; (ii) the sex ratio in the different sea age groups of the migrating stock and (iii) the age structure of the salmon stock. The Table 20.5 shows this variation for a salmon stock migrating through the Adour estuary during the fishing season (March to July).

In Table 20.5 we have the values for each year (1985–2000) of the reproductive potential for a fish of a given type of salmon and in the column "Whole stock", the reproductive potential of an average individual spawner $R_{POT}$for a given yearaccording to (Eq. 1):

$$R_{POT} = \sum_{T=1}^{n} (P_T * (W_T * SR_T * 2000)) \text{ with}$$

P_T: % of salmon type T in the landings; $P_T \in [0, 1]$
W_T: mean weight of the salmon type expressed in kg
SR_T: Sex ratio for a salmon type T; $SR_T \in [0, 1]$
n: number of salmon types

From Eq. 1 we can estimate the mean on the period 1985–2000, $\mu_{1985\text{–}2000} = 4750$ ova per salmon[9] with a σ of 1630 and a $(\frac{\sigma}{\mu})$ equal to 34%. This variation of the

---

[8]This is commonly observed in the most northerly salmon rivers (Power 1969; Lee and Power 1976; Beall 1994)

[9]Combined with the escapement value, this indicator provides an estimate of the total potential number of eggs available for the river basin. For example, according to MIGRADOUR (www.migradour.com), technical structure in charge of the Atlantic salmon counting in the Adour basin, 4483 salmons escaped the fisheries in 2010, so the egg deposition for that year could be estimated to 21 millions of eggs for the whole salmon basin.

Table 20.3 Mean size (fork length and weight) of Atlantic salmon sampled on different French rivers

River	Salmon type	Fork Length in mm	Weight in g	Sampling period	River Length in km	References
Elorn	1SW	[a]591 [b]29 [c]18	1868 360 17	March–June	58	Prouzet and Jézéquel (1983)
	2SW	740 34 432	4013 563 425			
	3SW	794 42 40	4863 813 40			
Elorn	1SW	632 51 189	2173 600 126	May-December	58	Prouzet and Gaignon (1985)
Aulne	1SW	576 – 3	1745 – 2	March–June	144	Prouzet 1984
	2SW	738 37 334	3832 605 351			
	3SW	806 42 31	5003 900 34			
Adour-Gaves	1SW	647 37 957	2801 527 957	March–July	350	This study
	2SW	784 31 747	5167 685 747			
	3SW	865 21 90	6168 697 90			
Allier	1SW	670 – 4	2910 – 3	March–July	960	Baglinière and Porcher (1994) Fork Length is estimated from Total Length ($L_F = L_T/1{,}022$)
	2SW	784 – 314	4533 – 314			
	3SW	941 – 851	7690 – 802			

[a]mean estimate; [b]standard deviation; [c]number sampled—SW: Sea Winter

reproductive potential of an average potential spawner is strongly influenced by the sea age structure of the migrating stock: more the grilse component is high and more R_{POT} is low.

Table 20.4 Sex ratio of different types of Atlantic salmon sampled on different French rivers

River	Type of salmon	Sex ratio (sample size)	References
Elorn	Grilse (early run)	0,78 (54 M/69F)	Prouzet and Gaignon (1985)
	2SW	0,25 (40 M/161F)	Prouzet and Jézéquel (1983)
	3SW	0,78 (7 M/9F)	
Saint-Jean (Elorn tributary)	Grilse (late run)	0,38 (15 M/40F)	Prouzet and Gaignon (1985)
Adour—Gave	Grilse (early run)	1,56 (8462 M/5439F)	Prouzet et al. (2001)
	2SW	0,35 (1846 M/5272F)	
	3SW	0,36 (90 M/252F)	

M: male; F: female; SW: Sea Winter

Table 20.5 Variation of the mean reproductive potential per type of salmon and per spawner in the whole stock (from Prouzet et al. 2001)

	Type of spawners				
	Grilse (1SWF)	Small spring salmon (2SWF)	Small summer salmon (2^{+}SWF)	Large spring and summer salmon (3 and 3 + SWF)	Whole stock
1985	2640	6468	7640	12500	5550
1986	2400	8085	8190	7520	7030
1987	2140	7450	8500	14275	2660
1988	2450	7415	8580	17200	7010
1989	2200	7280	6450	15540	4350
1990	2365	3970	6330		3200
1991	1560	8055	6550	4920	3350
1992	3085	7958	7246		4599
1993	2210	9380	8247	15840	3053
1994	1882	4810	10225	14700	4345
1995	1496	4143	7845	12339	4232
1996	1566	7140	8582	17600	5950
1997	1680	5376	7844	11200	4800
1998	1512	4603	5592		2947
1999	2291	8456	8103	14200	7903
2000	933		5601		5013
Mean	*2025*	*6705*	*7595*	*13430*	*4750*

20.2.3.5 Migratory Behavior

The migratory behavior of the salmon is characterized by the "Homing" phenomenon (return to its home river).[10] From the mid of the 20th century, different tagging campaigns have been undertaken in French rivers or on the feeding grounds in Greenland to know where the French salmon migrated in the Ocean and in which feeding areas (Vibert 1994). From recapture of the tag fishes, three areas were identified: one in the South of Greenland and the others in the vicinity of the Faroe Islands and in the Norwegian Sea (Davaine and Prouzet 1994). For the French salmon, it is a sea migration of 4 thousands nautical miles to reach the Greenland and to come back in the Bay of Biscay in the vicinity of the Adour river for example. From 1999 to 2001, experiments with radio and acoustic tags (Bégout-Anras et al. 2001; Mahaut and Prouzet 2010) were undertaken in the Adour estuary in order to have a better knowledge of the migration behavior of salmon and a precise idea of the time for the salmon to pass through the fishing zone into the estuary.[11] These experiments were undertaken with the technical assistance of the professional fishers (to catch the fish alive and to track the fishes equipped with radio or acoustic transmitters). The main results showed that the swimming depth increases progressively as the salmon migrates upstream; the progression upstream occurs mainly during strong water movements (rising or descending tides); the effect of the tide on the behavior of individuals seems important in the lower part of the estuary but this effect diminishes after the first ten kilometers from the mouth. The average moving speed is estimated to 0.6 m/s[12] (roughly a body length per second). For all the salmon monitored by acoustics, the observations showed that an average of 37 h is needed to cross the professional fishing area and to escape the drift nets in the estuary. These observations fully justified the total hauling of nets after 48 h with a delay for a hauling period between the lower and upper parts of the estuary. The regulation which arose from this work have been fully accepted because they were collectively discussed and deduced from observations forged in partnership.

20.2.3.6 Genetic Structure

Preliminary investigations on the genetic structure of the North Atlantic salmon population showed a genetic difference between the North American and European sub-populations (Guyomard 1994). It was the reason why geneticists advised not to use eggs from the North American sub-population to restock the European rivers. More recent investigations undertaken on the genetic diversity of European

[10]It's also the case for others migratory fishes such as sea trout or shads.

[11]It was important for the managers to know the optimal duration of the total net hauling zone in the estuary and to see if a staggered hauling regime between the lower and upper estuary enabled the harvesting rate of the salmon population migrating into the estuary to be minimized for the same number of days of hauling.

[12]Speeds between 2 and 4 m/s during short periods of few seconds are not rare (Mahaut and Prouzet 2010).

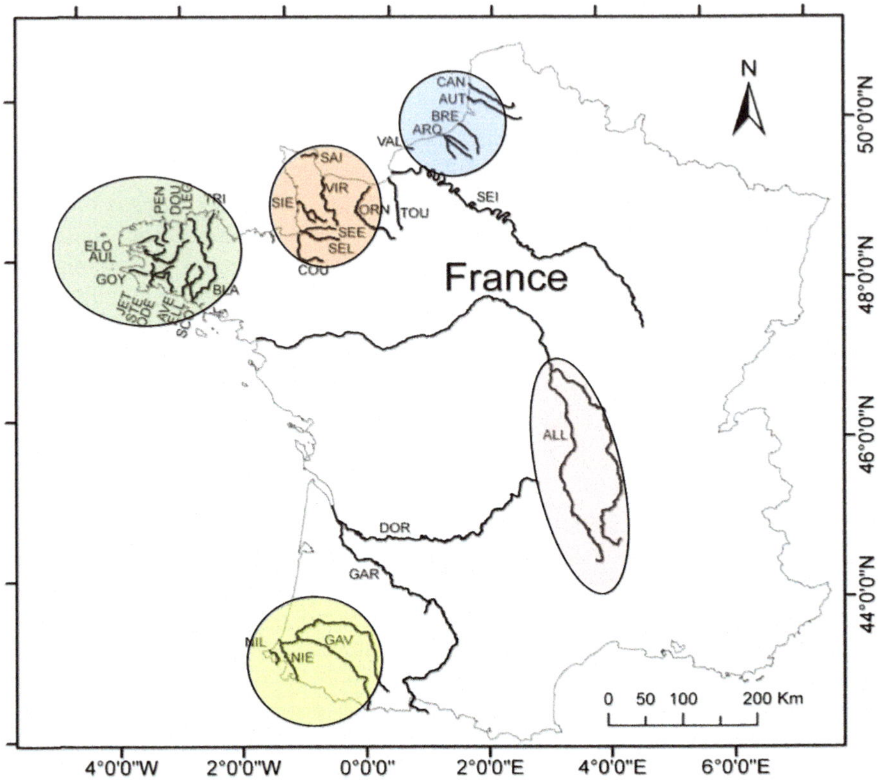

Fig. 20.6 Geographical location of the 5 genetically distinct groups of salmon (after Perrier 2010)

sub-population (King 2001) from microsatellite data suggest three geographical groupings in the context of isolation by distance: Iceland, Finland, Atlantic Europe (Western Norway, Ireland, Scotland and Spain). Thus, at a regional scale, we can consider that the European Atlantic salmon population is constituted of 3 groups genetically differentiated: Iceland, Baltic Sea and Atlantic Europe. In France, recent studies (Perrier 2010) made on fishes sampled on 34 rivers, revealed the existence of five genetically and geographically distinct groups; distance among estuaries and river length were some strong predictors of population structure (Fig. 20.6).

20.3 Trends in Catch Abundances

20.3.1 In the North Atlantic Area

Since 2000, no fishing for salmon has been prosecuted at Faroes. Since 1998, with the exception of 2001, the export of Atlantic salmon has been banned at Greenland.

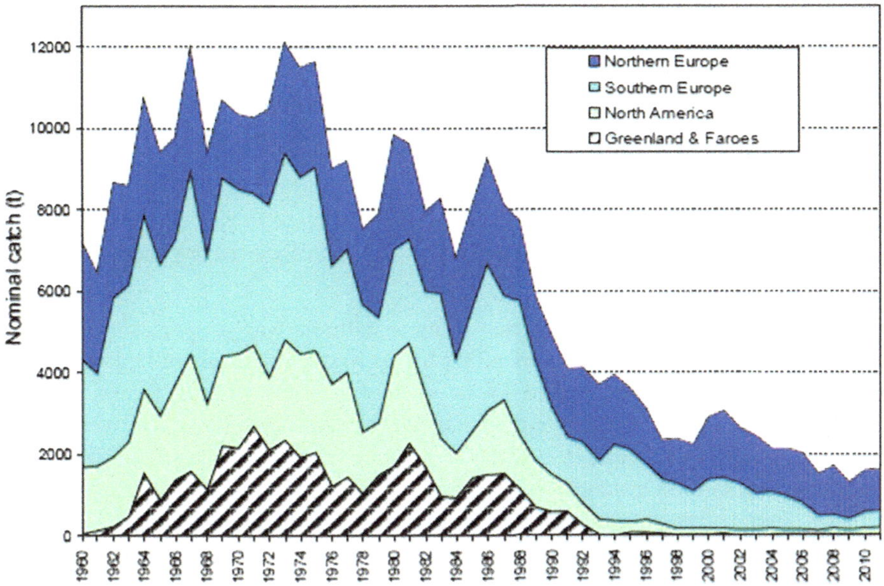

Fig. 20.7 Trends in abundance of Atlantic salmon catches in the different geographical areas colonized by Atlantic salmon stocks. (After ICES 2012)

So, since 2002, the salmon catches are sold on the local market or used for the private consumption. For the 2014 fishery, 57.8 tons was reported (ICES 2015).[13] For the whole area of catches, since the end of the eighties, we observed a continuous decline of the salmon landings partly due to the fishery constraints but mostly to the decrease of salmon abundances in its distribution area link to increasing anthropogenic pressure on its freshwater habitat see (Sect. 20.4) and more recently to the effects of climate change on the freshwater and marine ecosystems (Friedland et al. 2000; Peyronnet et al. 2008) (Fig. 20.7).

20.3.2 In France

For the period 1986–2013 (Fig. 20.8), the sport fishery is responsible in average for 56.2% of the total catches (3600 catches in average) and the rest is landed by the professional fishery in estuary and in the coastal area.[14] The professional fishery for salmon has been prohibited in large estuaries such as the Loire-Allier river system in 1994. It was not allowed during that period in the Gironde estuary and the

[13]11.2 tons of which for private human consumption.

[14]Mainly the Adour estuary and the Bay of Mont Saint-Michel in Normandy.

Garonne and Dordogne rivers to facilitate the restoration of the two salmon stocks in those river systems after the implementation of the French Salmon Plan in 1975.

Except the year 1987 characterized by the run of a large grilse number in the Adour estuary,[15] the catches are ranged between 3000 and 5000 thousands fish weighing between 12 and 20 tons approximately during the last 28 years.

20.4 Main Reasons of Salmon Stocks Decline in France

At the end of the 19th century, the Atlantic salmon was a species commonly exploited in a great number of rivers either by net in the lower part of large estuaries or by the "baro fisheries" in the lower course of some salmon rivers such as Dordogne or Gaves or by rod and line in the medium and upper courses of the salmon river systems. Even it is difficult to have a precise idea of the historical importance of salmon catches (Rainelli and Thibault 1980) some investigations made on some rivers from account books kept by the owners of the "baro" fisheries or by some important fishmongers have shown that salmon was a very valuable species for the fish economy at the end of the 19th or even just before the Second World War (Prouzet 2010). For example, in 1906, 42 tons of salmon catches were registered from the net and the "baro" fisheries in the Adour basin.[16] From 1896 to 1912, 27 tons of salmon were reported in average, annually, to the Maritime Affairs (Anonymous 1987) by the marine fishermen operating in the Adour estuary.[17] At that landing we have to add the catches made by the "baro" fisheries. The account books from two of the most important "baro" fisheries[18] indicated that, in average, 11000 pounds of salmon were caught annually at the same period. If we add the rod and line catches, we can estimate the average total catches between 40 to 50 tons annually in the Adour basin, at least before the First World War. That represented more than ten times the salmon catches declared by the professional and sportive fishers on the Adour-Gaves basin at the beginning of the 21st century.[19] The account books of fishmongers analyzed just before the Second World War showed that the catches remained abundant: 14,8 tons (1860 salmon) were sold by a single fishmonger from 1930 to 1933 (3.7 tons annually).

[15]Difficult to explain as it was not linked to an increase of the salmon run in all the others French salmon rivers perhaps a possible consequence of a successful restocking.

[16]Without taking into account the numerous catches from the rod and line fishery that were sold also but not recorded.

[17]The marine fisheries statistics gave for the period 1901–1925 an average landing of 58.8 tons of salmon per year for the maritime French fishery; statistics not taking into account the salmon caught by the "baro" fisheries and by anglers. (after de Drouin de Bouville 1943)

[18]30 «baros» were registered at the end of the 19th century, 4 located on the Gaves caught a large number of salmon (Cuende and Prouzet 1992):Cauneille, Lahontan, Puyoo and Sordes.

[19]Around 4.5 tons, in 2013, for the Adour estuary.

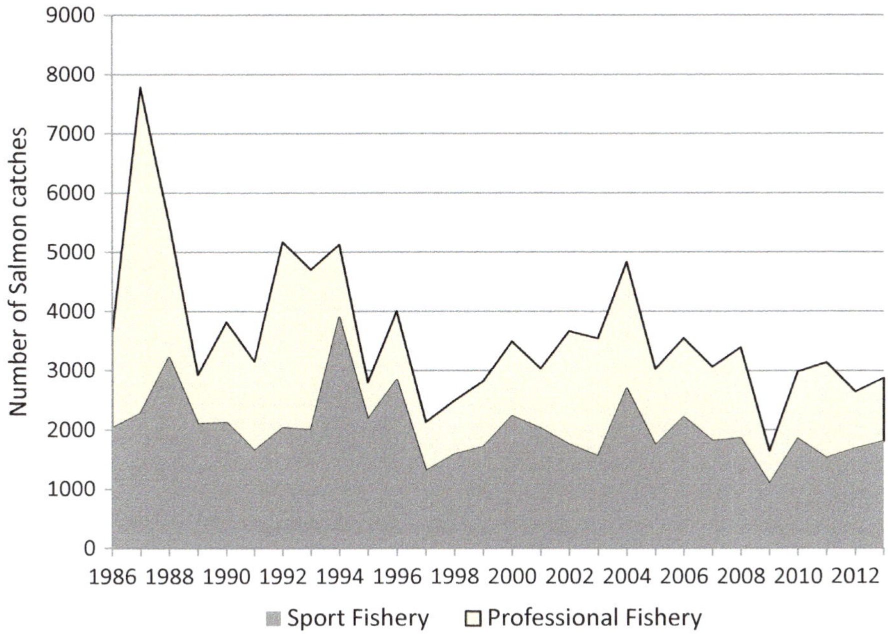

Fig. 20.8 Fluctuations of Atlantic salmon catches in France from 1986 to 2013—from Ifremer and Onema data bases

At the beginning of the 20th century and in the inter-war period, some major hydraulic works led to the eradication of migratory fish (Rabic, Allardi and Prouzet 2010). For example, the buildings of dams not equipped of fish ladders prevent the salmon to join the upper course of the rivers where the habitat is suitable for its spawning activity. This artificially shortened the river length and could have been, on some large salmon river systems, contributed to a decrease in mean reproduction age with the consequence, as mentioned previously in Sect. 20.2.3.1, of a strong reduction of the reproductive potential of the salmon population. This is clearly described by Cuende and Prouzet (1992) for the Adour-Gaves Basin. Prouzet (2013) estimated, from biological data collected on salmon catches, the decrease of the number of eggs laid by spawner two times less at the end of 20th century (see Table 20.5) than at the end of the 19th century.[20] Simultaneously at the development of the hydraulic infrastructures,[21] the inland water shipping may have led to the development of important towns in the vicinity of rivers. That increased the

[20]8,400 eggs per fish.

[21]We count 2,250 hydroelectric power stations in France and 75,000 dams of different kinds. This number was higher in the 19th century, around 100,000, but at that time watercourses formed the principal means of communication and transport and the rivers were maintained and the migratory species abundant due to the maintenance of the river continuity (Rabic et al. 2010).

organic pollution which became considerable at some points. Since 1930, Sabatier de Lachadenède warned the disruptive effect of the installation of sewer systems at Oloron, a town located just upstream one of the main salmon spawning area of the Adour-Gave river system (de Lachadenède 1939). The consequence was the clogging of spawning grounds and the increase in egg mortality. From the middle of the 20th century, the agriculture activities become more and more efficient to produce a great quantity of food for human consumption but with a great number of negative impacts on the aquatic environment: increasing use of chemical products, increasing water demand for plant irrigation, disappearance of vegetation on river banks, hedges removal with a significant increase of the siltation and the turbidity of water even in the upper course of the river. The combined effect of the reduction of accessible habitat for salmon and the degradation of environment quality, particularly in the medium course of the salmon rivers, is responsible for the decrease of the productivity of these salmon stocks in a great number of French salmon rivers and probably in its distribution area as a whole. This has led to a strong decrease of the maximum sustainable yield of many salmon stocks that are no longer to sustain a low or moderate fishing exploitation (Dumas and Prouzet 2003; Prouzet 2010).

20.5 Present Salmon Management in France

20.5.1 The Fishery

At the end of the 19th century and the beginning of the 20th century, we counted around 90,000 marine fishermen working on board and 58,600 people fishing from the shore.[22] For salmon and other migratory species, there were in average, from 1900 to 1912,[23] each year, 2,060 marine fishermen working on board in estuaries or in the lower part of the rivers of the Atlantic French coast and the salmon landings were estimated at 51 tons annually, in average for that period. At that time, we had in the maritime part of the Adour estuary 926 marine fishermen on fishing boats equipped of seine (to catch sturgeon, salmon and trout, shads, sea lampreys) or sometimes small trammel nets (to catch mullet mainly). The mean salmon landing was estimated to 25 tons. During all the 20th century, we observe a decrease of the total number of marine fishermen and it was very significant and continuous after the Second World War.

In 2014, it remains 215 marine fishermen that hold a fishing license to practice the net fishery in estuaries, a tenfold decrease in comparison with the number recorded at the beginning of the 20th century (Michelet 2014). For the Adour

[22]Annexe 3—rapport CREBS—INRA—sur les captures de poissons migrateurs dans les estuaires français depuis le milieu du 19ème siècle. 1987.

[23]Annexe 8—rapport CREBS—INRA—sur les captures de poissons migrateurs dans les estuaires français depuis le milieu du 19ème siècle. 1987.

Fig. 20.9 Migratory salmonids caught in the Adour estuary and tagged before the landing (Copyright N. Prouzet)

salmon fishery, the only French estuarian fishery for salmon, the situation is similar. Presently, it remains a small population of marine professional fishers: 20,[24] a reduction of almost fifty times in comparison with the population we had at the beginning of the 20th. If there existed at the end of the 19th and the first part of the 20th a great diversity of fishing gear: "baro", purse seine, trammel net, rod and line,[25] to catch the migratory species, currently the drift netting with gill or entangling nets from a boat is the most common fishing technique used close to the surface or the bottom.[26]

The salmon fishing in estuary is well regulated. To catch the salmon, you must purchase a fishing license the number of which is restricted at the national and regional scales. Some technical measures are added: length of the fishing vessel inferior to 12 m and an engine power not higher than 100 horse power (or 73 kW). Immediately upon capture, the migratory salmonid (salmon or trout) have to be tag (Fig. 20.9) with a special tag bearing a unique number allocated to the salmon license owner. The fishing season extents from mid-March to the end of July and

[24]Less than 30 fishers catch salmon if we add the inland professional fishers with a social status of farmers and who have the right to fish in the fluvial part of the estuary (39 fishing licenses have been attributed in 2015).

[25]The "baro" was banned in 1926 on the Adour-Gaves basin; the purse seine was banned in 1978; the rod and line is now used only for the recreational fishery and the salmon sale is prohibited for the leisure fishery.

[26]To catch mullets, melgrim, migratory salmonids, shads, sea lamprey, sea bass, spotted sea bass.

the number of salmon catches declared in 2013 was estimated to 1100 for the whole Adour professional fishery.[27] It was on average 1500 fishes for the period 1986 to 2013.

On salmon rivers, there exists also a recreational fishery and the number of salmon fishing fees is on average 2664 for the period 1986 to 2012; 1674 salmons have been recorded in 2012. The salmon angling season takes place from mid-March to mid-September and the number of catches is regulated by a TAC (Total Available Catches) or by a maximum number of salmon per angler (in the Adour—Gaves basin). The average number of catches on the period 1987 to 2012 is estimated to 2032 fishes.

20.5.2 Salmon Restocking

A restocking program should be a transitional operation the end of which is defined by the achievement of the salmon stock restoration. Sometimes as on the Elorn River, the restocking program is kept because the release of salmon fry and smolt is considered as a compensation of an irreversible loss of spawning areas. For others salmon systems such as the Loire-Allier, the Aulne river, the Gave de Pau or the Garonne-Dordogne river system, the restocking programs are kept to sustain some salmon stocks considered as endangered mainly due to impeded migration towards their best spawning areas.[28] When the situation of the salmon stock is considered in equilibrium with a sufficient healthy productive fish habitat, the restocking program is stop as it is the case on the Gave d'Oloron (a tributary of the Adour river system) (Table 20.6).

Some experiments of "Sea Ranching" were undertaken by Ifremer/Cnexo at the beginning of the 80's from a tidal pond located in Northern Brittany (Le Conquet). From 1980 to 1985, 45,000 smolts were released after an imprinting phase with morpholine. Some salmon returns were observed in the tidal pond: 51 tagged adult fish during the period 1983–1985 from the release of 36,594 tagged smolts (0.14% for the rate of return to the site of release[29]). However due to the great difficulty to know the number of tagged salmon caught at sea in the vicinity of the site of release, the experiment was stop in 1986.

---

[27]The number of salmon incidental caught by gill nets along the coast is difficult to estimate, perhaps a few hundred.

[28]In 2002, a report on the salmon migration on the Aulne river showed that only 5% of the spawners were able to migrate on the upper part of this river where 73% of the production areas for juvenile were located (Croze et al. 2002).

[29]Prouzet—non published data. The total number of salmon caught in the tidal pond from 1983 to 1985 was 61 fish from a release of 40,614 smolts (rate of return 0.15%).

Table 20.6 Important restocking programs undertaken in France since the beginning of the eighties

River system	Origin of eggs	Number of fry	Number of smolts	Rate of return or additional comments	References
Elorn	Foreign strain up to the beginning of the 80's (Scottish and Icelandic eggs), and use of eggs from local spawners after.	Release of unfed fry in nursery streams from foreign strains (Scotland, Ireland or Iceland) from 1976 to 1981.	Between 2,000 and 10,000 per year since the beginning of the 80's	Estimated between 0.9 and 3.6% from smolt release for the period 2007–2009. For the period 2008–2012 the proportion of fish with adipose fin clipped fish in the stock was ranged between 13% and 34%. Return rate from nursery streams estimated to 5.8%.	Dartiguelongue (2013) Prouzet and Gaignon (1982)
Aulne-Douffine	Since 1986, mainly with spawners caught on the Douffine river, a tributary flowing in the estuary of the Aulne river or in the Aulne river.	For the period 1995–2001: 76,000 parrs annually In 2002, release of parr (6 months old) only, around 200,000.	For the period 1995–2001: 30,000 smolts annually.	In 2009, 22% of the salmon stock of the Aulne river came from restocking program.	Croguennec (2011)
Loire-Allier	Local strain	1 million per year since 2001 and 200,000 parr 1+	200,000 smolts per year	Low level: estimated to 0.6% but it seems that the restocking could constitute 80% of the salmon stock of the Allier river.	Valadou (2014)
Garonne-Dordogne	Non local strain but issued from French stocks	15 millions of fry introduced during the last 15 years (1 million annually)		Low level. 0.05–0.08% from the fry 0.2–1% from the smolts	Valadou (2014)
Adour-Gaves	Foreign strain at the beginning and local strain in the 90's.	4.6 millions of fry from 1983 to 2010 (end of the restocking program on the Gave d'Oloron and Nive 10.8 millions of fry from 1983 to 2012 on the Gave de Pau		Not recorded but the salmon population on the Gave de Pau comes mainly from restocking program: 3137 salmons were counted at the Artix counting fence from the release of 4,760 millions of fry during the period 2005–2014; an estimated rate of return of 0.07%.	Valadou (2014) Aravelo (2013)

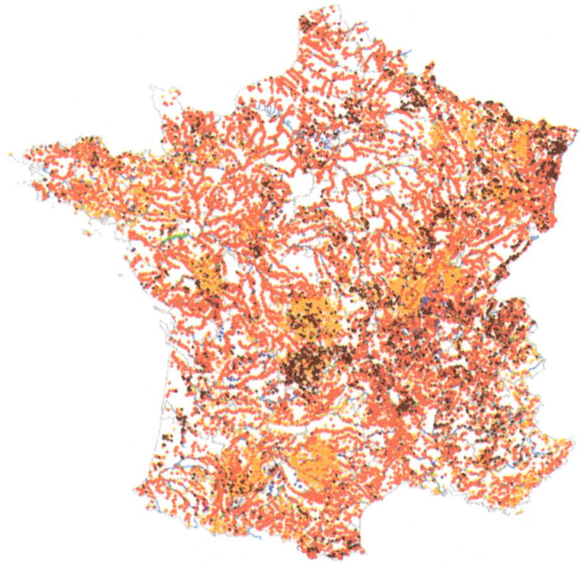

Fig. 20.10 Map showing the location and the density of dams on the French river systems. From CARMEN website

20.5.3 *Evolution of Quality and Fragmentation of Salmon Environment*

As mentioned previously in Sect. 20.4, the productivity of salmon environment was largely affected by many anthropogenic effects, particularly after the Second World War. One of the most important pressures on these amphihaline species such as shad, salmon, trout or eel was the development of the production of electricity from hydroelectric power stations. This development led to the edification of a great number of dams not equipped of fish ladders or equipped of facilities not really efficient to make easier the passage of migratory fishes. Moreover, the decrease of the use of waterways for the transport of goods and people with the development of railways and road networks during the first half of the 20th century made the crossing of numerous dams or obstacles (water mill, derivation of water for fish ponds, storage reservoir,..) already existing in the 19th century more difficult (see footnote 20) (Fig. 20.10).

This led to keep the salmon spawners in the medium course of rivers in less productive areas for the juvenile production or to slow down the salmon migration with some physiological consequences.[30]

So, since the middle of the 20th century, some technical researches have been made to improve the design of the fish ladders (Larinier and Gosset 1994) and to

[30]Over maturation for the spawner migrating towards spawning areas located far from the mouth (e.g. Allier river) or delayed of the downstream migration of smolts with an increase mortality during the sea water transfer.

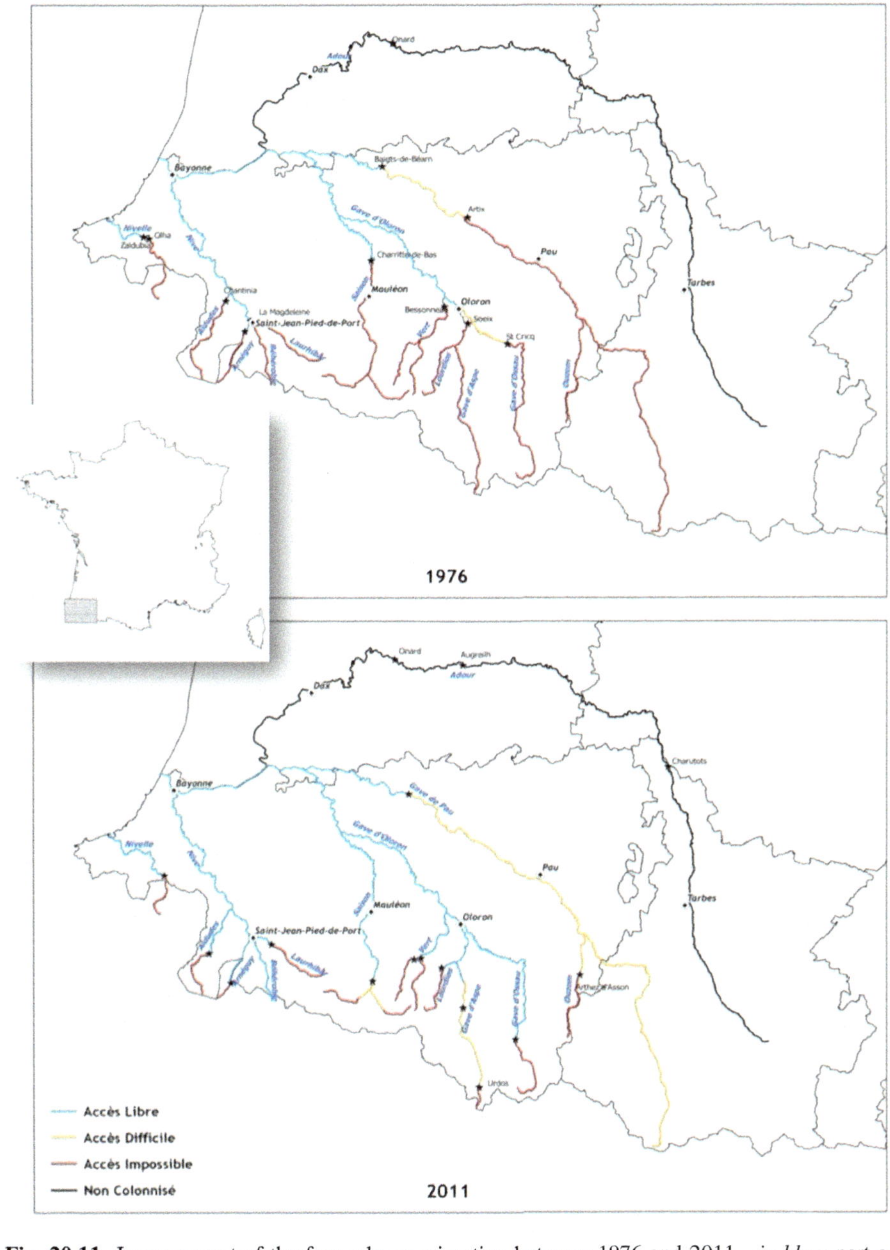

Fig. 20.11 Improvement of the free salmon migration between 1976 and 2011—*in blue:* part of river accessible to spawners—*in yellow or red*: part of river hardly or not accessible to spawners. (From MIGRADOUR 2012)

equip at least some impassable dams that prevented the salmon to reach on some important migratory fish axes. It is the case, for example of the Adour-Gaves basin as shown in Fig. 20.11.

As shown in Fig. 20.11, it remains an important effort to improve the free migration on one of the main salmon tributary of the Adour-Gaves basin: the Gave de Pau to restore the situation we knew at the beginning of the 20th century. For the others main Salmon river systems in France, the situation is worse despite some significant efforts in recent years. It is the case on the Loire–Allier axis where 91 dams have to be equipped to allow or facilitate the access of the best spawning and nursery areas located on the upper courses. The situation is not better on the Dordogne and Garonne rivers where the free migration of amphihaline species should be enhanced in the next years (Valadou 2014).

As mentioned previously by Lachadenède in 1930 for the salmon status in the Gaves river system the edification of dams and the decrease of the quality of water on the medium course of rivers due to urban concentrations were the major reason for the salmon decrease in the Adour-Gaves basin. During all the 20th century and particularly during the second half, the quality of our rivers and more generally of our aquatic ecosystems has significantly deteriorated. Farming, untreated urban seawage have a huge impact on the water quality and the clogging of spawning grounds. After the Second World War, in addition to the organic pollution, we have had gradually the pollution from more and more synthetic products that Nature does not know how to assimilate (Rabic et al. 2010). According to Loizeau and Tusseau-Vuillemin (2014), "around 100,000 chemical substances are produced, imported and used on the European market, and 5,000 of them are considered to be dangerous for mankind and the environment". Many chemical molecules are dispersed in aquatic environment: aromatic polycyclic hydrocarbons, the most worrying for aquatic environments; pesticides with 900 types currently used and 80,000 tons applied each year; biocides (copper for example for the treatment of fruits or organic such as atrazine for the treatment of vegetables or grains); organic synthetic substances such as PCB that we found now in significant concentrations in the adipose tissue of the eel. In addition, the increasing water demand for drinking or for irrigation is a major concern for the respect of the minimum of water that must be retained in the stream, especially in the context of climate change.

20.6 The Adour—Gave System: An Attempt to Implement an Ecosystemic Approach

Clearly, as mentioned above, the fishing activities should not be used as an adjustment variable intended to compensate the negative effects from all the human-induced activities, if the fishing exploitation is not the primarily cause of this decline as it seems to be the case. Unfortunately, the salmon management, at least in the latter half of the 20th century, has mainly consisted of fishing restrictions rather than a broader vision of the salmon environment and the ecological continuity.

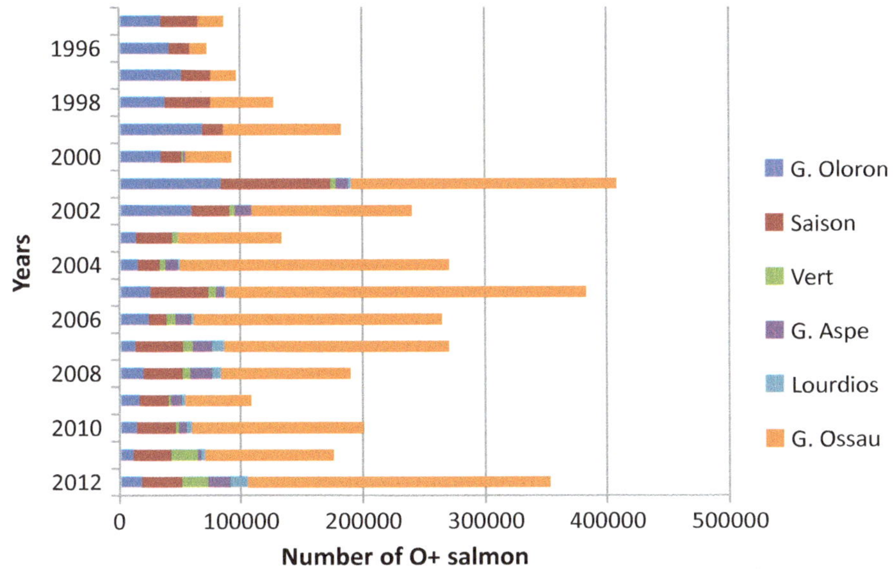

Fig. 20.12 Production of 0 + salmon on the gave d'Oloron (from MIGRADOUR data—www.migradour.com) see Fig. 20.11 for the geographical location of the salmon tributaries

Considering, the poor results obtained by the only constraint of the fishing pressure, it was decided in 1995 by the different stakeholders involved in the management and restoration of the species to implement a more systemic approach on the Adour basin. As mentioned above in Fig. 20.11, the Atlantic salmon was confined in 1976 to the spawning grounds of the lower stretches of the Gaves which are of inferior quality for reproduction. The situation was quite similar to that observed just after the Second World War. At the end of nineties, and due to a significant effort made on a period of 20 years, the free migration of salmon was greatly improved as well as access to the superior spawning grounds of the upper stretches where eggs have a higher rate of survival under the gravel of the river bed and where spawners can scatter their eggs on a larger suitable area. In these conditions, any fishing restriction enables to increase significantly the contribution for reproduction. Reductions in net fishing time were calculating after studying the upstream migration behavior of salmon in the Adour estuary (Mahaut and Prouzet 2010). From tracking data gathered during two years (1999 and 2000), it has been estimated that salmon took, on average, 2 days to cross the estuary (33 kms). This has enabled a statutory net hauling regime to be set up in order to allow a large proportion of salmon to avoid the nets. Thus, a rapid increase of the juvenile production was obtained as mentioned in Fig. 20.12. This Figure mentions not only the increase of production of juveniles from 2001 but also a better distribution of salmon on the best spawning areas.

Before this date, the spawning is located mainly below Oloron city, a part of the river where the quality of the spawning beds is poor (see the 1976 map in

Fig. 20.11). After this period the opening to the upper courses of the salmon tributaries allows a better scattering of spawners and a better egg survival (see the 2011 map in Fig. 20.11). The final result of this salmon management is a rapid increase of the production of juveniles enabling to keep a self sustaining salmon population with a significant professional and leisure exploitation.

This systemic approach highlights the importance of (i) identifying and making available information on the condition of resources and habitats as they were before; (ii) proceed methodically; salmon restoration was only able to begin in the Gaves system after prior efforts were made in the improvement of the free migration.[31] In these conditions, the fishing restrictions which have been approved by the fishers may quickly bear fruit; (iii) don't forget that regulations defined and discussed collectively among implied actors are generally fully accepted.

References

Anonymous (1987) Les captures de poissons migrateurs dans les estuaries français depuis le milieu du XIXème siècle (Tableau B Annexe 7). Rapport CREBS-INRA Rennes Contrat DPMA, Novembre 1987. 48 pages + annexes

Aravelo E (2013) Bilan des connaissances et propositions d'actions visant à restaurer la population de saumons atlantiques (Salmo salar L.) sur le gave de Pau. Mémoire M1 Dyn. Syst. Aqua., UPPA, contrat ONEZMA/ Agence de l'Eau Adour-Garonne, 98 p

Baglinière J-L, Porcher J-P (1994) Caractéristiques des stocks de reproducteurs et comportement lors de la migration génésique. In: Guéguen J, Prouzet P (eds), Le Saumon Atlantique, 101–122

Baglinière JL (2013) Les populations de saumon atlantique (Salmo salar) en France: Caractéristiques écologiques et migratoires, évolution face au changement global. Communication orale, 65 planches. Document INRA—Interreg

Beall E (1994) Chapter 6 - Les phases de la reproduction. In Le saumon atlantique—Biologie et gestion de la ressource. In: Gueguen J, Prouzet P (eds), Edition Ifremer, 123–140

Bégout-Anras M-L, Lagardère J-P, Gosset C, Rives J, Prouzet P (2001) Comportement migratoire et structures hydrodynamiques: suivi individuel par pistage des saumons atlantiques dans l'Adour. In: d'Elbée et P (eds) Océanographie du golfe de Gascogne, VIIème colloque international de Biarritz 4–6 avril 2000, Edition Ifremer, vol 31, 340–345

Boeuf G (1994) La phase de préadaptation à la vie en mer: La smoltification. In: Guéguen J, Prouzet P (eds), Le Saumon Atlantique, 47–63

Caudal L, Prévost E (2014) Etat du stock de saumon atlantique (*Salmo salar*) du Scorff (Bretagne Sud—France) en 2013. Rapport FDP 56—INRA, 26 p

Chadwick EMP (1985) The influence of spawning stock on production and yield of Atlantic salmon, *Salmo salar* L., in Canadian rivers. Aquaculture and Fisheries Management 1:111–119

Croguennec E (2011) Rapport sur le suivi par vidéo-comptage de l'activité ichtyologique sur l'Aulne. Observatoire aquatique de Châteaulin. Rapport SMAT Aulne et Hyères, Contrat de Plan Etat—Région, 60 p

Croze O, Senecal A, Woillez M (2002) Suivi par radiopistage de la migration anadrome du Saumon atlantique sur l'Aulne—Campagne 2000. *GHAAPPE*, RA03.01, 135 p. + annexes

[31]It is the reason of the failure of the salmon restoration in the large salmon system in France.

Cuende F-X, Prouzet P (1992) Comparison of Atlantic salmon, Salmo salar L., stocks exploited by commercial fisheries in the Adour Basin, France at the end of the 19 and 20th centuries. Aquac Fish Manag 23:373–383

Cuende F-X, Prouzet P (1994) Le saumon en France: Exploitation, gestion et avenir. In: Guéguen J, Prouzet P (eds) Le Saumon Atlantique, 155–267

Dartiguelongue J (2013) Contrôle du fonctionnement de la passe à poissons installée à Kerhamon sur la rivière Elorn. Suivi de l'activité ichtyologique en 2012. Rapport SCEA, 78 p

Davaine P, Prouzet P (1994) La vie marine du saumon atlantique dans son aire de répartition géographique. In: Gueguen J, Prouzet P (eds), 64–86

de Drouin de Bouville M (1943) Le saumon revient-il à sa rivière natale ou tout au moins à sa rivière d'origine. Bull Fr Piscic 123:10–27

De Pontual H, Prouzet P (1987) Atlantic salmon, *Salmo salar* L., stock discriminationby scale-shape analysis. Aquac Fish Manag 18:277–289

Dumas J, Prouzet P (2003) Variability of demographic parameters and population dynamics of Atlantic salmon (*Salmo salar L.*) in a southwest French river. ICES J Mar Sci 60:356–370

Friedland KD et al (2000) Linkage between ocean climate, post-smolt growth and survival of Atlantic salmon (*Salmo salar*) in the North Sea area. ICES J Mar Sci 57:419–429

Gueguen J, Prouzet P (1994) Le Saumon Atlantique. Edition Ifremer, 330 p. ISBN: 2-905434-48-1

Guyomard R (1994) La diversité génétique des populations de saumon atlantique. In: Gueguen J, Prouzet P (eds) Le Saumon Atlantique, 141–152

Hasler AD (1960) Homing orientation in migrating fish. Ergebn Biol 23:94–115

ICES (2015) ICES WGNAS report 2015. ICES advisory committee, ICES CM 2015/ACOM: 09, 332 p

Kermarrec J-Y, Le Maout G (2013) Reconquête de l'Elorn et Pêche de loisir: Valeurs culturelle, sociologique et économique. Communication orale «Colloque des Saumons et des Hommes» Brioude octobre 2013, 34 planches

King TL, Kalinowski ST, Schill WB, Spidle AP, Lubinski BA (2001) Population structure of Atlantic salmon (*Salmo salar* L.): a range-wide perspective from microsatellite DNA variation. Mol Ecol 10:807–821

Lange F, Carlut E, Prévost E, Servanty S (2015) Les populations de saumon, truites de mer et grandes aloses de la Nivelle en 2013. Rapport INRA—AQUAPOLE, 29p

Larinier M, Gosset C (1994) La libre circulation des migrateurs. In: Guéguen J, Prouzet P (eds) Le Saumon Atlantique, chapitre 11, 197–217

Lee RLG, Power G (1976) Atlantic salmon (*Salmo salar*) of the Leaf River, Ungava Bay. J Fish Res Board Can 33:2616–2621

Logrami (2013) Recueil de données biologiques sur les poissons migrateurs du bassin de la Loire. Rédacteurs. In: Bach JM, Parouty T, Léon C, Senecal A, Portafaix P, Guerit A, Le Peru Y, Baisez A (eds). Rapport d'activité du 01/01/12 au 31/12/12, Mar 2013, p 359

Loizeau V, Tusseau-Vuillemin MH (2014) Marine ecosystems under toxic pressure. In: Vulnerability of coastal ecosystems and adaptation, Chapter 1, 1–64. In: Monaco A, Prouzet P (eds) ISTE—Wiley, Paris

Michelet N (2014) Projet de règlement prévoyant une interdiction de la pêche au filet dérivant. Cas des pêcheries maritimes estuariennes françaises des façades atlantiques, Manche et mer du Nord. Note CNPMEM, juillet 2014, 13 p

Perrier C (2010) Structure génétique des populations de saumon atlnatique en France. Thèse Université de Caen, 181 p

Peyronnet A et al (2008) Different ocean climate factors control the marine survival of wild and hatchery Atlantic salmon *Salmo salar* in the north-east Atlantic Ocean. J Fish Biol 73:945–962

Power G (1969) The salmon of Ungava Bay. Arct Inst North Am Tech Pap 22:72

Prouzet P, Gaignon J-L (1982) Production de saumon atlantique (*Salmo salar* L.) juvéniles et adultes sur un ruisseau pépinière de Bretagne Nord (France) à partir d'une souche irlandaise. Rev Trav Inst Pêches marit, 454(2):155–174

Prouzet P, Jézéquel M (1983) Caractéristiques des populations de saumon atlantique (Salmo salar L.) capturées à la ligne sur l'Elorn (rivière de Bretagne Nord) durant la période 1974–1981. Bull Fr Piscic 289:94–111
Prouzet P (1984) Caractéristiques du stock de saumon atlantique (Salmo salar L.) capturé à la ligne sur l'Aulne (rivère de Bretagne-Nord) durant la période 1973–1981). Rev Trav Inst Pêches Marit 46(4):285–298
Prouzet P, Gaignon J-L (1985) Caractéristiques du stock de saumon atlantique d'un hiver de mer (Salmo salar L.) capturé sur l'Elorn de 1974 à 1984. Rev Trav Inst Pêches Marit 47(3 et 4):167–168
Prouzet P, Martinet J-P (1989) Fécondité des saumons de plusieurs hivers de mer de la Nive. Rapport DRV/RH/St-Pée sur Nivelle, 10
Prouzet P (1990) Stock characteristics of Atlantic salmon (*Salmo salar*) in France: a review. Aquat Living Resourc 3:85–97
Prouzet P, Martinet J-P, Cuende F-X (1997) Rapport sur la pêche des marins pêcheurs dans l'estuaire de l'Adour en 1996. Rapport Ifremer, 27 p
Prouzet P, Lissardy M, Morandeau G, Cuende F-X (2001) La pêche maritime professionnelle dans l'estuaire de l'Adour en 2000: Importance économique et caractéristiques des captures de poissons migrateurs amphihalins. DRV/RST/RH/2001-08, 35 p
Prouzet P (2010) Environnement et fonctionnement de populations halieutiques: Nécessité d'une approche systémique pour une gestion intégrée de ces ressources. Editions EUE, 138 p, ISBN 978-613-1-50462-4
Prouzet P (2013) Le Saumon de l'Adour: Son voyage initiatique, sa pêche. Edition Pleine Page, 120 p. ISBN 978-2-36042-018-6
Putman NF, Lohman KJ, Putman KJ, Quinn TP, Klimley AP, Noakes DLG (2013) Evidence for geomagnetic imprinting as a homing mechanism in Pacific Salmon. Current biology 23:312–316. https://doi.org/10.1016/j.cub.2012.12.041
Quinn TP (1982) Intraspecific differences in sockeye salmon by compass orientation mechanisms. Proceedings of the salmon and trout migratory behavior symposium, Seattle, pp 79–85
Rainelli P, Thibault M (1980) La fabuleuse richesse en saumons des rivières bretonnes d'autrefois, mythe ou réalité. Annales de Bretagne et des Pays de l'Ouest. Tome 87 4:697–713
Rabic J, Allardi J, Prouzet P (2010) Goods and services offered by coastal, estuarine and inland ecosystems. In: International meeting of small scale professional sea and inshore fishers. The official records. (CNAPPED Publication), pp 22–26
Reddin D, de Pontual H, Prouzet P (1992) A comparison of two techniques to discriminate continental origin of Atlantic salmon (*Salmo salar*) off West Greenland. Aquat Living Resour 5:81–88
Sabatier de Lachadenède M (1939) Travaux piscicoles dans le bassin des Gaves. Bull Fr Piscic 116:97–109
Schaffer WM, Elson PF (1975) The adaptative significance of variations in life history among local populations of Atlantic salmon in North America. Ecology 56(3):577–590
Saunders RL and Bailey JK (1980). The importance of genetics in Atlantic Salmon management. NASRC report, 1, 13 p
Valadou B (2014) Le saumon atlantique en France: vue d'ensemble. Présentation powerpoint, communication orale, 3rd june 2014, 3&ème assemblée générale de 'OCSAN, France, 28 diapositives
Vibert R (1994) Le saumon atlantique: Origine et caractéristiques essentielles. In: Guéguen J, Prouzet P (eds) Le Saumon Atlantique, pp 11–26

Chapter 21
Challenges to Harmonize Sustainable Fishery with Environmental Conservation in the Coastal Ecosystems Under Oligotrophication

Masakazu Hori, Masahito Hirota, Franck Lagarde, Sandrine Vaz, Masami Hamaguchi, Naoaki Tezuka, Mitsutaku Makino and Ryo Kimura

Abstract Coastal environments of the world have been exposed to eutrophication for several decades. Recently the quality of coastal waters has been gradually and successfully improved, however the improvement has caused another issue in ecoastal ecosystem services, called oligotrophication. Local stakefolders have suggested that oligotrophication reduces pelagic productivity and moreover fishery production in coastal ecosystems, while oligotrophication with high transparency has recovered benthic macrophyte vegetation which have been depressed by phytoplankton derived from eutrophication. In particular, seagrass species is one of the most important coastal vegetation for climate change mitigation and adaptation, which has been welcomed by another stakefolders. Therefore, harmonizing coastal fishery with environmental conservation is now essential for the sustainable use of ecosystem services. Here, we just started some practice in field based on the interdisciplinary approach including ecological actions, socio-economical actions and moreover psychological actions to find the integrative coastal management

M. Hori (✉) · M. Hamaguchi · N. Tezuka
National Research Institute of Fisheries and Environment of Inland Sea,
Japan Fisheries Research and Education Agency, 2-17-5 Maruishi, Hatsukaichi,
Hiroshima 739-0452, Japan
e-mail: mhori@affrc.go.jp

M. Hirota · M. Makino
National Research Institute of Fisheries Science,
Japan Fisheries Research and Education Agency, Yokohama, Japan

F. Lagarde · S. Vaz
Laboratoire Environnement-Ressources du Languedoc-Rousillon, IFREMER/UMR
MARBEC, 32400 Sete, France

R. Kimura
Japan Fisheries Research and Education Agency, Headquarters, Yokohama, Japan

T. Komatsu et al. (eds.), *Oceanography Challenges to Future Earth*,
https://doi.org/10.1007/978-3-030-00138-4_21

maximizing well-beings of various stakefolders, which is essential to harmonize environmental conservation with sustainable fishery and aquaculture. Now we are focusing on the interaction between oyster aquaculture and seagrass vegetation as an ecological action.

Keywords Oyster aquaculture · Seagrass · Indigenous and local knowledge · Integrated coastal management

21.1 Introduction

Marine coastal area is one of the most important ecosystem providing a wide variety of ecosystem services to human community. It has been suggested that 23% of the world human population live within 100 km distance of the coastal line (about three times higher than the global average: IPCC 2007) and utilize these ecosystem services, suggesting that coastal ecosystems has been exposed to the influence of over-use and pollution caused by the human systems. Actually coastal environments of the world have been influenced by artificial eutrophication with environmental pollution such as red tide bloom and hypoxia for several decades (e.g., Selman and Greenhalgh 2009; Yanagi 2015). Recently as public awareness of coastal pollution has become higher, the quality of coastal waters has been gradually and successfully improved in some regions (Matsuda 2015). However, some coastal stakeholders suggest that the improvement of water quality is now causing another issue in coastal ecosystem services called oligotrophication (Collos et al. 2009; Yamamoto and Hanazato 2015).

It has been suggested that oligotrophication has reduced pelagic productivity in coastal ecosystems, sometimes resulting in the decline of the fishery catch and the harvest of seaweed and bivalve aquaculture (Yamamoto and Hanazato 2015; Yanagi 2015). This is presumably because of the following reasons. First, the coastal fishermen have shifted their fishing technique, gears and moreover target species as an adaptation to the eutrophication of coastal waters for several decades. For example in Seto Inland Sea, Japan, most of the recent fishery catch are derived from pelagic production, although the previous most dominant fish catch was another species derived from benthic production before the 1960s when the eutrophication had not occurred yet (Tsurita et al. 2016). Second, eutrophication generally increases pelagic productivity and contrarily decreases benthic productivity due to lower transparency by the increased dense phytoplankton. The eutrophication has urged fishermen to change their occupation from coastal fishing to seaweed or bivalve aquaculture using enriched nutrients and abundant phytoplankton, due to its stable benefits (Tsurita et al. 2016). Therefore their recent occupation cannot be held without eutrophic environment. Third, oligotrophication decreases total amount of nutrients in coastal area from higher to lower level, indicating a possibility that total coastal productivity itself had decreased with oligotrophication.

In contrast, oligotrophication generally improves water quality, in particular transparency higher by decreasing massive pelagic phytoplankton (Scheffer et al. 2001). The higher transparency by oligotrophication has recovered various important ecosystem functioning and services in coastal areas, especially benthic primary production including seagrass vegetation (Burkholder et al. 2007; Hori and Tarutani 2015). Seagrass bed has been suggested as one of the most important coastal vegetation for climate change mitigation and adaptation, such as blue carbon storage and protection from sea-level rise and storm surges (Duarte et al. 2013; Arkema et al. 2013). These ecosystem services are generally welcomed by another stakeholders concerned with environmental issues, in particular globally well-considered by several international organizations such as IPCC, UNFCCC, UNEP, FAO and IPBES. These organizations aim to make some ocean agenda and national/international climate change initiatives for climate mitigation and adaptation, which include effective conservation and restoration of the coastal vegetation. Therefore, harmonizing coastal fishery with environmental conservation is now essential for the sustainable wise-use of ecosystem services.

Here, we just started some practice in field to make a harmony between sustainable coastal fishery and continuous water-quality improvement under oligotrophication, consequently resulting in both food security and climate change mitigation in future. Thau lagoon and Seto Inland Sea located at southern France and western Japan, respectively, are our target sites well-known as typical coastal areas with oligotrophication (Collos et al. 2009; Yanagi 2015). In both study sites, the aquaculture of pacific oyster (*Crassostrea gigas*) is an economically important fishery for the human community, and moreover eelgrass (*Zostera marina*) recovery by water-quality improvement is also apparent now.

Our practice is based on the interdisciplinary management approach including ecological actions to clarify the relationship among nutrient, ecosystem functioning and ecosystem services, socio-economical actions to estimate both the interface and the pathways between ecosystem services and the recipient human community, and moreover socio-psychological actions to estimate human-well beings of various stakeholders in the community. The aim of our approach is to identify the effective actions to maximize well-beings of various stakefolders, which is an essential key to harmonize coastal fishery with water-quality improvement under oligotrophication.

21.2 Materials and Methods

Seto Inland Sea (coordinates at the centre: 34.1667 N, 133.3333 E) is located at the southwestern part of main island of Japanese archipelago (Fig. 21.1a).

More than two thousand islands are interspersed in the Seto Inland Sea, so that the complicated coastal lines are formed in a local seascape which offers better calm areas for oyster rafted-aquaculture as well as habitats for eelgrass vegetation by preventing there from havy winds and wave actions. Oyster aquaculture using natural spats of native pacific oyster (*Crassostrea gigas*) at offshore area is

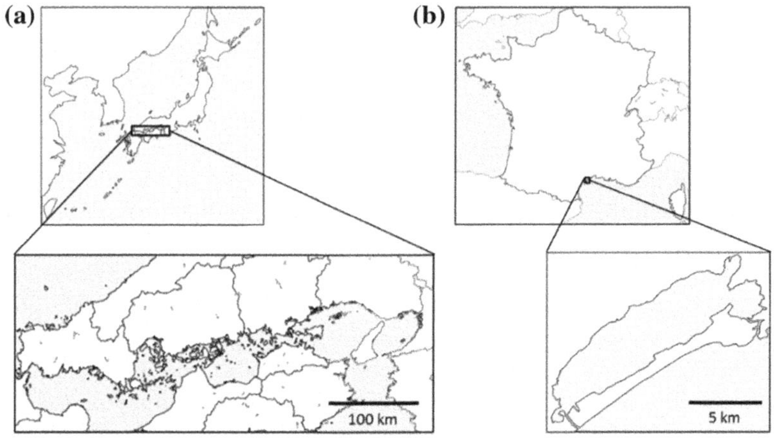

Fig. 21.1 Study sites of this research, **a** Seto Inland Sea in Japan, **b** Thau lagoon in France (modified from Hori et al. 2018)

flourishing in many areas of the Seto Inland Sea, so that the annual production amount in the Seto Inland Sea accounts for more than 60% of the national production of Japan. Seagrass recovery in the Seto Inland Sea also becomes apparent in recent 10 years with oligotrophication because of legal restriction of nutrient input from the water sheds. It has been estimated that the area of seagrass meadows increased from 6,000 ha to about 10,000 ha (Hori and Tarutani 2015).

Thau lagoon (coordinates at the centre: 43.41 N, 3.6241 E) is the largest lagoon located at southern French coast of the Mediterranean Sea (Fig. 21.1b). The lagoon is famous of oyster farming using non-native pacific oyster spats attached on longlines. The longlines with the spats are hung on oyster tables established in nearshore zone. About 10% of French national production of oyster is cultivated there, which is the largest oyster farming area in the Mediterranean Sea. It has been suggested that the recovery of seagrass beds is still proceeding, and that now the area of seagrass distribution extended up to 1,000 ha (Hori, personal communication with Syndicat mixte du bassin de Thau). The expansion of eelgrass meadows was observed even under oyster tables within oyster farming areas in June 2016.

As a first step to make a harmony between sustainable oyster aquaculture and seagrass conservation under oligotrophication in both Thau lagoon and Seto Inland Sea, we made a management strategy based on the interdisciplinarily approach, which consisted of ecological actions, socio-economical actions and socio-psychological actions. First, ecological actions aimed to improve or maintain the ecosystem functioning and ecosystem services of a target ecosystem. The ecological actions also include various ecological researches and managements to understand the status of the target ecosystem, in particular the investigation of the relationship among nutrient, phytoplankton production, seagrass production and oyster production, and the interactions between oyster aquaculture and eelgrass bed.

Socio-economical section is the important part to convey the change of ecosystem state and ecosystem services to the recipient human community. The socio-economical actions aimed to clarify the commodity and value chains of the human community around oyster and recreational business as well as the interface between ecosystem services and socio-economical activities in the target ecosystem. Moreover, the actions aim to identify the effect of the changes in ecosystem functioning and ecosystem services on the structure of these chains.

The fundamental purpose of socio-psychological actions was to identify the potential stakeholders and their well-beings in the recipient community, and to influence their view of nature and values. Some of ecosystem functioning and services cannot be appreciated by financial aspects, so that we need to develop the psychological method to directly identify well-beings. In our approach with analyses using structural equation modeling, we will try to identify some better management strategies by clarifying the conflict among ecosystem services, what ecological action is the most important for the development of the community and environmental conservation, which combination of ecological action and socio-economical actions is the most effective to maximize well-beings of the recipient community, and which is the key pathway from the ecosystem services to the well-beings via the value chain to improve or maintain both fishery and the coastal environment under oligotrophication. We showed the schematic of an ongoing ecological research by this approach here, clarifying several interactions between oyster farming and seagrass bed to identify better ecological actions for the management of ecosystem services in Thau lagoon and Seto Inland Sea.

21.3 Results and Discussion

Our interdisciplinary approach for the coastal management under oligotrophication was constructed as simple as possible, which consisted of three sections and three actions in each section (Fig. 21.2).

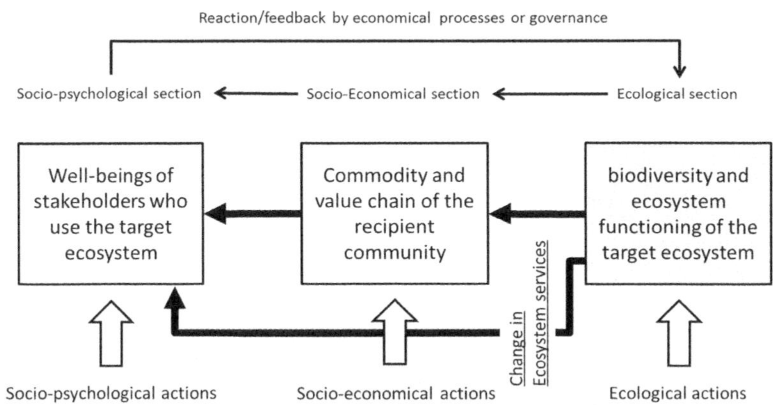

Fig. 21.2 Schematic explanation of the approach adopted in this study (modified from Hori et al. 2018)

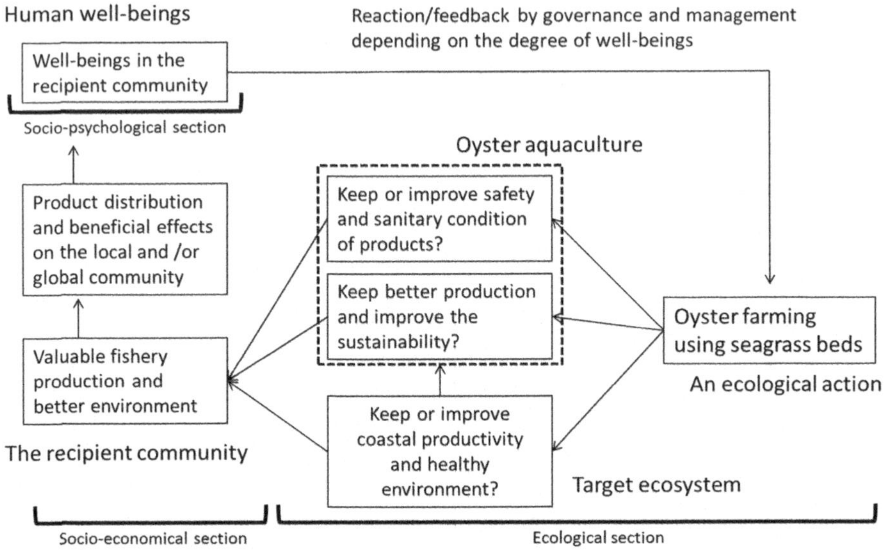

Fig. 21.3 The possible pathway that the oyster-seagrass interaction as an ecological action positively influences the recipient community via the change in the ecosystem functioning and ecosystem services (modified from Hori et al. 2018)

Based on this approach, we are now proceeding the research on clarifying the interactions between oyster farming and eelgrass bed to estimate the possibility of the oyster farming using seagrass beds as an ecological action in our management (Fig. 21.3). We have three working hypotheses at the moment as follows. First is whether the oyster farming using seagrass bed can keep or improve coastal productivity and healthy environments even under oligotrophication, meaning the effect of the oyster-seagrass interaction on ecosystem functioning of the target ecosystem. Second is whether the seagrass bed can keep better oyster production and improve the sustainability. This is a trophic effect of the change in the ecosystem functioning by seagrass bed on the oyster production, as a dominant ecosystem service in our study sites. The last is whether the seagrass bed can keep or improve the safety and sanitary condition of oysters. This is an effect of the change in environmental condition by seagrass bed on the quality of oysters, which is the possibility to give the added value to oyster products. If these hypotheses were successfully demonstrated, the recipient human community in socio-economical section of our approach can get valuable products and better environment. The change in the community by the ripple effect of the ecological action would cause the change in well-beings of the stakeholders of the target ecosystem.

In the original habitat of pacific oyster in Japan, the interaction between native eelgrass and native pacific oyster is common and an important component of the ecosystem functioning in coastal ecosystems, because pacific oyster reef is often formed adjacent to seagrass beds. Unfortunately, the natural mixed-landscape of oyster reef with seagrass bed has been lost due to coastal developments by reclamation and enbankment in many regions of Japan. In some regions, however, the aquaculture area has been established within or adjacent to seagrass beds. The oysters have been grown using this original oyster-seagrass interactions including facilitation of spats recruitment and spats survival rate, and especially trophic support from eelgrass bed to oysters for long time (Kasim and Mukai 2006; Tanaka 2014). This would be a typical indigenous and local knowledge of Japanese oyster aquaculture.

One of our interests in this research is also whether these oyster-eelgrass interactions in the original habitats of Japan has appeared in Thau lagoon ecosystem with native eelgrass bed and non-native pacific oysters, and whether the oyster-eelgrass interactions can work well to facillite the ecosystem functioning there. It was revealed by the analysis of both carbon isotope ratio and fatty acid signatures that the food resource of pacific oysters seasonally varied among phytoplankton, diatoms, bacterias and terrestrial organic matters in Thau lagoon (Pernet et al. 2012). This result suggests oysters in Thau lagoon have a potential to consume various kind of food resources derived from different origins, indicating that the oysters can also consume food resource subsidized from eelgrass beds, as well as the oysters in original habitat of Japan.

In oligotrophic environment, our first ecological action aims to facilitate total productivity based on not only pelagic production by increasing nutrient level, but also various benthic productions including seagrass beds. Seagrass-oyster interaction would become the key factor to make bio-resource cycling that increases turnover efficiency of the ecosystem functioning in the target ecosystem. Interactive resource subsidy between eelgrass and oyster can supply epiphytes and detritus as food resources for oysters, and nutrient and POM as resources for eelgrass and the eelgrass-associated organisms. The oyster-eelgrass interaction would keep high water-transparency, which is also ideal for recreational use. Increased eelgrass beds can absorb more carbon dioxide from the atmosphere and storage them as organic carbons, which can offset the carbon emission from oyster aquaculture and the recreational activities. This kind of local offset system of carbon emission can contribute promotion of Paris Agreement adopted at UNFCC-COP21. Our study is just starting out, so we have to make steady progress to identify wise-use and better management for oligotrophic coastal ecosystems through these ecological actions, socio-economical actions and socio-psychological actions in future.

Acknowledgements We would like to thank our French colleague, Yves Henocque for giving various suggestions and much help to us. This research could not be performed without his considerable support. We are also grateful to the Japanese–French Oceanographic Society for giving this opportunity to present our research. This French–Japanese collaboration was supported by Ministry of Agriculture, Forestry and Fisheries, Japan, and the ongoing research is supported by FRA and Ifremer.

References

Arkema KK, Guannel G, Verutes G, Wood SA, Guerry A, Ruckelshaus M, Kareiva P, Lacayo M, Silver JM (2013) Coastal habitats shield people and property from sea-level rise and storms. Nat Clim Chang 3:913–918

Burkholder JM, Tomasko DA, Touchette BW (2007) Seagrasses and eutrophicaton. J Exp Mar Biol Ecol 350:46–72

Collos Y, Bec B, Jauzein C, Abadie E, Laugier T, Lautier J, Pastoureaud A, Souchu P, Vaquer A (2009) Oligotrophication and emergence of picocyanobacteria and a toxic dinoflagellate in Thau Lagoon, southern France. J Sea Res 61:68–75

Duarte CM, Losada IJ, Hendriks IE, Mazarrasa I, Marba N (2013) The role of coastal plant communities for climate change mitigation and adaptation. Nat Clim Chang 3:961–968

Hori M, Tarutani K (2015) Changes in the distribution of seagrass vegetation with relation to the possible regime shift from pelagic-dominant to benthic-dominant system in Seto Inland Sea. In: Yamamoto T, Hanazato T (eds) Issues of oligotrophication in ocean and lakes. Chijinshokan, Tokyo, pp 129–148 (in Japanese)

Hori M, Hamaoka H, Hirota M, Lagarde F, Vaz S, Hamaguchi M, Hori J, Makino M (2018) Application of the coastal ecosystem complex concept toward integrated management for sustainable coastal fisheries under oligotrophication. Fish Sci 84 (2):283–292

IPCC (2007) Climate change 2007. Impact, adaptation and vullnerability. Contribution of working group II to the fourth assessment. In: Parry ML, Canziani OF, Palutikof JP, Van Der linden PJ, Hanson CE (eds) Report of the intergovernmental panel on climate change. Cambridge University Press, Cambridge, p 976

Kasim M, Mukai H (2006) Contribution of benthic and epiphytic diatoms to clam and oyster production in the Akkeshi-ko estuary. J Oceanogr 62:267–281

Matsuda O (2015) Towards rich Seto Inland Sea, considerable change of the basic plan on seto inland sea approved by the cabinet. Yutakana Umi 36:7–12 (in Japanese)

Pernet F, Malet N, Pastoureaud A, Vaquer A, Quere C, Dubroca L (2012) Marine diatoms sustain growth of bivalves in a Mediterranean lagoon. J Sea Res 68:20–32

Scheffer M, Carpenter S, Foley JA, Folke C, Walker B (2001) Catastrophic shifts in ecosystems. Nature 413:591–596

Selman, M. and S. Greenhalgh (2009). Eutrophication: policies, actions, and strategies to address nutrient pollution. WRI Policy Note, vol 3, pp 1–16

Tanaka T (2014) Satoumi with eelgrass and oyster beds, 'Hinase sengen-ryoshimachi' (Hinase, Okayama). Nippon Suisan Gakkaishi 80:72–75 (in Japanese)

Tsurita I, Hori M, Makino M (2016) Fishermen and conservation: sharing the case study of Hinase, Japan. In: Westlund L, Charles A, Garcia SM, Sanders J (eds) FAO-FA Technical Paper 603: marine protected areas: interactions with fishery livelihoods and food security, FAO, Rome, pp 43–50

Yamamoto T, Hanazato T (2015) Issues of oligotrophication in ocean and lakes. Chijinshokan, Tokyo, 195 p (in Japanese)

Yanagi T (2015) Eutrophication and oligotrophication in Japanese estuaries: the present status and future tasks. Springer, Dorderecht, p 97

Chapter 22
One-Year Colonization by Zoobenthic Species on an Eco-Friendly Artificial Reef in the English Channel Intertidal Zone

Jean-Claude Dauvin and Aurélie Foveau

Abstract The RECIF project aims to enhance the use of shellfish by-products, i.e. crushed shells of the queen scallop *Aequipecten opercularis*, in the development of an eco-friendly material for artificial reefs. The short-term colonization of three different types of artificial structure (blocks made of two rough-surface concretes, each with a different porosity, and blocks of ordinary concrete) is investigated to highlight possible differences between substrate materials and observe the succession of colonizing species. A total of 75 blocks were emplaced in March 2014 and monitored until February 2015 in the intertidal zone of Luc-sur-Mer (Calvados coast on the French side of the English Channel). The abundance and species richness of the whole community were recorded. Analyses show significant temporal differences in species abundance between blocks, but no variations in species richness, while also revealing differences in the settlement of species depending on their position on the blocks, i.e. on the face exposed to the main tidal currents and to the light, or on the other faces. These differences are explained by environmental parameters around the blocks and by larval behaviours. This unique experimental study focuses on all species found on the blocks, showing that it is important to take account of all the fauna to estimate the benthic production and the functional role of such artificial reefs.

Keywords Eco-reef · Short-term colonization · English channel · Intertidal zone · Monitoring

J.-C. Dauvin (✉) · A. Foveau
NormandieUniv – UNICAEN, Laboratoire Morphodynamique Continentale et Côtière, UMR 6143 M2C, 24 Rue Des Tilleuls, 14000 Caen, France
e-mail: jean-claude.dauvin@unicaen.fr

J.-C. Dauvin · A. Foveau
CREC, Station Marine de l'Université de Caen Normandie, 54, Rue Du Docteur Charcot, BP 49, 14530 Luc-Sur-Mer, France

A. Foveau
Ifremer Laboratoire Environnement Ressources Bretagne Nord, 38 Rue Du Port Blanc, 35800 Dinard, France

T. Komatsu et al. (eds.), *Oceanography Challenges to Future Earth*,
https://doi.org/10.1007/978-3-030-00138-4_22

22.1 Introduction

Submerged structures corresponding to artificial reefs are nowadays present in many urbanized coastal zones worldwide, but mainly in areas with high anthropogenic pressure such as the Mediterranean Sea and around Japan. Two main types of artificial reef have been deployed, with the aim of increasing local fisheries production, or protecting the coastal zone from the effects of trawling (Andersson et al. 2009). However, the presence of such immersed structures in the marine environment modifies seabed habitats by adding areas of hard substrate to soft-bottom habitats, which, in response, leads to changes of local flora and fauna (Perkol-Finkel et al. 2006; Petersen and Malm 2006).

The essential issue of any programme concerning artificial reefs is to carry out an inventory of the species colonizing a reef, i.e. the species richness, the temporal succession of species during the immersion phase and the growth of existing species.

While most artificial reefs are constructed to increase the catches of fishes and invertebrates of commercial interest, surveying the epifaunal community colonizing the reef is an important aspect since these species control the local diversity and are also potential prey for other species such as commercial crabs and fishes (Figley 2003).

Indeed, the main taxa settling on the surface of the reef consist of sessile epifauna as well as microflora and macroflora. Algae represent the pioneer fouling colonizers on marine reefs (Fletcher 1988), making up a diverse group of photoautotrophic organisms of various shapes (filamentous, ribbonlike, or platelike) and size (Fletcher 1988; Miller et al. 2012; Hughes et al. 2013). Moreover, the composition and heterogeneous nature of the immersed artificial substratum play important roles during the early phases of colonization by fixed organisms. Rough textures induce greater microhabitat diversity than smooth surfaces (Anderson and Underwood 1994; Guillitte 1995; De Muynck et al. 2009). Moreover, the settlement and development of organisms on the surface of the reef (Anderson and Underwood 1994) is favoured by the water retention properties of the porous artificial reef material in intertidal zones (Guillitte 1995), while the crevices and pits protect small organisms from predators. Many vagile taxa are also associated with this fouling community, which increases the attractiveness of the reef for predators (Moreau et al. 2008).

The RECIF project, co-financed by the ERDF, was sponsored from 2013 to 2015 within the framework of the European INTERREG IVA Programme for cross-border cooperation between France and the United Kingdom in adjacent regions around the English Channel (*La Manche* in French). This project proposes the incorporation of crushed seashells of the queen scallop *Aequipecten opercularis* (Linnaeus 1758) into the substrate of concrete blocks (Cuadrado et al. 2014), through the development of innovative building materials for artificial reefs, allowing the recycling of waste while offering a favourable solution to opening up services provided by the marine ecosystem. Another main objective of this project is to assess the role of reefs in the improvement and development of marine biodiversity in the case of the English Channel ecosystem.

The main aims of the present study are: (1) to compile an inventory of the sessile and vagile macrofauna recorded over a period of one year during a survey of blocks positioned on oyster culture tables in the intertidal zone near Luc-sur-Mer area on the Calvados coast in the eastern basin of the English Channel, (2) to describe the biological succession on these blocks over time, and (3) to compare the colonization of fauna between different types of artificial reef material, and (4) to assess the role of environment parameters in controlling the eventual differences in colonization of the substrates.

22.2 Materials and Methods

Three types of concrete substrate material were used to make the 20 × 20 × 40 cm blocks deployed in the experiments: an ordinary concrete made from natural aggregates and two rough-surface concrete mixes incorporating 40% crushed queen scallop shells and showing different porosities (Cuadrado et al. 2015) (Fig. 22.1).

The experimental structure, i.e. 75 blocks (i.e. 9–10 blocks per oyster culture table corresponding to the three block types, with 30 blocks of eco-friendly material for each of the two porosity values and 15 blocks of ordinary concrete), were fixed onto eight oyster culture tables in the intertidal zone of Luc-sur-Mer (49° 19′ 15″ N, 0° 20′ 55″ W), on the southern shore of the Bay of Seine, in the eastern basin of the English Channel, on 19 and 20 March 2014 (Figs. 22.2 and 22.3).

The experimental site is about 350 m from the Luc-sur-Mer seawall, lying in a slight depression in a zone mainly composed of sand and rocks, which is only accessible once a fortnight at low water during spring tides. At high tide, the water depth is 6.5 m. We estimate that, during the period of the experiment (March 2014–February 2015), the concrete blocks were emerged 44% of the time. The main

Fig. 22.1 Queen scallop *Aequipecten opercularis* (Linnaeus 1758) and crushed queen scallop shells used to make up the substrate of concrete blocks (after Cuadrado et al. 2015)

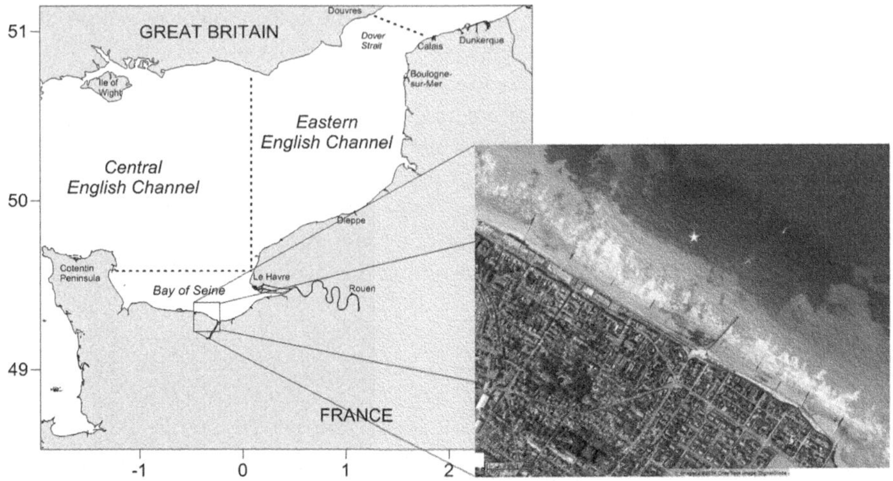

Fig. 22.2 Location of experimental site on the intertidal zone at Luc-sur-Mer (Calvados coast, southern shore of the Bay of Seine)

Fig. 22.3 In situ experimental structure

environmental factors controlling settlement, colonization and competition between species are: speed of tidal currents, wind force and direction, sunshine hours and rainfall during emergence. The blocks were collected about every 15 days between 1 April 2014 and 4 February 2015, making a total of 22 sampling dates (Fig. 22.4).

The two types of blocks containing 40% seashell by-products were collected for analysis at each sampling date. The third type of block, i.e. control concrete without seashells, was sampled once a month. The blocks were replaced by the same type of block at each occasion to assess the seasonal cycle of colonization.

The collected blocks were analysed under a binocular microscope at the Marine Station of Luc-sur-Mer. Sub-sampling was carried out over unit areas of 25 cm^2 delimited by a mask, and analyses covered approximately one quarter of the total surface-area of each face. A series of operations was performed on these sub-sampling units: photographs of each sub-sample and the entire face were stored in an image library; all of the animal sessile species were identified and counted, including highly abundant species such as barnacles. Then, the whole surface of the studied face was examined to identify the colonizing species. The seawater used to keep the fauna alive during the laboratory observations was then filtered on a 0.5-mm mesh sieve to collect the captured biota, fixed with 4% formaldehyde, and examined at a later stage to identify and count the vagile fauna associated with the blocks.

Statistical analyses (Kruskal-Wallis and Tukey tests) were performed on the collected fauna using tools within the R software.

Fig. 22.4 Sampling strategy

22.3 Main Results

During the period of the survey, a total of 153 macrozoobenthic taxa were identified, i.e. 49 sessile epifauna and 104 associated fauna mainly composed of crustaceans and polychaetes. There is a succession of colonization with time, with the blocks being firstly colonized by barnacles, then hydrozoans, simple ascidians, fixed worms, colonial ascidians, bryozoans, entoprocta and finally porifera (Foveau et al. 2015). At the beginning of the summer, the dominance of tunicates was accompanied by a significant colonization of hydroids. This tunicate/hydroid community persists until the end of the summer, before gradually being replaced by a third dominant sponge community. For about five months, these three communities co-exist on the different faces of the blocks, according to the ecological preference of the different species (Foveau et al. 2015).

Among the epifauna, 28% are tunicates (simple compound ascidians), followed by cnidarians (22%), bryozoans (18%) and sponges (11%). In order of decreasing abundance, the ten most common taxa are: the crustacean *Balanus crenatus* (up to 189,000 ind.m^{-2}), the simple ascidian *Corella eumyota* (until 6,850 ind.m^{-2}), the mollusc *Mytilus edulis* (until 2,000 ind.m^{-2}), and the annelid *Spirobranchus lamarcki* (until 1,350 ind.m^{-2}), followed by the complex ascidian *Botryllus schlosseri*, the simple ascidian *Molgula*, the complex ascidian *Polyclinum*, the cnidarian *Kirchenpaueria pinnata*, and the sponges *Halichondria* and *Sycon*. Two tunicates, *Corella eumyota* and *Perophora japonica*, originate from Asian waters and are Non-Native and Invasive species.

Statistical analyses on the epifauna species richness show that there are no significant differences between the three types of blocks, or between the various faces of a block. Nevertheless, there is a significant difference when comparing various faces as against various dates (the non-parametric Kruskal-Wallis chi-square and p-values are, respectively: 253.3666 and $<2.2e^{-16}$ vs. 258.9103 and $<2.2e^{-16}$). While there are no significant differences of epifaunal abundance between the three types of blocks or between the sampling dates, differences are observed between the faces of the blocks (the non-parametric Kruskal-Wallis chi-square and p-values are, respectively: 53.1332 and $1.103e^{-9}$). The Tukey test reveals that the lower face accounts for the difference, in particular when compared with the upper, landward and along-shore oriented faces.

Mobile fauna associated with the sessile fauna account for 104 taxa, with annelids and arthropods showing high and approximately equal species richness (respectively 40 and 39% of the total diversity). The five most abundant *taxa* are all crustaceans: the amphipods *Corophium acutum* and *Jassa falcata*, the tanaid *Zeuxo holdichi*, and the decapods *Carcinus maenas* and *Pisidia longicornis*. Among the crustaceans, a new species of tanaidacean (*Zeuxo holdichi* Bamber 1990) is recorded for the first time along the Calvados coast. This species lives mainly in the seaweed which settles on the blocks or directly in crevices present at the surface of the blocks, but always in mud tubes like those built by amphipods of the genus *Corophium* spp.

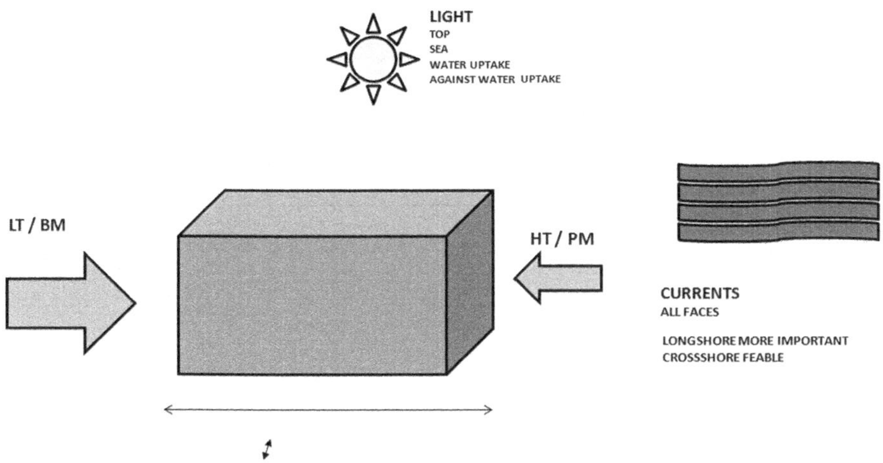

Fig. 22.5 Potential environmental parameters controlling the colonization of blocks in the intertidal zone at Luc-sur-Mer (southern shore of the Bay of Seine)

Figure 22.5 illustrates the importance of the main environmental factors in the colonization of the different faces of the blocks. The various 'longshore' and 'cross-shore' currents influence the arrival and distribution of larvae. Light is another important parameter in the colonization of the different block faces, with large differences between the face exposed to light and the lower face. The colonisation of photophilic or sciaphilic algae also plays a role in the interaction with animals.

22.4 Discussion

The blocks are colonized by a highly diversified fauna (153 taxa) including 2/3 of mobile fauna and 1/3 of sessile fauna, including annelids, tunicates, hydrozoans, bryozoans, barnacles and mussels. The succession observed during the period of the survey (one year) is comparable to the successions observed in shallow waters (8–12 m depth) by Svensson et al. (2007) and Andersson et al. (2009) along the Swedish coast, and by Hatcher (1998) in UK waters. The succession corresponds to temporal changes in species composition and community structure (Pickett 1976). A biological succession over time is observed on the studied blocks: the community is initially colonized by abundant barnacles, and then becomes dominated by ascidians. These changes may be the consequence of direct interactions between the colonising species and may also arise from indirect effects, i.e. pioneer species renew the environment, making it more or less favourable for subsequent colonizing species or non-interactive processes, linked to the probability of arrival of available larvae in the zooplankton or differences in growth and longevity among the species present on the reef (Drury and Nisbet 1973; Connell and Slatyer 1977). Moreover, the biological succession is often the result of a complex combination of these different mechanisms.

Some authors suggest that the composition of the substrate and the heterogeneity of the surface are also important factors in the establishment of organisms (Mullineaux and Garland 1993). For example, the cypris larvae of barnacles show preferences in terms of substrate surface (Mullineaux and Butman 1991) or structural complexity of the basal substrate (Lemire and Bourget 1996). Nevertheless, our results show no significant difference in colonization according to the block composition, i.e. with or without crushed queen scallop shells used in the concrete mix, and with two distinct porosity values, probably because the different block surfaces show a similar degree of heterogeneity which favours a high level of colonization. In an estuarine zone at Ijmuiden (the Netherlands), Paalvast (2015) observed large differences of colonization on substrates with fine and coarse texture compared to smooth texture, as well as between structures with holes and with vertical pits.

Differences in colonization may also be caused by environmental factors, such as wave exposure and level on the intertidal zone (Paalvast 2015). At the site studied here, 'longshore' and 'cross-shore' currents influence the arrival and distribution of larvae on the blocks. Moreover, the availability of larvae is linked to the spawning season of the different colonising species. Accordingly, the pioneer species may be different depending on the period when the artificial structures are deployed (Anderson and Underwood 1994). As a result, the succession will be governed by the interaction between resident species and the new colonising species, and will vary with the time of year of the experiment (Connell and Slatyer 1977). Light appears to be an important parameter, and differences mainly concerning the algae have been observed in the colonisation of the block faces and, in particular, the bottom face (Foveau et al. 2015). The bottom face, which is sheltered from light, is colonized predominantly by sciaphilic species. By contrast, the upper face, which is turned towards the light, is colonized by different photophilic algal species, with the exception of small red sciaphilic algae.

In summary, numerous results obtained with this survey strategy involve a high financial cost. Such a short frequency of observation (every 15 days) is not feasible in a subtidal zone with strong tidal currents, high turbidity, and high meteorological variability in a temperate region under more or less permanent westerly winds. Nevertheless, by focusing on all species found on the blocks, this unique experimental study in the intertidal zone shows that it is important to take account of all the fauna to estimate the benthic diversity. Such an approach will then allow us to assess the production and the functional role of artificial reefs. This kind of survey also appears to be a good starting point for setting up an observatory of Non-Native Species; in our survey lasting for one year, three NNS were recorded. The survey was continued for a further year until March 2016, but with a three-month period between removals of blocks.

Acknowledgements Aurélie Foveau received a post-doctoral grant from the collaborative project, RECIF, which was selected within the framework of the European INTERREG IVA Programme for cross-border cooperation between France and the United Kingdom in adjacent regions around the English Channel, and co-funded by the ERDF. The authors wish to thank the co-financiers and all project partners for their support. The authors also thank M.S.N. Carpenter for the English revision. Jean-Claude Dauvin is grateful to the Japanese-French Oceanographic Society in Japan and its President, Professor Teruhisa Komatsu (SFJO-J), for his invitation to take part in the 16th Japanese-French Oceanography Symposium, held in Tohoku/Tokyo, Japan, from 17–21 November 2015, on the topic 'The sea under human and natural impacts: challenge of oceanography to the future Earth'.

References

Anderson MJ, Underwood AJ (1994) Effect of substratum on recruitment and development of an intertidal estuarine fouling assemblage. J Exp Mar Biol Ecol 184:217–236

Andersson MH, Berggren M, Wilhelmsson D, Öhman MC (2009) Epibenthic colonization of concrete and steel pilings in a cold-temperate embayment: a field experiment. Helgol Mar Res 63:249–260

Bamber RN (1990) A new species of *Zeuxo* (Crustacea: Tanaidacea) from the French Atlantic coast. J Nat Hist 24:1587–1596

Connell JH, Slatyer RO (1977) Mechanisms of succession in natural communities and their role in community stability and organization'. Am Nat 111:1119–1144

Cuadrado H, Sebaibi N, Boutouil M, Boudart B (2014) Physical properties and mechanical behavior of concrete made with crushed queen scallop shells. In: Proceedings of the international symposium on environmentally friendly concrete—ECO-Crete, 13–15 August 2014, Reykjavik (Iceland), p 21–28

Cuadrado H, Sebaibi N, Boutouil M, Boudart B (2015) Properties of concretes incorporating crushed queen scallops for artificial reefs. In: Proceedings of congress on artificial reefs: from materials to ecosystems, Caen, France, p 9–18

De Muynck W, Maury-Ramirez A, De Belie N, Verstraete W (2009) Evaluation of strategies to prevent algal fouling on white architectural and cellular concrete. Int Biodeterio Biodegradat 63:679–689

Drury WH, Nisbet ICT (1973) Succession. J Arnold Arbor 54:331–368

Figley B (2003) Marine life colonization of experimental reef habitat in temperate ocean waters of New Jersey. New Jersey Department of Environmental Protection Division of Fish and Wildlife Report, 64 p

Fletcher RL (1988) Brief review of the role of marine algae in biodeterioration. Int Biodeterio Biodegradat 24:141–152

Foveau A, Dauvin JC, Rusig AM, Mussio I, Claquin P (2015) Colonisation à court terme par le benthos sur un éco-récif artificiel. In: Proceedings of congress on artificial reefs: from materials to ecosystems, Caen, France, p 119–126

Guillitte O (1995) Bioreceptivity: a new concept for building ecology studies. Sci Total Environ 167:215–220

Hatcher AM (1998) Epibenthic colonisation patterns on slabs of stabilized coal-waste in Pools Bay, UK. Hydrobiologia 367:153–162

Hughes P, Fairhurst D, Sherrington I, Renevier N, Morton LHG, Robery PC, Cunningham L (2013) Microscopic examination of a new mechanism for accelerated degradation of synthetic fibre reinforced marine concrete. Construct Build Mater 41:498–504

Lemire M, Bourget E (1996) Substratum heterogeneity and complexity influence micro-habitat selection of *Balanus* sp. and *Tubularia crocea* larvae. Mar Ecol Prog Ser 135:77–87

Miller AZ, Sanmartin P, Pareira-Pardo L, Dionisio A, Saiz-Jimenez C, Macedo MF, Prieto B (2012) Bioreceptivity of building stones: a review. Sci Total Environ 426:1–12
Moreau S, Péroni C, Pitt KA, Connolly RM, Lee SY, Meziane T (2008) Opportunistic predation by small fish on epibiota of jetty pilings in urban waterways. J Fish Biol 72:205–217
Mullineaux LS, Butman CA (1991) Initial contact, exploration, and attachment of barnacle (*Balanus amphitrite*) cyprids settling in flow. Mar Biol 110:93–103
Mullineaux LS, Garland ED (1993) Larval recruitment in response to manipulative field flows. Mar Biol 116:667–683
Paalvast P (2015) The role of geometric structure and texture on concrete for algal and macrofaunal colonization in marine and estuarine intertidal zone. In: Proceedings of congress on artificial reefs: from materials to ecosystems, Caen, France, p 77–84
Perkol-Finkel S, Shashar N, Benayahu Y (2006) Can artificial reefs mimic natural reef communities? The role of structural features and age. Mar Environ Res 61:121–135
Petersen JK, Malm T (2006) Offshore windmill farms: threats to or possibilities for the marine environment. Ambio 35:75–80
Pickett STA (1976) Succession: an evolutionary interpretation. Am Nat 110:102–119
Svensson JR, Lindegarth M, Siccha M, Lentz M, Molis M, Wahl M, Pavia H (2007) Maximum species richness at intermediate frequencies of disturbance: consistency among level of productivity. Ecology 88:830–838

Chapter 23
New Installations of Artificial Reefs Along the Coast of the Landes (South–West Atlantic Coast of France)

Gérard Fourneau, Florence Dufour, Aurélie Penne, Nelly Ferrou, Thomas Scourzic, Aurore Laborde and Elodie Zaccari

Abstract The « Aquitaine Landes Récifs » (Aquitaine Landes Reefs Association) was founded in 1996. Its aim is to create, immerse, manage and scientifically monitor stands of artificial reefs off the coast of the Landes (South–West Atlantic Coast of France). In 1999, this association first made the immersed artificial reefs by companies, then in 2007 the association built the reefs and immersed them. These are the most important projects that have been installed on the coast of southwestern France, with an installed volume of 2,000 m^3 immersed in three sites of 16 hectares. Its main objective is to establish habitats for marine animals, in order to increase the number of species present in this area, including local commercial species. To this end, it has built several new types of artificial reefs. The detailed observation of the structures in place, after their immersion, made it possible to study their mechanical and physical behavior, with respect to the specific conditions of the Landes sandy coast. Scientific monitoring, conducted since the first immersions in 2010, demonstrated the attractiveness of the various modules for many species, including commercial species. A third objective is to observe marine species that settle in and around artificial reefs in the Bay of Biscay, allowing scientists to have new fields of experience for their research, in a context of climate change. This possibility can extend to incursions of recreational divers, even to amateur biologists and schools. Existing reef improvement projects are planned by this association, as well as tests of immersion of commercial molluscs in deep water, opening an experimental way to shellfish farming.

Keywords Artificial reefs · Fishery resources · Fishing · Oyster-farming

G. Fourneau (✉) · F. Dufour · A. Penne · N. Ferrou · T. Scourzic · A. Laborde · E. Zaccari
Aquitaine Landes Récifs, 2128 avenue de la résistance, 40990 Saint-Paul Lès Dax, France
e-mail: contact@aquitaine-landes-recifs.fr

T. Komatsu et al. (eds.), *Oceanography Challenges to Future Earth*,
https://doi.org/10.1007/978-3-030-00138-4_23

23.1 Introduction

The global fishery for marine natural resources has been steadily declining for many years (Jackson et al. 2001; FAO 2016). The establishment of artificial reefs makes it possible, to a certain extent, to improve the presence of marine organisms in coastal areas accessible to fishing (Baine 2001). In order to overcome the steady decline in marine resources in France and Aquitaine, to support artisanal fishing and to combat trawling within the three-mile zone and after studies on the recruitment of marine species in this area (Bouchet 1984), Meetings were held locally with professional and amateur fishermen, biologists and various institutions to find and implement sustainable solutions.

The objective of this publication is to present the different phases of development of the marine concessions in which the experiments took place. In order to carry out effective actions at the regional level, the association Aquitaine Landes Récifs was created in 1996 to Create, immerse, manage and scientifically control the artificial reefs off the coast of Aquitaine, which is mostly sandy, and also experience the immersion of molluscs in deep water.

It is important to confirm that marine species habitually living in rocks can settle and live within or on the periphery of artificial reefs on large sandy bottoms.

It is also necessary to know, under these conditions, whether the provision of habitats of various shapes makes it possible to obtain the presence of many species which are not usually found in this zone. Previous experiments carried out in the Adriatic Sea (Bombacce et al. 1994) have shown that this is generally the case with varying degrees of success. It was therefore necessary to obtain new data in the situation of the coasts of the Landes.

Finally, it was also desirable to obtain new concrete evidence for the production of molluscs species consumed by mankind on artificial structures immersed in open water. Such trials have been conducted in other countries (DeWitt et al. 1999), and it was interesting to try new approaches in this field.

23.2 Site Selection

The Association decided to submerge artificial reefs in three sites along the south–west coast (France): Capbreton, Soustons/Vieux-Boucau, Messanges/Moliets and Maâ. Site selection was based on four key principles:

- Proximity to the Port of Capbreton (only port of the Landes) to facilitate diving expeditions.
- Proximity to the coast to ensure that the depth of immersion is not too deep (between 18 and 25 m) to allow for sufficiently long dives to collect scientific data.
- A shallow gradient and appropriate grading to help the reefs maintain their position on the seabed.

- Proximity to an alluvial tributary: Lake Hossegor for the Capbreton concession, Port D'Albret, Soustons Lake/Vieux-Boucau and Courant d'Huchet for Messanges/Moliets and Maâ. Another advantage of these sites is that they are used as nurseries and shelters by species visiting artificial reefs (Fig. 23.1).

Preliminary studies (ALR et BIO-SUB 1999) were conducted prior to immersing artificial reefs to study the physical and biological characteristics of potential immersion sites and to acquire new and reliable mean depth data, Tides, currents, swell frequency and height, substrate composition, average underwater visibility, and species abundance at the surface and within sediments. The sites studied and selected according to these criteria were considered suitable for the immersion of artificial reefs and their monitoring by divers under optimal conditions. It is important that they have good visibility for their observations. It is also essential that constructed reefs possess sufficient characteristics to avoid the silting of structures and the impacts caused by tides and tides.

The French official bodies granted ALR, in partnership with the multi-community intercommunal trade union (SIVOM Côte Cote Sud), three maritime concessions each with an area of 16 hectares for a renewable six-year period, for a total of 15 years. These concessions are considered as pilot tests and, as such, are granted free of charge. As the Association conducts scientific monitoring, it therefore has a permanent authorization from the Departmental Delegation to the Sea and the Littoral.

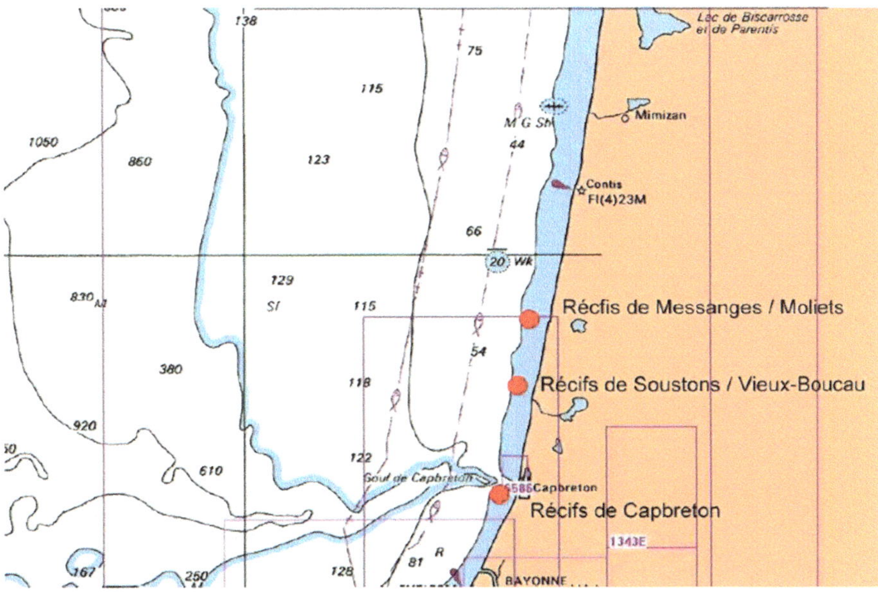

Fig. 23.1 Locations of the different sites of reef implantations SHOM 1978 modified by Dufour et al. 2011)

23.3 First Immersions (1999–2004)

The first submerged cylindrical structures were made of concrete. Honeycomb modules of different sizes were used weighing between 0.9 and 1.6 tonnes with an average diameter of 90 cm and a length of 1 m. Each duct has a volume of about 1 m^3.

- The first phase of the immersion took place in 1999 in the Capbreton concession. 800 m^2 of recovered honeycomb structures were deposited individually from a barge in a relatively random pattern (Fig. 23.2).
- Between 2001 and 2002, 800 m^2 of honeycomb structures were immersed in the Soustons/Vieux-Boucau site. To be more efficient and to obtain a larger volume, five modules were interconnected before being deposited on the seabed by the vessel Aquitaine Explorer. Seven clusters of about 120 m^3 were deposited in a circle around a central point (Fig. 23.3).
- Between 2003 and 2004, with the help of Aquitaine Explorer, 600 m^3 of artificial reefs were immersed on the Messanges/Moliets and Maâ site. The modules, specially made for the Association, were assembled together by cables and deposited in three clusters (Fig. 23.4).

This initial phase allowed the dumping of 2,200 m^3 of artificial reefs.

Scientific monitoring by divers was carried out after initial immersion in order to follow the development of the structure and its colonization over time. A follow-up protocol has been developed for this purpose.

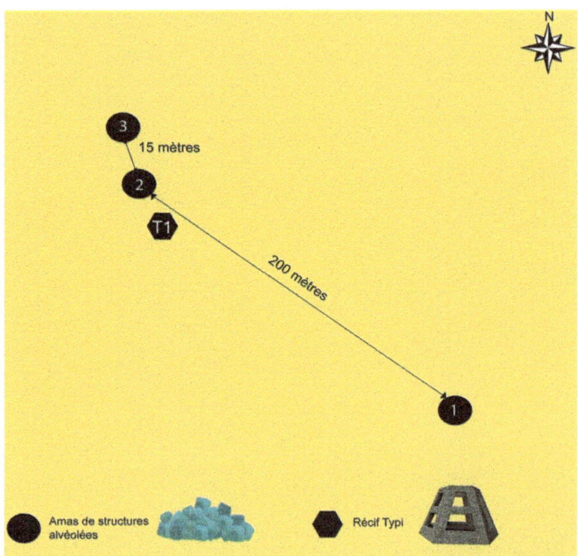

Fig. 23.2 Design of the Capbreton concession (Dufour et al. 2011)

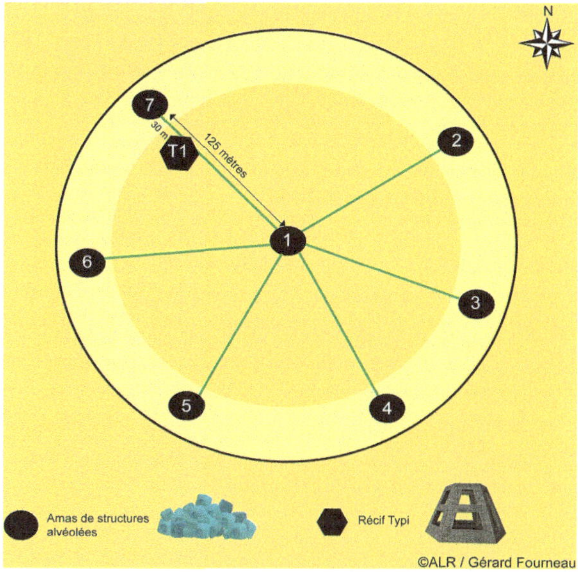

Fig. 23.3 Design of the Soustons/Vieux Boucau concession (Dufour et al. 2011)

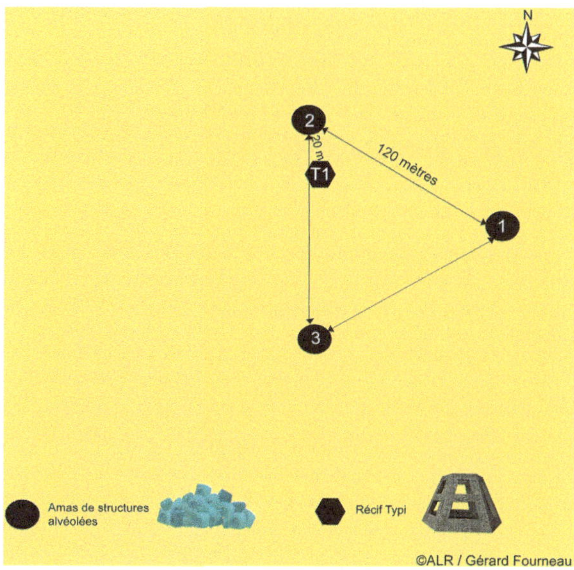

Fig. 23.4 Design of the Messanges/Moliets and Maâ concession (Dufour et al. 2011)

23.4 Reef Construction by the Association

From 2007, the association decided to develop its own structures, using a design based on the shape of a stable, broad-based pyramid made up of six parts. This reef was named "Typi" because of its shape reminiscent of Indian tents.

The association acquired a special steel mold, to produce concrete elements. It has thus become a manufacturer of artificial reefs. Weighing 13 tonnes, measuring 2.60 m and a diameter of 4.60 m, it is a real obstacle for trawlers. Three examples of this type of structure were immersed in 2010 on each of the three concession sites of the Association using Buoys, Gascogne and Lighthouses and Beacons of Verdon sur Mer (France) in order to compare the state of the structures, their attractiveness and their rate of colonization.

Observations have shown a high stability of this reef on sandy bottoms. Moreover, it has not been eroded by the sand movements at the bottom.

In order to increase the reef complexity and to provide new habitats for marine animals (Sherman et al. 2002; Charbonnel et al. 2002; Taniguchi and Tokeshi 2004), a new model was developed from the previous reef design, which makes it possible to use the same mold. Named "Babel", due to its high cylinder shape, this new reef could be made of different plates with feet of different heights, and central openings with different diameters. These slabs could be installed inside the "Typi reef".

23.5 Reef Populations

Stand studies have been carried out several times by specialized companies (ALR et Océanide 2008; ALR et Duchassin 2013; ALR et Maslies 2015). A standardized protocol for biological and ecological monitoring of the structure was developed following the methodology used by Castege et al. (2016) between May and September. These observations led to the identification of many marine species between 2010 and 2015 on "Typi reefs". Among the invertebrates, a large number of Anthozoa, Alcyonaria, Actinarians, as well as Polychaetes, Cephalopod Molluscs, Gastropods and Lamellibranchs were observed.

The presence of many species of fish in the larval, post-larval and juvenile state demonstrates the attractiveness of "Typi reefs" for species living in the bottom (Carr 1991) (Fig. 23.5). This new type of structure also attracts many species of fish, including commercial species (*Sparus aurata*, *Mullus surmuletus*, *Scorpaena porcus*, *Diplodus vulgaris*, *Spondyliosoma cantharus* and *Zeus faber*). Cuttlefish (*Sepia officinalis*) and squid (*Loligo vulgaris*) find shelters and spawning areas, thus supporting local fisheries.

These observations must necessarily be pursued for a longer period (Charbonnel and Bachet 2010), so that the state of equilibrium of the local ecosystem is reached (Fig. 23.6).

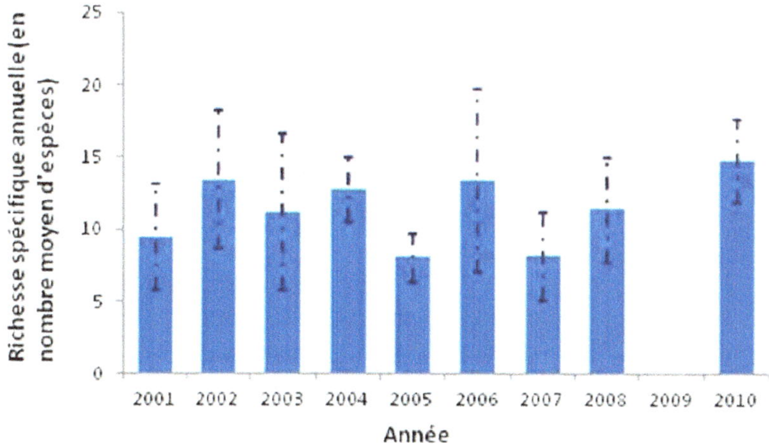

Fig. 23.5 Annual abundance of species (average number of species) on the Capbreton artificial reef between 2001 and 2010; the standard deviations are indicated by the dots. Indeterminate species, which could represent duplicates, are not shown here

Fig. 23.6 Colonisation of the Typi reef one year after immersion in 2011

23.6 Outlook and Prospects

The new projects aim to further improve the shapes and characteristics of the reefs, such as the “Typi” and “Babel” models, so that they have more habitats of varying shapes and sizes (Charbonnel et al. 2002), and using original techniques (Ceccaldi and Nakagawa 2002). It is envisaged to install measuring equipment in the interior of the “Typi reef” in order to establish correlations between environmental conditions and the presence of certain species.

On the other hand, three copies of the “Babel reef”, consisting of 4 plates weighing 10 tons, 2.5 m high and 2.77 m wide, were immersed in each of the three concession sites in 2015 and the evolution of their stands will be followed as the previous ones.

In addition, a project to create new structures adapted to the culture of oysters (*Crassostrea gigas*) and scallops (Pectinidae) in deep water. A steel structure fixed on a heavy concrete base was made and immersed in the site of Capbreton. Weighing 5 tons, a height of 3 m, a width of 2.5 m and a length of 3.32 m, this new model is equipped with Australian fish traps (65 and 85 cm) with different meshes. This project will be carried out in collaboration with the Regional Conchyliculture Committee of Arcachon Aquitaine and the oysters of Hossegor (France).

The association Aquitaine Landes Récifs wishes to further develop its scientific management by setting up a monitoring system using new means of observation (mini-ROV) in addition to the dives by scientific teams.

In addition, the Association, as a manufacturer of artificial reefs, is considering the establishment of an underwater space equipped with artificial reefs in the Bay of Biscay to test reef models adapted to a particular environment and which encourage the presence of certain target species.

23.7 Conclusions

2,600 m^3 of artificial reefs have been immersed by the Aquitaine Landes Recifs Association since 1999, making this project the fourth most important of its kind in France in terms of submerged volume, surface area of the concession and the first along the French Atlantic coast.

Today, at least 134 species have been identified on the three concessions, showing that the presence of artificial reefs on sandy bottoms allows for the establishment of new ecosystems of limited size, but which contribute both to the preservation of biodiversity and taking advantage, to some extent, of local artisanal fisheries.

Lastly, a final objective is to provide training in marine biology, particularly for universities, by organizing diving expeditions on immersed artificial reefs in order to observe the colonizing species in situ.

Acknowledgements For the implementation of these projects, the association is supported by various partners who have materially and financially supported its activities: the Regional Council of New-Aquitania, the Departmental Council of Landes, Action Pin, SIVOM South Coast, the Intercommunal Syndicate of Port d'Albret, the Caisse d'Epargne Foundation, the Macif Foundation and the Crédit Coopératif Foundation.

They also thank Mrs Nelly Ferrou, who presented these results at the 16th French-Japanese Symposium of Oceanography in Shiogama.

Finally, they thank Professor Hubert-Jean Ceccaldi, Drs. Patrick Prouzet and Yves Hénocque for their assistance in the correcting and improving this manuscript.

References

ALR et BIO-SUB (1999) Étude préliminaire à l'implantation de récifs sur la Côte Landaise (Capbreton, Vieux-Boucau, Moliets), 24 p (diffusion restreinte)

ALR et Duchassin M (2013) Suivi scientifique des récifs artificiels d'Aquitaine Landes Récifs, Capbreton, Soustons/ Vieux-Boucau, Messanges/ Azur/ Moliets & Maâ, Année: 2013, 50 p (diffusion restreinte)

ALR et Maslies M (2015) Suivi scientifique du récif Typi de Capbreton, 2010–2015, 10 p (diffusion restreinte)

ALR et Océanide (2008) Suivi scientifique des récifs artificiels Capbreton, Soustons/ Vieux-Boucau & Messanges/ Azur/ Moliets, 78 p (diffusion restreinte)

Baine M (2001) Artificial reefs: a review of their design, application, management and performance. Ocean Coast Manag 44(3):241–259

Bombacce G, Fabi G, Fiorentini L, Speranza S (1994) Analysis of the efficacy of artificial reefs located in five different areas of the Adriatic Sea. Bull Mar Sci 55(2,3):559–580

Bouchet JM (1984) Etude du recrutement des espèces fixées et vagiles sur un récif artificiel au large d'Arcachon. Actes Séminaire Déterminisme du Recrutement, I. S. T. P. M., Nantes, 2–4 juillet 1984, contr. no 40, 6 p

Carr MH (1991) Habitat selection and recruitment of an assemblage of temperate zone reef fishes. J Exper Mar Biol Ecol 146(1):113–137

Castege I, Milon E, Fourneau G, Tauzia A (2016) First results of fauna community structure and dynamics on two artificial reefs in the south of the Bay of Biscay (France). Estuar Coast Shelf Sci 179:172–180

Ceccaldi HJ, Nakagawa H (2002) Le progrès technologique au service de l'exploitation durable du littoral: 189–195. In: IV ème Conf. internat. « Droits de Propriété, Economie et Environnement: le Littoral », Aix-en-Provence, 26-28 juin 2002 Bruylant édit. Bruxelles

Charbonnel E, Bachet F (2010) Artificial Reefs in the Côte Bleue Marine Park: Assessement after 25 years of experiments and scientific monitoring: 73–79. In: Ceccaldi H-J et al (eds) Global change: mankind-marine environment interactions, Proceedings of the 13th French-Japanese oceanography symposium. Springer International Publishing, 339 p

Charbonnel E, Serre C, Ruitton S, Harmelin JG, Jensen A (2002) Effects of increase habitat complexity on fish assemblages associated with large artificial reef units (French Mediterranean coast). ICES J Mar Sci 59:S208–S213

DeWitt OM, Alspach GS, Parks W Jr, Speir C (1999) Comparative growth and mortality of American oysters, *Crassostrea virginica* on artificial reefs and natural substrates in the Chesapeake bay (Vi. USA). In: Proceedings of the 7th international congress on artificial reefs. San Remo, Italy, 7–11 Oct. 1999

Dufour F, Penne A, Ferrou N, Scourzic T, Rubio E, Laborde A, Fourneau G (2011) Artificial reefs development on Aquitaine coast. International Symposium Vulnerability of Coastal ecosystems to global change and extreme events (18–21 october 2011), Biarritz

FAO (2016) La situation mondiale des pêches et de l'aquaculture 2016. Contribuer à la sécurité alimentaire et à la nutrition de tous. Rome. 224 pages. ISBN 978-92-5-209185-1

Jackson JBC, Hughes TP, 16 co-authors (2001) Historical overfishing and the recent collapse of coastal ecosystems. Science 293:629–638

Sherman RL, Gilliam DS, Spieler RE (2002) Artificial reef design: void space, complexity, and attractants. ICES Journal of Marine Science 59:196–200

SHOM (1978) Carte de Biscarosse à. San Sébastian

Taniguchi H, Tokeshi M (2004) Effects of habitat complexity on benthic assemblages in a variable environment. Freshw Biol 49(9):1164–1178

Chapter 24
Marine Ecosystem Services: Perception of Residents from Remote Islands, Taketomi Town

Kazumi Wakita, Keiyu Kohama, Takako Masuda, Katsumi Yoshida, Taro Oishi, Zhonghua Shen, Nobuyuki Yagi, Hisahi Kurokura, Ken Furuya and Yasuwo Fukuyo

Abstract Marine ecosystem services provide various benefits to people. In order to receive those benefits sustainably, conservation of marine environment is an important measure, and how to motivate people to marine conservation would be one of the keys to secure sustainable receipt of marine ecosystem services. This study explores perception of marine ecosystem services by residents of remote islands, namely Taketomi Town in Japan and how the perception would influence their behavioural intentions for marine conservation. A questionnaire survey was administered to the residents, and factor analysis and Structural Equation Model were applied to analyse data from 344 respondents. The results show that respondents perceive marine ecosystem services in four categories, namely "Benefits closely related to daily lives", "Benefits from supporting services", "Benefits from regulating services", and "Benefits irrelevant to daily lives". Among the four categories, "Benefits from regulating services" is the most influential to enhance behavioural intentions for marine conservation. The perception of marine ecosystem services by respondents of Taketomi Town and their influence on

K. Wakita (✉)
School of Marine Science and Technology, Tokai University, 3-20-1 Orido, Shimizu-Ku, Shizuoka 424-8610, Japan
e-mail: kazumiw@tokai-u.jp

K. Kohama
Planning and Finance Division, Taketomi Town 11-1 Misaki-Cho, Ishigaki, Okinawa 907-0012, Japan

T. Masuda · Z. Shen · N. Yagi · H. Kurokura · K. Furuya · Y. Fukuyo
Graduate School of Agricultural and Life Sciences, The University of Tokyo, 1-1-1 Yayoi, Bunkyo-Ku, Tokyo 113-8567, Japan

K. Yoshida
Laboratory of Aquatic Science Consultant Co., LTD, 2-30-17 Higashi-Kamata, Ota-Ku, Tokyo 144-0031, Japan

T. Oishi
Faculty of Social and Environmental Studies, Fukuoka Institute of Technology, 3-30-1 Wajiro-Higashi, Higashi-Ku, Fukuoka 811-0295, Japan

T. Komatsu et al. (eds.), *Oceanography Challenges to Future Earth*,
https://doi.org/10.1007/978-3-030-00138-4_24

behavioural intentions for marine conservation are different from the results of previous studies administered to residents in the main island Honshu, Japan. This shows possibility that perception of marine ecosystem services and motivation for behavioural intention for marine conservation would relate to their connectednesspossibility to the sea.

Keywords Marine ecosystem services · Perception · Marine conservation · Connectedness

24.1 Introduction

24.1.1 Background

Marine ecosystem services provide various benefits to people. One of the most prevailing classifications of marine ecosystem services is the four categories provided by the Millennium Ecosystem Assessment (2003), namely Provisioning Services, Regulating Services, Cultural Services, and Supporting Services. However, how the value of ecosystem is viewed varies depending on cultural norms among others (Goulder and Kennedy 1997). Although number of studies have been conducted to analyse perception of marine ecosystem services by people (Blasiak et al. 2015; Wakita et al. 2014), they are still not enough to understand their variety. Wakita et al. (2014) provided an example of perception of marine ecosystem services by residents in the main island Honshu, Japan, however, analysis of difference of perception of marine ecosystem services according to lifestyle has not been explored. Hence, taking Taketomi Town as a distinguished site which is composed of multiple remote islands, this study aims at analysing perception of marine ecosystem services by remote islands' residents and their influences on behavioural intentions for marine conservation to compare with those by residents in the main island Honshu, Japan.

24.1.2 Study Site

Taketomi Town is the most southern located municipality of Okinawa prefecture in Japan, composed of 16 islands.[1] It has subtropical climate and is rich in both terrestrial and marine biodiversity (Naha Natural Environment Office of Ministry of Environment 2010). Not only whole terrestrial area of Irimote Island which is one of the 16 islands of Taketomi Town is designated as national park, but also a part of its coastal area is designated as the only natural environment protection area in the

[1]http://www.town.taketomi.lg.jp/town/index.php?content_id=8.

sea in Japan because of variety of coral reefs and endemic species.[2] Precious natural environment of Taketomi Town is one of the causes of many tourists visiting there. Taketomi Town attracts more than one million tourists per year from 2013 to 2015[3] and the number of annual tourists is over 250 times larger than residents' population: 4,161 in March 2016.[4]

24.2 Methods

A questionnaire survey regarding perception of marine ecosystem services and behavioural intentions for marine conservation was conducted from February 15 to March 8, 2013 with a letter from Taketomi Town asking for cooperation for the survey. Questionnaire items followed the items developed by Wakita et al. (2014) to examine differences between perception of marine ecosystem services by remote islands' residents and those by residents in the main island Honshu, Japan. Eighteen questionnaire items on perception of marine ecosystem services and five items regarding behavioural intentions for marine conservation were administered to the respondents (Table 24.1).

The questionnaires were distributed by community leaders to all homes in Taketomi Town, i.e. 2,186 family units face-to-face. Collection of responses was via postal mails. Data from the survey were analysed in two stages. At the first stage, a factor analysis was conducted using SPSS Statistics Ver. 21 to examine categorisation of marine ecosystem services by the respondents. Then, SPSS AMOS Ver. 20 was used to assess goodness-of-fit for a proposed Structural Equation Model which hypothesizes relationships between the perception of marine ecosystem services and behavioural intentions for marine conservation.

24.3 Results and Discussion

A total of 393 responses were collected, and after omission of incomplete responses, 344 responses were used for the analysis. By the factor analysis, four underlying factors of perception of marine ecosystem services were identified (Promax with Kaiser normalisation, principal component analysis as the extracting method). Variables were retained if their rotated loadings were above the threshold value of 0.4. The final results are shown in Table 24.2.

Table 24.3 shows results of reliability analysis to assess stability and consistency of the measured items of each latent construct.

---

[2]http://www.env.go.jp/press/100392.html.

[3]http://www.town.taketomi.lg.jp/town/index.php?content_id=53.

[4]http://www.town.taketomi.lg.jp/uploads/fckeditor/uid000005_20160405180559401fa423.pdf.

Table 24.1 Questionnaire items to measure perception of marine ecosystem services and behavioural intentions for marine conservation

Component	Abbreviation	Items
Provisioning services	P_{food}	Without foodstuffs like fish and seaweed provided by the sea, our diet would be extremely affected
	$P_{oranament}$	Without corals and beautiful shells, which can be used as ornaments, our lives would be extremely colorless
	P_{med}	Without marine resources, which can be utilized as medicine, our health would be extremely endangered in the future
	$P_{mineral}$	Without mineral resources such as cobalt and nickel in the seabed, high-tech industries would be extremely hampered
	P_{energy}	Without energy resources such as natural gas and methane hydrate in the seabed, supplies of energy would be severely limited
	P_{water}	Without water for human consumption and irrigation produced through desalination of seawater, our lives would be extremely inconvenient
Regulating services	R_{beach}	Without sandy beaches to reduce waves, we would be extremely vulnerable to high waves
	$R_{reef\text{-}mang}$	Without coral reefs and mangroves to calm waves, we would be extremely vulnerable to high waves
	R_{tidal}	Without clams and other sea creatures living in tidal flats to purify the water, water quality would experience severe deterioration
	R_{cd}	Without the sea to contribute to carbon dioxide absorption, there would be severe advancement of global warming
Cultural services	$C_{religion}$	Without the sea to be utilized for religious and traditional events, our culture would be extremely impoverished
	C_{rec}	Without recreational opportunities such as swimming, diving, and surfing, our recreation opportunities would be far less interesting
	C_{health}	Without opportunities to spend time by the sea, our health would be considerably worsened
	$C_{culture}$	Without the sea, our coastal cultures would be far less attractive and far more monotonous
	$C_{scenery}$	Without white sandy beaches, pine trees, and night views along the coasts, we would have far fewer opportunities to be moved by coastal scenery
Supporting services	S_{life}	Because the sea exists, life continues and nature is sustained
	S_{ncycle}	Because the sea exists, the nutrient cycle of the earth is well regulated and nature is sustained
	S_{place}	Without the sea, there would be no place for marine organisms to live, causing fatal damage to the earth

(continued)

Table 24.1 (continued)

Component	Abbreviation	Items
Behavioural intentions for marine conservation	MC_{tax}	I would accept a tax increase for marine conservation
	$MC_{donation}$	I would donate money for marine conservation
	$MC_{volunteer}$	I would volunteer for marine conservation
	MC_{supcom}	I would support companies that contribute to marine conservation
	$MC_{envgoods}$	I would purchase pro-environmental goods for marine conservation even at higher prices

Source Wakita et al. (2014)

Table 24.2 Results of factor analysis for the variable representing the indispensability of marine ecosystem services

Component	Variable	Factor 1	Factor 2	Factor 3	Factor 4
		Explained variance			
		35.78	9.99	4.93	4.13
		Rotated loadings			
Provisioning services	P_{food}	**0.49**	−0.08	0.18	−0.03
	$P_{ornament}$	**0.56**	−0.17	0.10	0.16
	$P_{mineral}$	−0.01	0.03	0.00	**0.98**
	P_{energy}	0.04	0.03	0.03	**0.77**
Regulating services	R_{beach}	0.09	−0.18	**0.67**	0.05
	$R_{reef\text{-}mang}$	−0.01	0.05	**0.62**	−0.02
	R_{tidal}	0.03	0.21	**0.60**	0.00
	R_{cd}	−0.06	0.17	**0.61**	0.02
Cultural services	$C_{religion}$	**0.49**	0.01	0.20	−0.05
	C_{rec}	**0.79**	−0.03	−0.20	0.10
	C_{health}	**0.72**	0.02	0.10	−0.08
	$C_{culture}$	**0.60**	0.25	0.02	−0.06
	$C_{scenery}$	**0.58**	0.24	−0.07	−0.01
Supporting services	S_{life}	0.03	**0.84**	−0.01	0.01
	S_{ncycle}	0.04	**0.93**	−0.08	0.02
	S_{place}	−0.11	**0.63**	0.18	0.03

All the Cronbach Alpha Coefficient of latent construct exceed Nunnally and Bernstein's (1994) recommendation of 0.70, which supports the use of these variables. All corrected item-total correlations surpasses 0.30 which is the threshold value for elimination of items (Parasuraman 1988), hence all variables are retained.

Based on the characteristics of observed variables of respective latent constructs, the authors named the four values as (i) Benefits closely related to daily lives, (ii) Benefits from supporting services, (iii) Benefits from regulating services, (iv) Benefits irrelevant to daily lives. This categorization is different from the three

Table 24.3 Reliability analysis of observed variables

Latent constructs and description of observed variables		Corrected items-total correlation	Alpha value (for item deletion)
Benefits closely related to daily lives			**0.83**
P_{food}	Food provision	0.51	0.81
$P_{ornament}$	Ornament provision	0.51	0.82
$C_{religion}$	Religious usage	0.55	0.81
C_{rec}	Recreation provision	0.60	0.80
C_{health}	Health provision	0.69	0.78
$C_{culture}$	Culture nurturing	0.66	0.80
$C_{scenery}$	Scenery provision	0.60	0.80
Benefits from supporting services			**0.85**
S_{life}	Life source	0.77	0.74
S_{ncycle}	Nutrient cycle	0.80	0.71
S_{place}	Living place for organisms	0.61	0.89
Benefits from regulating services			**0.75**
R_{beach}	Wave reduction by beach	0.52	0.72
$R_{reef\text{-}mang}$	Wave reduction by reefs and mangroves	0.53	0.70
R_{tidal}	Water purification by tidal flats	0.60	0.67
R_{cd}	Carbon dioxide absorption	0.58	0.67
Benefits irrelevant to daily lives			**0.88**
$P_{mineral}$	Mineral resources provision	0.79	–
P_{energy}	Energy resources provision	0.79	–
Behavioural intentions for marine conservation			**0.81**
MC_{tax}	Support level for tax increase for marine conservation	0.61	0.77
$MC_{donation}$	Support level for donation for marine conservation	0.66	0.75
$MC_{volunteer}$	Support level for volunteering for marine conservation	0.54	0.79
MC_{supcom}	Support level for private companies which contribute to marine conservation	0.55	0.79
$MC_{envgoods}$	Support level for buying goods which have a low-impact on marine environment	0.63	0.76

categorizations of marine ecosystem services perceived by residents in the main island Honshu, Japan, namely Indirect Benefits, Essential Benefits, and Cultural Benefits (Wakita et al. 2014). Including Taketomi Town, Okinawa is known that residents and sea have close relationships, i.e., they use inner reef area for their daily food (Kamimura and Kakuma 2012, http://ourworld.unu.edu/jp/satoumi-in-an-okinawan-coral-reef-system). Also, Taketomi Town has variety of sea-related cultural heritages and festivals (Taketomi Town 2011). At the same time, Taketomi Town is frequently hit by typhoon and natural disasters because of its location (Taketomi Town 2011). Considering these facts, the perception of marine ecosystem services by residents of Taketomi Town reflects their connectedness to the sea, which is no surprise of being greatly different from that by residents in the main island Honshu, Japan.

Figure 24.1 shows standardized estimate hypothetical model regarding relationships between perception of marine ecosystem services and behavioural intentions for marine conservation of the respondents. The goodness-of-fit index (GFI) and the adjusted goodness-of-fit (AGFI) are both acceptable at 0.886 and 0.852. The root mean square error of approximation (RMSEA) is 0.068 which is considered as good fit (Hooper et al. 2008). The explained variance was $R^2 = 0.41$. The strongest positive path coefficient (0.35) was calculated between "Benefits from regulating services" and behavioural intentions, which suggests that the islands' residents would place high importance on the function of the seas to protect their lives from natural disasters. This is also justified by the fact that only access to and from the islands of Taketomi Town is by ship, which would keep residents high

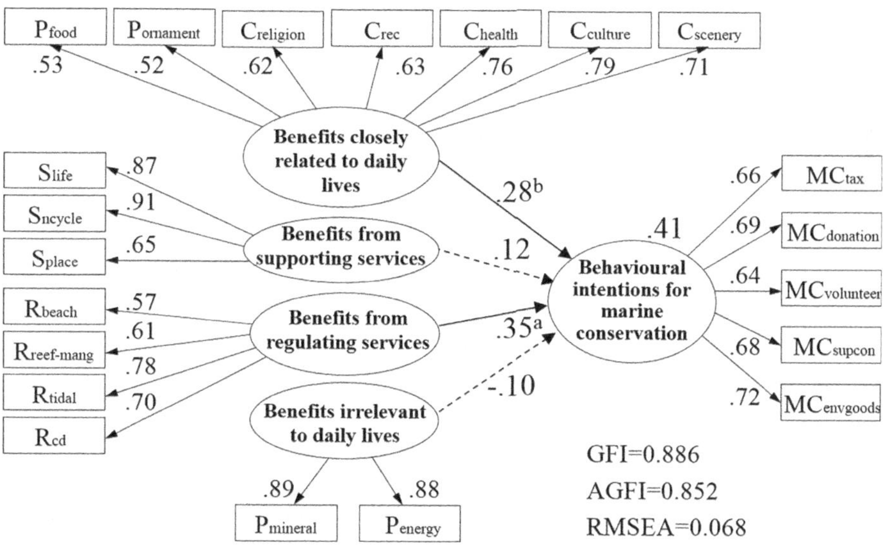

Fig. 24.1 Standardized estimate hypothetical model. **a** Significance at the 0.001 level. **b** Significance at the 0.01 level. Dashed line indicates path that is not significant at 0.05

awareness of natural disasters and daily condition of the sea. The above result of Taketomi Town is again different from the result of main island Honshu, Japan by Wakita et al. (2014) which shows strongest positive path coefficient between "Cultural Benefits" and behavioural intentions for marine conservation.

24.4 Conclusion

It was identified that respondents of islands' residents in Taketomi Town perceive marine ecosystem services in four categorized benefits, among which the most influential one for enhancing behavioural intentions for marine conservation is the "Benefits from regulating services". This suggests people's lifestyle and their connectedness to the sea would influence their value on marine ecosystem services and further influence behavioural intentions for marine conservation.

Acknowledgements The authors thank the Taketomi Town Government and community leaders for their support to the questionnaire survey. Appreciation goes to Nippon Foundation for financial support to this research through the Ocean Alliance Initiative Project, The University of Tokyo. This work was also supported by JSPS KAKENHI (Grant number 24121009).

References

Blasiak R et al (2015) Marine ecosystem services: perceptions of indispensability and pathways to engaging citizens in their sustainable use. Mar Policy 61:155–163

Goulder LH, Kennedy D (1997) Valuing ecosystem services: philosophical bases and empirical methods. In: Daily GC (ed) Nature's services: social dependence on natural ecosystems. Island Press, Washington, DC, pp 23–47

Hooper D et al (2008) Structural equation modelling: guidelines for determining model fit. Electron Bus Res Methods 26(1):53–60

Kamimura M, Kakuma S (2012) Conservation activities of coral reef ecosystems in Okinawa. United Nations University website. http://ourworld.unu.edu/jp/satoumi-in-an-okinawan-coral-reef-system. (In Japanese)

Millennium Ecosystem Assessment (2003) Ecosystems and human well-being: a framework for assessment. Island Press, Washington DC, p 245

Naha Natural Environment Office of Ministry of Environment (2010) Biodiversity of Amami-Ryukyu Islands. Naha natural environment office of ministry of environment, p 7. (In Japanese)

Nunnally JC, Bernstein IH (1994) Psychometric theory, 3rd edn. McGraw-Hill Education, New York, p 640

Parasuraman A, Zeithaml VA, Berry LL (1988) Servqual: a multiple-item scale for measuring consumer perceptions of service quality. J Retailing 64(1):12–40

Taketomi Town (2011) Taketomi Town Ocean Basic Plan, Taketomi Town, p 111. (In Japanese)

Wakita K et al (2014) Human utility of marine ecosystem services and behavioural intentions for marine conservation in Japan. Marine Policy 46:53–60

Chapter 25
Quantitative Mapping of Fish Habitat: From Knowledge to Spatialised Fishery Management

Sandrine Vaz and Olivier Le Pape

Abstract The delineation of essential fish habitats is necessary to identify, design and prioritize efficient marine protected area (MPA) networks with fishery objectives, capable, in addition to other possible objectives and functions of MPAs, of sustaining the renewal of marine living resources. Generally, the first step to obtain maps of essential fish habitats consists in choosing one of the numerous existing statistical approaches to build robust habitat suitability models linking relevant descriptors of the marine environment to the spatial distribution of fish presence or density. When these descriptors are exhaustively known, i.e. maps are available for each of them, geo-referenced predictions from these models and their related uncertainty may be imported into Geographic Information Systems for the quantitative identification and characterization of key sites for the marine living resources. The usefulness of such quantitative maps for management purposes is endless. These maps allow for the quantitative identification of the different habitats that are required for these marine resources to complete their life cycles and enable to measure their respective importance for population renewal and conservation. The consequences of anthropogenic pressures, not only fishing but also land reclamation, aggregate extractions or degradation of habitat quality (e.g. nutrient excess or xenobiotics loadings, invasive species or global change), on living resources, may also be simulated from such habitat models. These quantitative maps may serve as input in specific spatial planning software or to spatialise population or fishery dynamics, ecosystem or trophic models that may then be used

S. Vaz (✉)
MARBEC, Univ Montpellier, CNRS, Ifremer, IRD, Avenue Jean Monnet, 34203 Sète Cedex, France
e-mail: sandrine.vaz@ifremer.fr

O. Le Pape
AGROCAMPUS OUEST, UMR985 ESE Ecologie et Santé Des écosystèmes, 35042 Rennes, France
e-mail: olivier.le.pape@agrocampus-ouest.fr

T. Komatsu et al. (eds.), *Oceanography Challenges to Future Earth*,
https://doi.org/10.1007/978-3-030-00138-4_25

to simulate various scenarios. Fish habitat maps thus may help decision makers to select relevant protection areas and design coherent MPA networks and management levels which objectives are to sustain fishing resources and fisheries.

Keywords Habitat models · Fishery management

25.1 Introduction

Many populations, particularly of fishes, are only remnants of their original numbers due to direct or indirect human pressure. We believe that there are two major causes for living marine resources decline: overexploitation by fishery and degradation of habitats essential to the renewal of living resources. Such degradation may occur through two main processes: habitat loss (through artificialisation of coastal habitats, offshore structure installation, aggregate extraction, etc.) and decrease in habitat quality (through for example eutrophication or chemical contamination of water and sediments). The protection of living resources aiming to make their exploitation sustainable requires taking both these aspects into account.

Spatial regulations, including Marine Protected Areas (MPAs) are now increasingly introduced to regulate these perturbations (Leathwick et al. 2008; Sale et al. 2005; Spalding et al. 2008; Wood et al. 2008). Within the framework of an ecosystem approach (Brownman et al. 2004; de Jonge et al. 2012), the European Commission (EC) is promoting the concept of marine spatial planning to improve the management of marine activities. Accordingly, for management objectives to be efficiently achieved, information from habitat maps needs to represent habitats at a spatial scale relevant at the ecosystem and fish population level (Le Pape et al. 2003; Planque et al. 2011), to allow accurate estimates to be made of the potential effects of pressure and protection.

Marine fish populations are not randomly distributed, but exhibit distributions that are structured both in space and time (Mello and Rose 2005). A species distribution results from the combined action of several forces (Planque et al. 2011), some of which are external (such as environmental conditions or food availability), whereas others are internal to the considered species, population or community (such as total population size). The set of conditions required for individual survival and reproduction constitute the "ecological niche" within which a species may indefinitely sustain itself. The geographical projection of this fundamental niche corresponds to the habitat of the considered species (Chase and Leibold 2003). The analysis of relationships between species and their habitats has always been a central issue in ecology and is used to investigate the role of the different factors that may affect a population and to characterize the mechanisms determining habitat suitability. For numerous marine organisms, habitat requirements change over the course of their development (Harden 1968), resulting in distinct distributions along the life cycle; the displacement between these zones being ensured by passive or active migration (Grüss et al. 2011a, b). Ontogenic habitat switching structures the spatial distribution of organisms from each phase to the entirety of their life cycles (van de Wolfshaar et al. 2011).

The successive steps of the general and most widespread approach to build quantitative maps of essential habitats are described here first. Then we focus on the usefulness of these maps to investigate and delineate fish habitat, to compare the respective importance of different habitats on population renewal and to estimate and/or to simulate the consequences of anthropogenic pressures on living resources. The use of these maps and quantitative information to prioritize protection areas and improve fisheries management systems is finally highlighted.

25.2 Developing Quantitative Mapping

25.2.1 Identifying Factors

Identifying factors that condition the spatial distribution of a given species represents the core of predictive geographical modelling in ecology. Habitat modelling (modelling species distribution) and Geographic Information Systems (GIS) are key tools that lead to a better understanding of species-environment relationships.

Habitat occupation results from the influence of a large number of constraints. These constraints may be of very different nature, and are linked either to the individuals' external environment, or to the individuals themselves, whether they belong to the studied population/species or other ones (Ashcroft et al. 2010; Grüss et al. 2011a; Planque et al. 2011). In order to identify factors controlling species spatial distribution at any given life stages, the response of a population to different control hypotheses, which can be external (geographic location, environment, associated fauna) or internal to the population (spatial autocorrelation, population size, age structure), must be studied.

Both abiotic (generally non-consumable habitat features) and biotic factors, including both consumable resources and non-consumable features, are potential driving environmental factors of habitat choice and species distribution (Hayes et al. 1996) at different scales (Meng et al. 2005; Grüss et al. 2011a). Therefore, environmental variables retained as potential descriptors in HS models must be selected in relation to these species and any life stage-specific requirements (Harden 1968; Le Pape et al. 2003; Loots et al. 2010). However the final aim is to develop HS maps by using models in a GIS to predict HS for a combination of habitat layers (Rubec et al. 1999). To achieve this goal, there is a need for an exhaustive knowledge of retained descriptors, i.e. obtaining static and/or temporal maps of the factors used in HS models. Therefore, the selection of environmental descriptors is a compromise between potential driving factors and available data. Exhaustive fine scale knowledge about relevant factors, in particular biotic factors, is presently seldom available and habitat mapping is predominantly based on abiotic factors only. With regard to available knowledge and data, HS models are often only based on hypotheses on how external environmental factors control species distribution and/or life stages (Guisan and Zimmermann 2000), although their capability to explain the observed distribution may sometimes be limited when they are used alone.

25.2.2 Statistical Modelling and Geographic Projection of Species Habitat

For the purpose of statistical modelling of species habitat, methods aim to define an HS for a species by providing a numerical estimate of species response (e.g. in terms of abundance and/or presence probability) to changes in one or more factor(s). There are numerous statistical modelling techniques to predict species distribution and several methods may be used to reach this objective. The first criterion driving the choice of an HS model is related to fish data. Indeed, fish occurrence and fish density are described by different methods in habitat models. A wide array of models has been developed to describe and predict habitat distribution, and the variety of available statistical techniques is growing (Guisan and Zimmermann 2000; Guisan and Thuiller 2005; Austin 2007; Guisan et al. 2006; Le Pape et al. 2014): Habitat modelling involves three steps: (i) choosing the type of model that is best adapted to the data, (ii) selecting the model, (iii) evaluating its adjustment, predictive capacity and uncertainty. This vast topic will not be further developed here.

To develop predictive habitat maps, the HS model parameter estimates are used to re-code (Eastwood et al. 2001) the environmental layers so as to produce maps reflecting HS. The same method could be used to produce maps of uncertainty in order to provide an overview of the prediction quality.

25.3 Predicting the Spatial Distribution of Marine Species and Their Dependence to Essential Habitats

25.3.1 Atlases

The vulnerability of living resources and their habitats often leads to create atlas of marine resource habitats in particular marine region so as to provide planners and decision-makers with the necessary information to help managing the use of its living and non-living resources. Atlases of fish spatial distributions and modelled habitats, may be used to investigate ontogenic and seasonal shifts in fish spatial distribution and habitat and may contribute to a better understanding of this species spatial ecology (Carpentier et al. 2009, Martin et al. 2009).

25.3.2 Essential Habitat Maps

If a policy ensuring sustainable use of marine resources is to be implemented in any given marine region, the knowledge on the location and functionality of essential fish habitats (such as spawning and nursery grounds) is a critical step for selecting a combination of sites with good spatial and ecological coherence.

The functional value of spawning habitats makes them critically important for the completion of fish life cycles. Inter-annual fluctuations in spawning ground distributions of fish can also be investigated using habitat modeling approach where egg spatial and temporal distributions are explored based on several environmental variables (Lelievre et al. 2014). This approach resulted in high resolution maps of spawning habitats, and new insights on their utilisation by spawners. Such study contributes the knowledge necessary to the spatial management of fisheries resources, and may also be used to identify marine areas with particular habitat features that need to be preserved.

25.3.3 Invasive Species

Over recent decades, invasive species have become a major ecological concern worldwide due to increased abundance in their native habitats that affects the local ecosystems (Mills 2001) and potential spreading to other areas either by natural expansions or by human activities. The invasive ctenophore *Mnemiopsis leidyi*, well-known for its dramatic disturbance of ecosystems, has been reported in various coastal and offshore locations in the southern North Sea in the past years and it became crucial to understand its distribution dynamics (David et al. 2015). Two modelling methods, habitat model and a particle tracking model, were used in conjunction (i) to identify habitats where the invasive ctenophore *M. leidyi* could survive the North Sea cold winters and (ii) to investigate the dispersal of individuals between these different habitats, emphasizing favorable areas where sustainable populations could have been established. This study, based on the agreement of habitat and dispersal models results, showed that *M. leidyi* has become established in areas of the North Sea where the environment conditions and retention processes allows overwintering and later blooms.

25.4 Analyzing Pressure Impacts (Hindcast or Forecast)

Habitat models have been used in different backward or forward predictive approaches to estimate quantitatively the past or future consequences of anthropogenic pressures on fish population renewal and related fisheries.

25.4.1 Habitat Loss

Rochette et al. (2010) developed quantitative maps of common sole nursery habitats in the English Channel by using a model based on bathymetry and sediment structure. Then historical maps of the highly degraded Seine estuary were used to build habitat maps of this estuary from 1850. This backward predictive approach showed that habitat loss in the Seine estuary has led to a more than 40% decrease in its nursery capacity. The loss related to both land reclamation and degradation of residual habitat quality was assessed at the Eastern channel population scale and for the related fisheries, at nearly 17% (15–32%). On the basis of these estimates, Cordier et al. (2011) addressed the fishery-related macroeconomic impact of this loss and that of the possible restoration of these nursery areas.

25.4.2 Climate Change

HS maps delineate geographic areas within which ranges of environmental factors define the presence or high abundance of a particular species or species assemblage. In this context, climate scenarios are of great interest to report correlations between the past or current geographical distribution of a species, as are some climate variables to extrapolate a future distribution. Such studies (e.g. Murrawski 1993; Heikkinen et al. 2006; ter Hofstede et al. 2010) generally assume that species distributions mirror climatic limitations. Alternatively, species may respond to shifting climate conditions by a shift in their realized niches and modelling strategies for predicting the potential impact of the natural system present limitations and should be used with caution. Nevertheless, the bioclimate envelope approach can provide a useful first approximation of the impact of climate change on biodiversity (Cheung et al. 2009) and on fisheries catch (Cheung et al. 2010). Other approaches combined basic regional climate change scenarios with spawning stock size scenarios to illustrate the relative impact of each component on the projected change on regional fish stock distributions (Le Pape et al. 2014).

25.5 Helping Decision Makers for Management

25.5.1 MPA Network Design

MPA networks have been recognized as an efficient spatial management tool for conserving marine biodiversity (Leathwick et al. 2008) but also for managing fishing resources and fisheries. HS maps are very powerful biological inputs in such processes (Elith and Leathwick 2009; de Jonge et al. 2012; Le Pape et al. 2014). The initial approach, compiling HS maps for different species and life stages, was to combine layers into aggregated HS indices (Rubec et al. 1999; Brown et al. 2000; Store and Jokimäki 2003; Hattab et al. 2013). This initial approach was progressively discarded as systematic conservation planning software became available (Margules and Pressey 2000). These tools integrating ecological as well as socioeconomic aspects may be based on HS maps (Delavenne et al. 2012) and use an spatial optimisation algorithm to identify priority areas for conserving biodiversity and ecosystem services (including fisheries), reducing fragmentation levels and minimizing socio-economic costs.

25.5.2 Marine Fisheries Spatial Planning

When applied to fisheries, systematic conservation planning fails to account for (i) changes in fleet dynamics induced by new conservation constraints and their associated feed-backs on conservation costs, or (ii) their influence on fish population dynamics and distributions, which may in turn alter the achievement of conservation targets. Such a static approach may therefore lead to short- or medium-term misestimates in forecasted costs and target achievements. In order to circumvent such limitations of systematic conservation planning we present a first attempt to couple a conservation planning tool (Marxan with Zones—MwZ) with a mixed-fisheries dynamics simulation model (ISIS-Fish) (Reecht et al. 2015). MwZ was used to provide near optimal management scenarios within the already planned MPA network. It highlighted combinations of different spatial management strategies, such as no-take and limited-take MPAs. The ISIS-Fish simulations enabled the evaluation of the management strategy under which the fishing fleets and fish stock will remain viable. Such model coupling allow testing what the dynamics of the main fleets and fish populations would be under different conservation scenarios, and how they would influence targets achievement. Similarly it becomes possible to test whether any proposed MPA network with a particular set of conservation measures is suitable to ensure medium to long-term viability of fleets and of the fish populations they harvest.

25.5.3 Marine Fishery Impact on Ecosystem

Similarly, it is difficult to determine the impacts that proposed MPAs could have on marine ecosystem health, fisheries and fisheries sustainability. Metcalfe et al. (2015) used MwZ to identify a series of MPA networks designed and coupled them with a spatially explicit ecosystem model developed in Ecopath with Ecosim. Using these existing software tools in combination provides a powerful policy-screening approach that could help inform marine spatial planning by identifying potential conflicts and designing new regulations that better balance conservation objectives and stakeholder interests. Such approach might be as effective as and less politically contentious than a network of no-take MPAs.

25.6 Conclusion

Some habitats are essentials to marine living resources to enable the completion of their life cycle and the renewal and sustainability of populations. However, habitat function is often neglected by fishery managers and as a consequence, habitat protection is insufficient. MPA network designation is a new opportunity for fishery management (protection of spawning ground or nursery of exploited species, spatialised management of fishery effort). Fish habitat maps are essential tools of knowledge to help decision in spatial management. These models and maps provide useful tools for managers faced with the multiple demands of resource users when considering strategies for conservation, resource management and spatial planning. Models and maps can be used to help decision making in bio-regionalization, conservation schemes and spatial management of marine systems, to prioritize protection areas. A large panel of data and methods is available and increasingly easy to implement to describe quantitatively the spatiotemporal distribution of marine ichthy ofauna and their essential habitats. More than a spatio-temporal representation, these approaches allow for simulating the consequences of disturbances and changes, based on realistic scenarios designed to assess the influence of the evolution of environmental descriptors, with regards to global change or management measures. They can also help defining the spatial structure of trophic or end-to-end complex models or be transferred into systematic conservation planning to allow the identification of both protection area and management strategies. Furthermore, they can be coupled to dynamic models (particle tracking, fishery or trophic network) to simulate impact of fish habitat change on other ecosystem components in the longer run.

Acknowledgements We wish to acknowledge the many colleagues who were involved in fruitful collaborations and helpful discussions and who generated many of the thoughts and results cited here.

References

Ashcroft M, French K, Chisholm L (2010) An evaluation of environmental factors affecting species distributions. Ecol Model 222(3):524–531

Austin MP (2007) Species distribution models and ecological theory: a critical assessment and some possible new approaches. Ecol Model 200:1–19

Brown K, Buja K, Jury S, Monaco M, Banner A (2000) Habitat suitability index models for eight fish and invertebrate species in Casco and sheepscot bays, Maine. N Am J Fish Manag 20:408–435

Brownman H, Stergiou K, Cury P, Hilborn R, Jennings S, Lotze H, Mace P, Murawski S, Pauly D, Sissenwine M, Stergiou KI, Zeller D (2004) Perspectives on ecosystem based approaches to the management of marine resources. Mar Ecol Prog Ser 274:269–303

Carpentier A, Martin CS, Vaz S (eds) (2009) Atlas des Habitats des resources Marines de la Manche Orientale—CHARM II/Channel Habitat Atlas for marine Resource Management—CHARM II. http://archimer.ifremer.fr/doc/00000/7377/

Chase JM, Leibold M (2003) Ecological niches: linking classical and contemporary approaches. University of Chicago Press, Chicago and London

Cheung W, Lam V, Sarmiento J, Kearney K, Watson R, Pauly D (2009) Projecting global marine biodiversity impacts under climate change scenarios. Fish Fish 10:235–251

Cheung WL, Lam VWY, Sarmiento JL, Kearney K, Watson R, Zeller D, Pauly D (2010) Large-scale redistribution of maximum fisheries catch potential in the global ocean under climate change. Glob Change Biol 16:24–35

Cordier M, Pérez Agúndez JA, O'Connor M, Rochette S, Hecq W (2011) Quantification of interdependencies between economic systems and ecosystem services: an input-output model applied to the Seine estuary. Ecol Econ 70(9):1660–1671

David C, Vaz S, Loots C, Antajan E, Van Der Molen J, Travers-Trolet M (2015) Understanding winter distribution and transport pathways of the invasive ctenophore *Mnemiopsis leidyi* in the North Sea: coupling habitat and dispersal modelling approaches. Biol Invasions 17(9):2605–2619

De Jonge VN, Pinto R, Turner RK (2012) Integrating ecological, economic and social aspects to generate useful management information under the EU Directives' 'ecosystem approach'. Ocean Coast Manag 68:169–188

Delavenne J, Metcalfe K, Smith RJ, Vaz S, Martin CS, Dupuis L, Coppin F, Carpentier A (2012) Systematic conservation planning in the eastern English Channel: comparing the Marxan and Zonation decision-support tools. ICES J Mar Sci 69:75–83

Eastwood PD, Meaden GJ, Grioche A (2001) Modelling spatial variations in spawning habitat suitability for the sole Solea solea using regression quantiles and GIS procedures. Mar Ecol Prog Ser 224:251–266

Elith J, Leathwick J (2009) The contribution of species distribution modelling to conservation prioritization. In: Moilanen A, Wilson KA, Possignham H (eds) Quantitative methods and computational tools. University Press, Oxford, UK, pp 196–210

Grüss A, Kaplan D, Guénette S, Roberts CM, Botsford L (2011a) Consequences of adult and juvenile movement for marine protected areas. Biol Conserv 144:692–702

Grüss A, Kaplan DM, Hart DR (2011b) Relative impacts of adult movement, larval dispersal and harvester movement on the effectiveness of reserve networks. PLOS One 6(5):e19960

Guisan A, Zimmermann NE (2000) Predictive habitat distribution models in ecology. Ecol Model 135:147–186

Guisan A, Thuiller W (2005) Predicting species distribution: offering more than simple habitat models. Ecol Lett 8:993–1009

Guisan A, Lehman A, Ferrier S, Austin M, Overton J, Aspinall R, Hastie T (2006) Making better biogeographical predictions of species distribution. J Appl Ecol 43:386–392

Harden J (1968) Fish migration. Edward Arnold, London, UK

Hayes DB, Ferreri CP, Taylor WW (1996) Linking fish habitat to their recruitment dynamics. Can J Fish Aquat Sci 53(1):383–390

Hattab T, Ben RaisLasram F, Albouy C, Sammari C, Romdhane MS, Cury P, Leprieur F, Le Loc'h F (2013) The use of a predictive habitat model and a fuzzy logic approach for marine management and planning. PLoS ONE 8(10):e76430. https://doi.org/10.1371/journal.pone.0076430

Heikkinen RK, Luoto M, Araujo MB, Virkkala R, Thuiller W, Sykes MT (2006) Methods and uncertainties in bioclimatic envelope modelling under climate change. Prog Phys Geog 30:751–777

Le Pape O, Chauvet F, Mahévas S, Lazure L, Guérault G, Désaunay Y (2003) Quantitative description of habitat suitability for the juvenile common sole (Soleasolea, L.) and contribution of different habitats to the adult population in the Bay of Biscay (France). J Sea Res 50(2–3):139–149

Le Pape O, Delavenne J, Vaz S (2014) Quantitative mapping of fish habitat: a useful tool to design spatialised management measures and Marine Protected Area with fishery objectives. Ocean Coast Manag 87:8–19

Leathwick J, Moilanen A, Francis M, Elith J, Taylor P, Julian K, Hastie T (2008) Novel methods for the design and evaluation of marine protected areas in offshore waters. Conserv Lett 1:91–102

Lelievre S, Vaz S, Martin CS, Loots C (2014) Delineating recurrent fish spawning habitats in the North Sea. J Sea Res 91:1–14

Loots C, Vaz S, Planque B, Koubbi P (2010) What controls the spatial distribution of North Sea plaice spawning population? Confronting ecological hypotheses through a model selection framework. ICES J Mar Sci 67:244–257

Margules CR, Pressey RL (2000) Systematic conservation planning. Nature 405:243–253

Martin CS, Carpentier A, Vaz S, Coppin F, Curet L, Dauvin JC, Delavenne J, Dewarumez JM, Dupuis L, Engelhard G, Ernande B, Foveau A, Garcia C, Gardel L, Harrop S, Just R, Koubbi P, Lauria V, Meaden GJ, Morin J, Ota Y, Rostiaux E, Smith R, Spilmont N, Vérin Y, Villanueva C, Warembourg C (2009) The Channel habitat atlas for marine resource management (CHARM): an aid for planning and decision-making in an area under strong anthropogenic pressure. Aquat Living Resour 22:499–508

Mello LGS, Rose GA (2005) Using geostatistics to quantify seasonal distribution and aggregation patterns of fishes: an example of Atlantic cod (*Gadusmorhua*). Can J Fish Aquat Sci 62:659–670

Meng L, Cicchetti G, Raciti S (2005) Relationships between juvenile winter flounder and multiple scale habitat variation in Narraganset Bay, Rhode Island. Trans Am Fish 134:1509–1519

Metcalfe K, Vaz S, Engelhard GH, Villanueva CM, Smith RJ, Mackinson S (2015) Evaluating conservation and fisheries management strategies by linking spatial prioritization software and ecosystem and fisheries modelling tools. J Appl Ecol 52(3):665–674

Mills CE (2001) Chapter 4 C Pelagic Cnidaria and Ctenophora. pp 23–38. In: Hines AH, Ruiz GM (eds) Marine invasive species and biodiversity of south central Alaska. Final project report, submitted to the regional citizens' advisory council of Prince William Sound, Anchorage, Alaska 74 p

Murawski SA (1993) Climate change and marine fish distributions: forecasting from analogy. Trans Am Fish Soc 122:647–658

Planque B, Loots C, Petitgas P, Lindstrom U, Vaz S (2011) Understanding what controls the spatial distribution of fish populations using a multi-model approach. Fish Oceanogr 20(1):1–17

Reecht Y, Gasche L, Lehuta S, Vaz S, Smith RJ, Mahévas S, Marchal P (2015) Toward a Dynamical approach for systematic conservation planning of Eastern English Channel fisheries. In: Marine productivity: perturbations and resilience of socio-ecosystems, pp 175–185. Springer International Publishing

Rochette S, Rivot E, Morin J, Mackinson S, Riou P, Le Pape O (2010) Effect of nursery habitat destruction on flatfish population renewal. Application to common sole (Soleasolea, L.) in the Eastern Channel (Western Europe). J Sea Res 64:34–44

Rubec PJ, Bexley JCW, Norris H, Coyne MS, Monaco ME, Smith SG, Ault JS (1999) Suitability modeling to delineate habitat essential to sustainable fisheries. Am Fish Soc Symp 22:108–133

Sale PF, Cowen RK, Danilowicz BS, Jones GP, Kritzer JP, Lindeman KC, Planes S, Polunin NV, Russ GR, Sadovy YJ (2005) Critical science gaps impede use of no-take fishery reserves. Trends Ecol Evol 20:74–80

Spalding MD, Fish L, Wood LJ (2008) Toward representative protection of the world's coasts and oceans—progress, gaps, and opportunities. Conserv Lett 1(5):217–226

Store R, Jokimäki J (2003) A Gis-based multi-scale approach to habitat suitability modeling. Ecol Model 169:1–15

Ter Hofstede R, Hiddink JG, Rijnsdorp AD (2010) Regional warming changes fish species richness in the eastern North Atlantic Ocean. Mar Ecol Prog Ser 414:1–9

Van de Wolfshaar K, HilleRisLambers R, Gardmark A (2011) Effect of habitat productivity and exploitation on populations with complex life cycles. Mar Ecol Prog Ser 438:175–184

Vaz S, Vermard Y, Gardel L (2011) Predicting species distributions: the impact of exploitation and climate change (on plaice). World Conference on Marine Biodiversity, Aberdeen, 26–30 Sep 2011

Wood LJ, Fish L, Laughren J, Pauly D (2008) Assessing progress towards global marine protection targets: shortfalls in information and action. Oryx 42(03):340–351

Chapter 26
Do Our Ocean Policies Make Any Difference in the Wellbeing of Coastal Communities?

Yves Henocque

Abstract Like many other countries, France and Japan now have their own ocean policy, though at different stage of development and in quite different context. On the European side, buzz words like 'Blue Growth', 'Maritime Spatial Planning', and others, are on the forefront and could make us feel that ocean policies are primarily focused beyond the coast, in offshore waters and their corresponding human activities, somewhat leaving coastal communities in the back seat. Through case studies, we will try to show that ocean policies should be coast-to-coast, across oceans, regional seas, or local well delineated water body, never forgetting that, beyond 'Blue growth', we should be heading towards a 'Blue society'.

Keywords Ocean policies · Ecosystemix approach · Blue growth · Blue society · Coastal communities

26.1 Ocean Policies Are Integrated, Ecosystem-Based and Co-implemented

Since coastal areas were first included in public policy, the number of concepts has multiplied; in the era of globalization, where the sea provides for all possible interconnections: human and non-human, universal and political, natural history interlinked with social histories, ebb and flow, ecology and economy, from sovereignty to world governance, sanctuary and network, enjoyment to catastrophe, etc. Oceans and coastal areas are hotspots of global phenomena and their consequences (climate change, bioinvasion, waste, pollution, piracy, migration, etc.). The answers to these problems are applied locally, but must be reflected on globally, thus requiring shared governance. This presupposes the coordination of state, interstate and supra-state actors, as well as cooperation with different actors in civil society. The growing awareness of the global issues, and the role of the seas and

Y. Henocque (✉)
Ifremer/JAMSTEC, Yokosuka, Japan
e-mail: yves.henocque@ifremer.fr

T. Komatsu et al. (eds.), *Oceanography Challenges to Future Earth*,
https://doi.org/10.1007/978-3-030-00138-4_26

shorelines, is rather recent and for the first time was high on the agenda of the Rio + 20 summit.

In the Rio + 20 outcome document, *The Future We Want* (which could have been called *The Ocean We Want*), oceans and their role in planetary survival and human wellbeing receive a central attention. Paragraph 158 of the Rio + 20 outcome documents highlights the importance of integrated, ecosystem-based ocean governance:

> We recognize that oceans, seas and coastal areas form an integrated and essential component of the Earth's ecosystem and are critical to sustaining it and that international law, as reflected in United Nations Convention on the Law of the Sea (UNCLOS), provides the legal framework for the conservation and the sustainable use of the oceans and their resources. We stress the importance of the conservation and sustainable use of the oceans and seas and of their resources for sustainable development, including through the contributions to poverty eradication, sustained economic growth, food security, creation of sustainable livelihoods and decent work, while at the same time protecting biodiversity and the marine environment and addressing the impacts of climate change. We therefore commit to protect, and restore, the health, productivity and resilience of oceans and marine ecosystems, and to maintain their biodiversity, enabling their conservation and sustainable use for present and future generations, and to effectively apply an ecosystem approach and the precautionary approach in the management, in accordance with international law, of activities impacting on the marine environment, to deliver on all three dimensions of sustainable development.

26.2 National Integrated Ocean Policies and Regional Sea Strategies

The principles and objectives as stated above underpin the national ocean policieslike there are now in many countries that have, or are taking concrete steps toward, cross-cutting and integrated national ocean policies (Cicin Sain et al. 2016).

In Japan, the "Bill for the Basic Ocean Law" took effect in July 2007. The contents of this new Law "define the basic principles of Japan on the ocean, clarify responsibilities of the national government, local governments, business operators, and citizens, specify the basic items concerning measures on the ocean, and stipulate the establishment of the Headquarters for Comprehensive Ocean Policy for the purpose of promoting these measures in a comprehensive and systematic manner, aiming at realizing a new ocean-oriented nation" (Kuribayashi 2008). As the basis of the Basic Ocean Plan, 12 basic measures have already been agreed upon. They include the promotion of development and conservation of the marine environment within the EEZ and other areas, securing the safety and security of the oceans including maritime transport, promotion of ocean science and technology, ocean industries and their international competitiveness, integrated coastal management and enhancement of citizen's understanding of the oceans, conservation of remote islands, and international cooperation.

Five years later (2013) and with the Great East japan Earthquake disaster in between, the Basic Plan on Ocean Policy was reviewed amid, among others, the ongoing national debate on future energies including marine renewable energies, exploration and future exploitation of deep sea mineral resources, and maritime safety and security. The new Basic Plan on Ocean Policy cites four targets for the stance on future efforts and the direction to be pursued: (1) contributing to international cooperation and the global community; (2) gaining wealth and prosperity by developing and utilizing the oceans; (3) changing from "a country protected by the sea" to "a country that protects the sea", and, 4) challenging unexplored frontiers (Yamamoto 2014).

In France, the recent national move (2009) for the management of the coast and ocean has been largely influenced by the new European Integrated Maritime Policy and its Marine Strategy Framework Directive (2008). It was initiated within the framework of the "Grenelle Environment" and then "Grenelle of the Sea" national consultations ending up with the publication of a Maritime Policy Blue Book giving the outlines of the national maritime strategy related to knowledge enhancement, governance arrangement, maritime activities development, education and training, and international commitments in metropolitan France and its overseas territories.

A first Basic Act (LoiGrenelle 1) enacted in 2009 was complemented in 2010 by a second more detailed one (LoiGrenelle 2) setting out the institutions and governance mechanism for implementation of the national maritime strategy. Although France does not yet have a full-fledged ocean policy as in the case of Japan, it started to actively create the conditions of its future development, more particularly through the national application of the EU Marine Strategy Framework Directive. From the Channel to the Mediterranean, consultative multi-stakeholders councils have been set up for each of the four 'ecoregions' designated within the Exclusive Economic Zone (EEZ). These consultative bodies are in charge of developing a coastal and ocean strategy and implementation plan for each of the four ecoregions. The ongoing development of 'programme of measures' to attain the 'Good Environmental Status' of the Directive, will represent the environmental pillar of each regional strategy.

In these respective national contexts, both countries have numerous ongoing local ICM-related initiatives, including Marine Protected Areas (MPAs), taking place in specific regional and local context. The question is therefore how to progressively build up from national to local and vice versa an adequate, viable and well supported governance and management process that can help each nation to consolidate their national framework hence promoting their ocean-state position in their respective maritime region and in the international arena.

Looking at both national maritime policies (though a formalized French maritime strategy has still to be drafted and enacted), respective goals and objectives look somewhat similar (Table 26.1) except the specifics of the French overseas territories, which are quite strategic since their EEZ represents around 97% of the French entire EEZ.

Table 26.1 National maritime policies France/Japan

France (Blue book on ocean policy)	Japan (Basic plan on ocean policy)
Invest into the Future 1. Generate the people's passion for the sea 2. Improved knowledge for improved management 3. Maritime education and professional training 4. Protect the coastal and marine environment 5. Develop coastal and ocean monitoring	Enhance the Knowledge of the Sea 1. Promotion of marine surveys 2. Promotion of marine science and technology R&D 3. Enhance citizens' understanding of the sea and foster human resources
Develop a maritime sustainable economy 1. Natural resources sustainable development 2. Sustainable fisheries and aquaculture development 3. Innovating and competitive shipbuilding industry 4. Rethink maritime transport 5. Develop ports international dimension 6. Strategy for leisure boats and aquatic sports development	Harmonize sea development and environmental protection 1. Promote development and use of marine resources 2. Preserve marine and coastal environment Sound development of maritime industries 1. Secure maritime transport 2. Promote maritime industries & strengthen international competitiveness
Promote maritime France in Europe and the world 1. France and international governance development 2. France and EU integrated maritime policy building up 3. Maintain sovereignty and fulfil responsibilities 4. Strengthen France's intervention capacity for defence and security	International partnership with regard to the sea 1. Secure international coordination and promote international cooperation
Set up a renewed governance 1. Coastal and ocean governance setting and policy instruments development 2. Foster the state operative capacity at sea 3. Foster operative efficiency at international level	Comprehensive governance of the sea 1. Promote development in EEZ and continental shelves 2. Comprehensive management of the coastal zones 3. Preserve the islands. Secure safety and security at sea 1. Secure safety and security at sea
Promote the French overseas territories' maritime dimension 1. Territorial governments, developers of the national policy 2. Marine environment: an asset as well as responsibilities 3. Marine resources as one of the key economic sectors	

26.3 The Millennium Ecosystem Assessment: A Common Inheritance

The 2005 Millennium Ecosystem Assessment (MEA) emphasises that 60% of ecosystem services are deteriorating. Among these, the renewal of fishery stocks and the production of freshwater seem to be the most threatened. This erosion has been more substantial during the last fifty years than in all of human history, and it will be even more substantial in the next fifty years hence the label of 'anthropocene' era.

Based on its assessments, the MEA has developed a table showing the dangers expected over the next hundred years in the form of four scenarios constructed using both the pooled opinions of experts on the "possible futures" of ecosystems, ecological services and human well-being, and global models which include the principal forces for change that impact ecosystem services, i.e. mainly habitat change (changes in land use, physical alteration of rivers or extraction of water from rivers); over-exploitation; invasive species; pollution; climate change.

Among these scenarios, one is very close to the current 'globalization' process, which envisages an increase in the liberalisation of trade, as well as stronger global interconnections and the emergence of a world governance (UN organisations). The approach to the management of environmental crises is still a reactive one. This scenario leads to the strongest economic growth and the weakest population growth, with an increase in environmental risks to human populations.

Its most wishable alternative would be the 'Adapting Mosaic', which refers to a vision of the world in which governance moves not towards the global but towards the local level. A great diversity of local styles of ecosystem management will then co-exist. Out of these local experiences networks are formed to improve the overall ecosystem management. However, there is no global-level governance. Economic growth is relatively weak at the beginning but increases after some time. Population growth is substantial.

Since the actual outcome will not be one but probably a combination of scenarios, the 'Adapting Mosaic' approach will need be supported by an ecological engineering approach and the integration of ecosystem services into the commercial sphere, in an approach that uses revolutionary technological change to reduce the use of physical resources and reach optimal management of ecological functions. It is the concept that underpins the EU "green/blue growth" policy which, at local level, is touching upon the matching of traditional ecological knowledge with scientific knowledge and new technologies, very much in the sense of the Japanese 'satoumi' concept that combines the use of ecological engineering with traditional coastal habitats and resources co-management (Fig. 26.1).

Such is the commonly accepted framework putting forward the importance of an integrated ecosystem- based management approach in managing biodiversity and ecosystem services, recognising the mosaic composition of ecosystem types and their inherent inter-linkages while pointing out the need for networking "unconnected and piecemeal" initiatives and setting up new forms of governance allowing

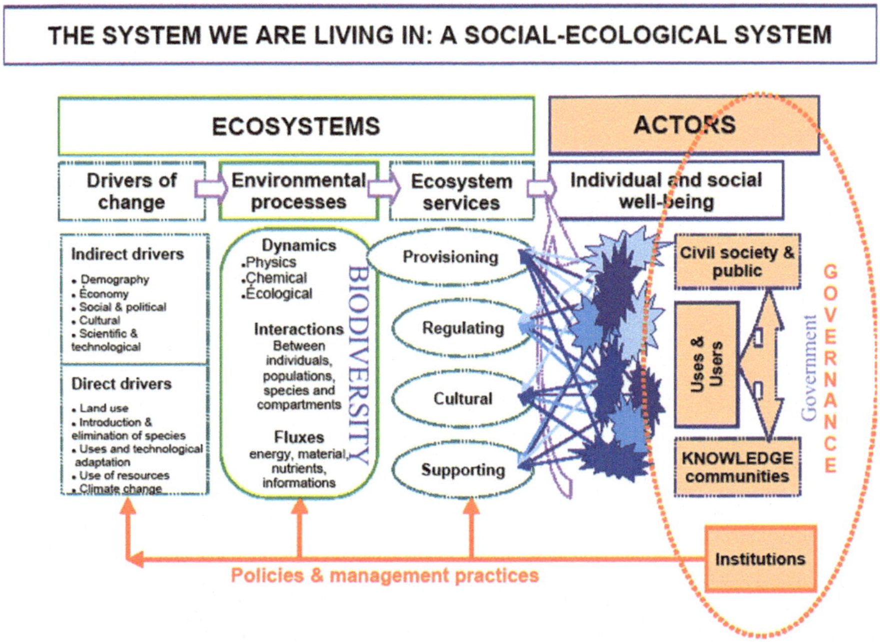

Fig. 26.1 A representation of the social-ecological system after the MEA model

the decentralised management of the "commons", which is particularly true for the marine areas. As regards drivers of change, the 11 March 2011 Great East Japan earthquake and tsunami correspond to an abrupt and "giant composite disaster" (Mimura et al. 2011) which led to the entire destruction of centuries-built social-ecological systems and their ecosystem services along the coast of Tohoku. More than never, reconstruction should take into consideration this fragile equilibrium between man and nature.

26.4 Respective Landscape and Seascape Cultural Values

Satoyama in Japan refers to a mosaic of ecosystems including wetlands, grasslands, woodlands, farmlands, paddy fields, and settlements. The interaction of humans with nature has led to the emergence of these ecosystems that provide significant habitats for a great variety of wild animals and plants. Thus, the interaction of humans with the satoyama social-ecological systems has played a vital role in biodiversity conservation, socio-economic progress, and the emergence of traditional knowledge on different components of these ecosystems.

When the newspaper *Asahi Shimbun* conducted a large public survey in 2008 (Iwata et al. 2011), the main conclusion was that among the nominated sites it was the Forest Type (areas covered by at least 90% forest) which came first with keywords such as 'nature', 'landscape beauty', or 'traditional lifestyle of local people'. The frequent use of 'genuine' or 'pristine' indicates that this type is regarded as being the ideal Japanese rural landscape by many people.

Next was the Mixed type (60% forest, 20% paddy field, 10% other fields) regarded as the place to conduct Satoyama, and the Paddy Field Type regarded as 'home village' (*Furusato*) indicating the strong emotional attachment to paddy field landscapes. The Coastal Type was generally ignored except by specialised organisations like Japan Fishery Associations, the National Institute for Islands, or NGOs like Marine Blue 21, the chosen sites being concentrated around the Seto Inland Sea and on the west coast of the Kyushu region where large inner bays are numerous.

In France, a Burgundy landscape with its vineyards in the foreground, the result of centuries of human interaction with vineyards and their "terroir", may be considered as a kind of Satoyama cultural landscape. Closer to the sea, marshlands and lagoons are characterized by centuries-old activities on the watershed, in the water body and along its shore at the interface between land and sea. Like in Japan, we traditionally have a continuum linking satoyama and satoumi practices.

26.5 NPOs, NGOs, and the Predominance of Coastal Fisheries in Japan

At this stage, a point about civil society and interest groups composition and activities should be made: although two countries like France and Japan function generally speaking on different types of governance though interest groups and the civil society in general present some similarities in their relation with the bureaucracy which remains central in structuring the political relationships of civil society organizations.

The difference is certainly that in Japan the type of group most emblematic of "civil society", business organisations have much greater resources and access to policy making. There is a well-known high level of communication and interaction (and the making process of the Ocean Basic Law and Ocean Basic Action Plan is a good illustration of it) between business organisations and the economic bureaucracy and policy-makers. In short, a "think tank" organisation like the Ocean Policy Research Institute (OPRI), its lobbying capacity based on research and networking activities, does not exist in France, at least specifically for the sea.[1] Interestingly, the French Maritime Cluster[2] is a federation of maritime industries which tend to play this role though it is specifically devoted to the private sector.

---

[1]The French think tank IDDRI (International Relations and Sustainable Development Institute) could sustain the comparison though it is devoted to all topics related to sustainable development including ocean.

[2]http://www.cluster-maritime.fr.

Soon after the March 11 devastating earthquake and tsunami, one could read the following editorial title in a regional newspaper[3]: "Japan earthquake underscores importance of social capital". It was said that "the conduct of the Japanese people after the most devastating disaster of the century showed the country's indestructible social capital". As a matter of fact, if considering one of the typical indicator of structural social capital, there has been a growing density of association (NPOs and NGOs) since the 50's in Japan, and this increase is considered to be a response to emerging social demands and problems; the recognition of the limitations of public and private-sector enterprises; and an increase in government subcontracts (Inoguchi 2002).

As stated in the Japan *Satoyama Satoumi* Assessment (2010), the origins of the latter (*satoumi*) "can be traced to the attempts of local communities to understand the relationship between human beings and the sea in the coastal areas of the Seto Inland Sea", hence very much dealing with the available social capital since these initiatives, though recently encouraged by governmental programmes, are much dependent on "citizens, non-profit organisations (NPOs), and non-governmental organisations (NGOs)".

Then, in most of the cases, the *satoumi* initiatives are related if not led by members of the Fishery Cooperative Association (FCA) and Fishery Management Organization (FMO), the functions of which facilitate the co-management regimes through the fishing rights "in response to declining harvest volume that threatened fishermen's economic well-being" (Uchida and Wilen 2004).

The strong sense of environmental ownership developed by the fishermen thanks to their fishing rights has also been instrumental in national programmes like the Fisheries Agency's "Promoting the multifunctionality of fishing community and fisheries", mainly focused on marine ecological engineering to restore nursery habitats (e.g. seagrass beds) and strengthened the productivity of locally important species.

In Japan, fisheries organisations and their networks at provincial and regional levels make them an important component of social capital for the benefit of *satoumi* initiatives provided they open up to the other coastal and maritime activities and their various private sector organisations including the NPOs.

In short, though *satoumi*is conceptually linked to the centuries-old *satoyama*in a very specific cultural context, it may be usefully compared to the adaptive co-management approach that has been developed in many places around the world while it is particularly focused on "enhancing biological productivity and biodiversity through human intervention" (Japan *Satoyama—Satoumi* Assessment 2010).

In France, after the French revolution abolishment of guilds and religious societies, there has been a new surge in associational life from the first half of the nineteenth century. It was most especially manifested in mutual aid societies, created to provide self-help insurance against the costs of sickness, accidents, old

[3]Want China Times. Editorial of 28 March 2011.

age, and burial. At the same time, cooperative organizations also began to spring up among both producers and consumers with two more and more entrenched camps: the state and the civil society organisations. More recently, a number of small associations devoted to the coastal areas and ocean sustainable development (e.g. LittOcean[4]) have started to develop their own strategy and activities in regard to public awareness, capacity building and expertise services.

In both cases, whatever each country respective context, civil society organisations should not be regarded as powerless or insignificant in politics and policy-making. As for any other group, they still have to come up as well-structured networks of local NPOs/NGOs in a same region in order that they can also counterbalance well organized international NGOs and their country representatives which tend to have a standardized approach, not always adapted to local cultures and needs.

Fishing communities, local stakeholders and authorities, both in Japan and in France, know very well what is and will be at stake tomorrow: the ecosystem health and its capacity to keep on providing the needed goods and services. Therefore, a lot of local initiatives, including marine protected areas or the like, have been developing in the last 20 years, but with great difficulties in interrelating them to pass at the next bigger scale, i.e. Department, Prefecture, and/or Region level.

26.6 Looking at Common Issues

26.6.1 Local Governments' and Stakeholders' Involvement

Beyond local initiatives, in regard to the implementation of a national ocean policy, what is the appropriate governance and planning system that may link the land and the sea, allowing local authorities to get into the process and to develop a sense of ownership?

Any ocean policy implementation requires local action ('Think global, act local'). Local government involvement is therefore essential to the successful implementation of the various economic development and environmental management policies and action plans forwarded by central government. This is practised through different arrangements in France like in Japan, but the question is totally new when it comes to the management of the marine and maritime area.

Generally speaking, there is an increasing shift in management responsibilities to local governments or coastal communities through specific institutional arrangement. In many of the known models, community sustainability issues often relate to traditional resource usage as a key management issue while overall socio-economic considerations have usually been a weak component of many of these coastal management models.

---

[4]http://www.littocean.fr.

At present, economic viability is the most pressing concern in sustaining coastal communities, particularly those linked to fisheries and any other kind of coastal and sea-related livelihood. Their sustainable development should therefore constitute an important component of any regional and national integrated maritime policy. To achieve this, any local initiative must be, among others, related to community development, regional/provincial land use and maritime spatial planning, as well as sector planning at the national and regional level.

26.6.2 Current State in France

In France, local authorities are increasingly creating their own institutional arrangement for allowing a cross-sectoral dialogue but they are doing so in a rather disorderly manner, without much consultation between them. One of the best example is the Sea Forum created by the Brittany Region under its Coastal Areas of Brittany Charter where the State is co-chairing the Forum. In other coastal regions, various kind of governance arrangement is gradually building up as well.

At a larger scale, the new national maritime policy legal framework has set a new consultative body at national level: the Coast and Sea National Board which, under the chairmanship of the Prime Minister, gathers representatives from the local governments (elected officials), private sector (all maritime activities) and the civil society (NGOs). The regional ramification of this national body is ensured through the setting up of inter-regions Coast and Sea Maritime Councils chaired by the State (Prefect of Region/Maritime prefect) in charge of preparing the corresponding interregional strategic plan. The latter will be articulated with the region (Sea and Coast regional strategy) and then with the local implementation arrangements between municipalities and departments.

Currently, the operational framework is given by the EU Maritime Strategy Framework Directive (MSFD), the environmental 'pillar' of the EU integrated maritime strategy toward the 'Good environmental status' or *"the environmental status of marine waters where these provide ecologically diverse and dynamic oceans and seas which are clean, healthy and productive" (MSFD, Article 3).* It is within the boundaries of pre-defined 'ecoregions' (Fig. 26.2) that the 11 descriptors of the 'Good Environmental Status' have to be stated and corrected when needed.

26.6.3 Current State in Japan

In Japan, there have been a series of almost simultaneous amendments of the River Act (1997), the Coast Act (1999), the Harbour Act (2000), and the Fishery Port Act (2001) towards the inclusion of environmental conservation. In 2000, these amendments were completed by the National Land Agency with the "Guidelines for Integrated Coastal Management Plan" addressed to Prefectures and Municipalities (Fig. 26.3).

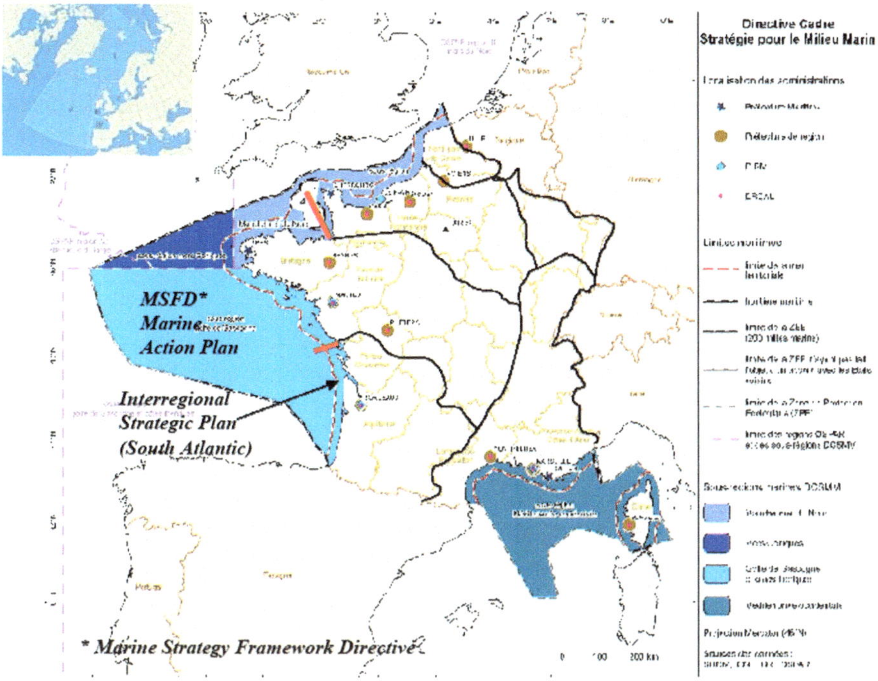

Fig. 26.2 MSFD marine eco-regions and interregional strategic plan boundaries within France EEZ

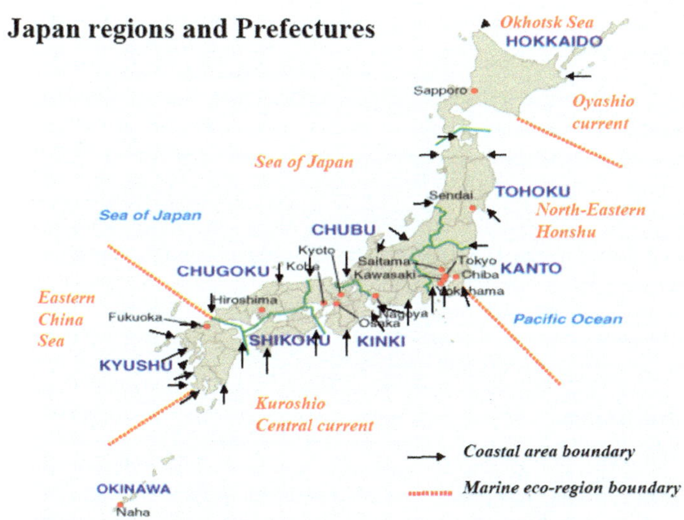

Fig. 26.3 Japan pre-defined 48 coastal areas and marine eco-regions (Spalding et al. 2007)

The main principles put forward by these guidelines were: (i) Participation and cooperation, concerning stakeholder groups such as the central and local governments (Prefecture and Municipalities), private sector, Non-Profit Organizations (NPO), fishermen and local communities; (ii) Wide overview, giving full consideration to entire bays, inland seas and river estuaries; (iii) Long-term view, setting a future vision of coastal areas following natural cycles analysis; (iv) Continuous implementation, based on the results of regular monitoring and evaluation.

- An Integrated Coastal Zone Management Commission was planned as well at the level of each coastal Prefecture and/or Municipality with a representation of all stakeholder groups.
- It seems that since then, there have been only a few initiatives from local governments, most of them coming from Municipalities and almost none from Prefectures. There may be a number of reasons to explain such a situation (Ebara 2000).
- The articulation between land use and urban planning (Municipal master plan) and the coastal zone management plans is uncertain. This has taken place mainly in the Seto Inland Sea, in regard to reclamations impacts assessment and marine environment recovery, while the development of non-coordinated sectoral policies remained the rule.
- The articulation between the 48 pre-defined coastal areas (which most of them correspond to one or two prefectures boundaries) and the bigger marine eco-regions that surround Japan is currently unknown as the respective role of local and central governments.
- The new coastal zone management plan has no statutory basis contrary to other administrative plans (e.g. Municipal Master Plans; Prefecture: Coastal Management Plan mainly related to the coastline defence).
- Although some municipal governments have tried to incorporate innovative methods for promoting public participation in the planning process, the average level of participation remains limited, often reduced to the use of passive channels such as written survey.
- There are many overlapping administrative statutory plans which make the integration process quite difficult to achieve and to enforce.
- Users other than fishers have a few legal channel to sue engineering projects like reclamation since most of them are considered of "public interest" (Kobutsu) and therefore entirely depend on the Governor's or administration's decision.

26.7 Conclusion

In both countries, there is a gap between top-down driven national ocean policies and bottom-up driven local coastal management initiatives, the meeting point of which could be at regional (France) and Prefecture (Japan) level. In France, an interesting exception is the region of Brittany which, a few years ago, set up a regional integrated coastal management strategy promoting and coordinating fourteen local initiatives representing a continuum along the whole coastline of Brittany.

Whatever the context, we need a 'nested governance system' or to consider a 'polycentric approach', i.e. mechanisms that build coherence of purpose and synergy of action at the varied scales from municipal to national layers of government. The dynamic interplay among local, regional and national levels is a common thread in both countries that need to be articulated on coastal and ocean matters.

In Europe, the situation is now changing in the frame of the EU Integrated Maritime Policy and new Directives like the Marine Strategy Framework Directive. Depending on countries, stocktaking analysis and demonstration programmes regarding their coast and marine waters led the way to legal and institutional arrangements where an interdepartmental committee may be replaced by an interdepartmental unit, and eventually an independent unit with overall responsibility for coastal and ocean management.

Emerging principles and practices of participation and subsidiarity both support the idea that decision making should be taken as near to the local level as possible, but a broader framework needs to be in place to ensure adequate perspective and coherence between different local initiatives, and to provide appropriate technical and financial support to make them sustainable.

The Region or other administrative level (e.g. Prefecture, in Japan) associated with strategic planning may hold the key to resolving the problem of territorial and sectoral integration in a manner which both reflects national and international policies and is adapted to local conditions. This is not necessarily the level at which detailed action planning and management should take place since they are essentially local tasks, but the Region (France)or the Prefecture (Japan) is potentially a critical enabling level in terms of ICM initiatives coherence in between the local and national level.

The whole process will be facilitated at national level by ensuring that policies relating to the coast and ocean are compatible, providing a national focal point for regional and local initiatives, and facilitating integrated approaches to coastal and ocean management.

References

Cicin Sain B, Vanderzwaag DL, Balgos MC (2016) Routlege Handbook of National and Regional Ocean Policies. Routledge Ed. 640 pp

Ebara A (2000) Coastal management system and ecological preservation in Japan. In: Proceedings of the international symposium on protection and management of coastal marine ecosystems, Bangkok, Thailand, 12–13 December 2000

Inoguchi T (2002) Broadening the basis of social capital in Japan. In: Putnam RD (ed) Democracies in flux: the evolution of social capital in contemporary society. Oxford University Press, New York

Iwata Y, Fukamachi K, Morimoto Y (2011) Public perception of the cultural value of Satoyama landscape types in Japan. Landscape Ecol Eng 7:173–184

Kuribayashi T (2008) Thinking about the basic ocean law: enactment of the Basic Ocean Law and its significance. Ship and Ocean Newsletter No. 10, September 2008

Mimura N, Yasuhara K, Kawagoe S, Yokoki H, Kazawa S (2011) Damage from the great East Japan earthquake and tsunami. A quick report. Mitig Adapt Strateg Glob Change 16:803–818

Spalding MD et al (2007) Marine ecoregions of the world: a bioregionalization of coastal and shelf areas. Bioscience 57(7)

Uchida H, Wilen JE (2004) Japanese coastal fisheries management and institutional designs: a descriptive analysis. In: IIFET 2004 Japan proceedings

Yamamoto I (2014) On the new revision of the basic plan on ocean policy. Ocean Newsletter No. 18. Ocean Policy Research Foundation

Part VI
Aquaculture

Chapter 27
Heterogeneity of Japanese Oyster (*Crassostrea Gigas*) Spat Collection in a Shellfish Farmed Mediterranean Lagoon

Franck Lagarde, Martin Ubertini, Serge Mortreux, Adeline Perignon, Axel Leurion, Patrik Le Gall, Claude Chiantella, Slem Meddah, Jean-Louis Guillou, Gregory Messiaen, Béatrice Bec, Cécile Roques, Delphine Bonnet, Hélène Cochet, Ismaël Bernard, Erika Gervasoni, Marion Richard, Gilles Miron, Annie Fiandrino, Stephane Pouvreau and Emmanuelle Roque D'orbcastel

Abstract Japanese oysters spatfields were recently discovered in the French Mediterranean Thau lagoon farmed for shellfish. This discovery led to interesting issues concerning research for a highly seashell exploited ecosystem. The analyses of various environmental parameters characterized favorable sites and periods for spat collection. Spat collection in Thau lagoon presented high temporal and spatial heterogeneities related to hydrodynamic, predation, competition and trophic resource variability. Here, the results highlight various explicative factors of the

F. Lagarde (✉) · M. Ubertini · S. Mortreux · A. Leurion
P. Le Gall · C. Chiantella · S. Meddah · J. L. Guillou · G. Messiaen
M. Richard · A. Fiandrino · E. Roque D'orbcastel
Ifremer, CNRS, IRD, Montpellier University, MARBEC, Sète, France
e-mail: Franck.lagarde@ifremer.fr

A. Perignon
CRCM, Mèze, France

B. Bec · C. Roques · D. Bonnet
Ifremer, CNRS, IRD, Montpellier University, MARBEC, Montpellier, France

H. Cochet
Cochet-Environnement, Locoal Mendon, France

E. Gervasoni
CEPRALMAR, Montpellier, France

I. Bernard
Eureka Modélisation, Lézardrieux, France

G. Miron
Université de Moncton, Moncton, NB, Canada

S. Pouvreau
Ifremer, UBO, CNRS, IRD, LEMAR, Brest, France

T. Komatsu et al. (eds.), *Oceanography Challenges to Future Earth*,
https://doi.org/10.1007/978-3-030-00138-4_27

Japanese oysters recruitment within Thau lagoon and underline the high ecological variability in the studied system under oligotrophication.

Keywords Spat collection · Japanese oyster · Crassostrea gigas · Thau lagoon · France

27.1 Introduction

Certain Mediterranean lagoons are used in the southwest of France for traditional shellfish farming activities. Though recognized for a long time for its oyster (*Crassostrea gigas*) growth vocation (Gangnery et al. 2001, 2004), lagoon oyster farming however depends on spat supplied from Atlantic nursery basins and hatcheries. Strong difficulties were reported by the industry with the observation in 2009 of mass oyster mortalities caused by the Ostreid herpesvirus (OsHV-1μvar) (Pernet et al. 2012; Pernet et al. 2014; Pernet et al. 2014). Until then, no intention to use local spat supplies was ever proposed by the oyster culture industry. Additionally, no scientific knowledge of the Japanese oyster recruitment process in Mediterranean lagoons was available. The epidemic crisis brought French Mediterranean producers to seriously consider a local source of seed. A research project was conducted between 2012 and 2014 with the aim to describe natural recruitment and oyster spat collection in Thau lagoon. The focus was placed on spat settlement to better understand the spatial and temporal heterogeneity in spat collection in the lagoon. This three years project led to the collaboration of several partners including business partners from the shellfish industry, research and academic organizations and consulting firms. This project shows a concrete example of local scientific development with the prospects of ending the oyster culture industry crisis.

27.2 Material and Method

The Thau lagoon is the largest nanotidal lagoon of the Languedoc Roussillon region in southern France. It covers an area of 7500 hectares (19 km × 4.5 km) on a north-east/south-west axis, with a mean depth of 3.5 m. It is supplied with seawater by channels linking the lagoon to the Mediterranean sea. Eight sampling spatfall stations were installed in the lagoon, Table 27.1), three inside the shellfish farming zones (Bouzigues, Mèze, Marseillan) and three to five (depending on the year) outside the shellfish farming zones (OFSZ): Marseillan_OSFZ, Listel_OSFZ, Meze_OSFZ, Bouzigues_OSFZ, Balaruc_OSFZ. The mooring buoy was attached to a 35 kg chain, the tension maintained by a subsurface float of 3 L and marked on the surface by a buoy and a pennant (Fig. 27.1).

Table 27.1 Depths, positions and sampling years of experimental stations

		Coordinates WGS84 (Decimal degree)		Year sampled		
Code station	Depth (m)	Long	Lat	2012	2013	2014
Bouzigues	8.0	3.64332690	43.4379778	X	X	X
Meze	3.2	3.58326056	43.3977765	X	X	X
Marseillan	3.8	3.57124851	43.3791309	X	X	X
Marseillan_OSFZ	3.6	3.54214175	43.3484724	X	X	X
Listel_OSFZ	6.9	3.61106341	43.3887515	X	X	X
Balaruc_OSFZ	5.2	3.69180552	43.4319240	X	X	X
Meze_OSFZ	5.0	3.607380	43.414427		X	X
Bouzigues_OSFZ	8.0	3.64766667	43.417500			X

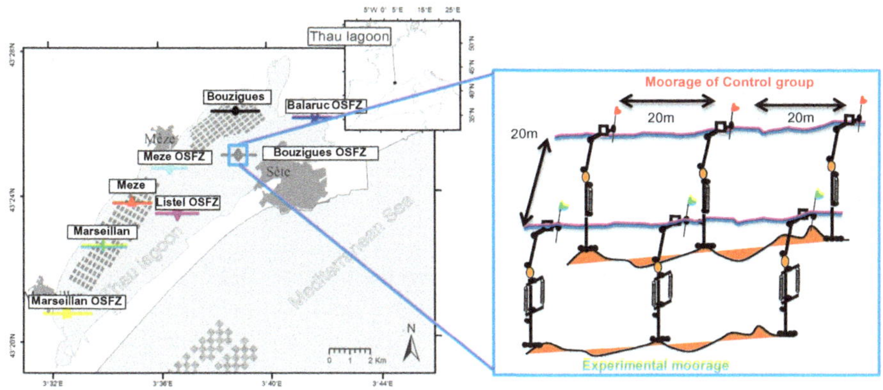

Fig. 27.1 Sampling sites located in Thau lagoon. Areas with grey boxes are bivalve farms. Coloured symbols represent experimental spatfall stations. Spatfall stations consisted of 3 control collector moorages and 3 spat sampling moorages outside shellfish farming zone. Collectors were simply suspended under the farming structures inside the shellfish farming zones

We monitored oyster spat recruitment between the months of June and October during the 2012–2014 period. At each sampling time (T_0, T_1, $T_2 \ldots T_9$), the stations were equipped with three control collectors set up for the entire duration of the experiment and two sets of three collectors immerged respectively for 2 and 4 weeks to estimate the quantities of pediveligers and post larvae (Fig. 27.2).

Collectors, made of 44 plastic cups attached together, were used to collect oyster larvae at different stages of their life-cycle (pediveliger, post larvae and spat). Collectors measured 110 cm long. The sample unit was represented by the cup. Cups were white and had a diameter of 15 cm. Each collector was sampled at

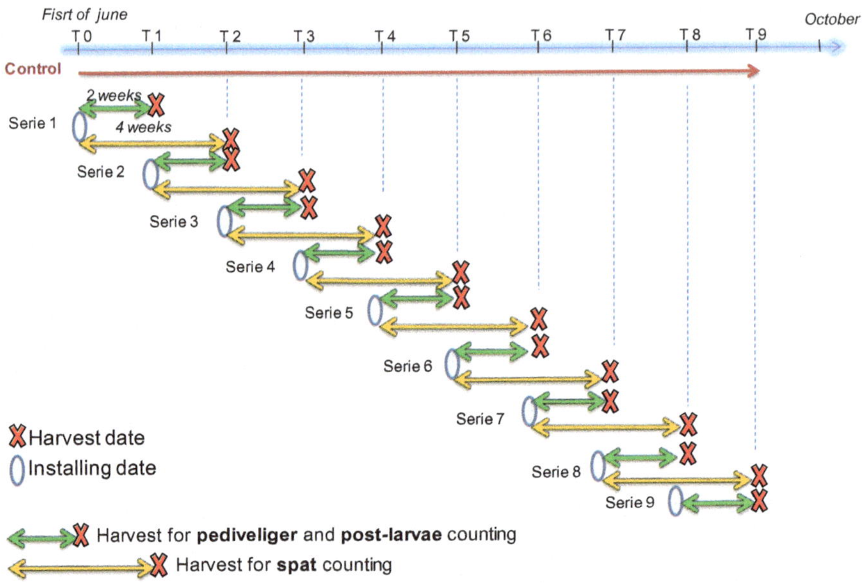

Fig. 27.2 Sampling strategy of collector deployment and collection of larvae and spat. Figure recreated from Lagarde et al (2015), http://archimer.ifremer.fr/doc/00279/39054

three depth levels (high cup = 39th cup from the bottom, mid cup = 22th, bottom cup = 5th). The cup presented two faces, "above" and "below", that were observed under a binocular microscope to count the oyster larvae. Each face was divided in 18 sub-surfaces with equal areas referred as "basic subunit". Larval densities were determined by counting individuals in subunits. The number of subunits used (1 to 4) was dependent on larval density to minimize the counting time.

The cup collectors were deployed vertically, submerged 2 meters below the surface and replicated three times at each experimental station. They were designed to support 2 sets of collectors, the collectors immersed for "2 weeks" to assess pediveliger supply and the collectors immersed for "4 weeks" to assess metamorphosis success and spat abundances.

The variability of pediveliger and spat densities per cup were categorize dusing 4 categories (Zero: 0individual percup; Low: 1 to 20 individuals; Medium: 21 to 200 individuals; High: 201 to 2000 individuals, Overabundant: >2000 individuals percup) (Stephane et al. 2012) (Fig. 27.3).

Temperature, salinity and plankton samples were associated to spat assessment weekly to characterize the environment and to define planktonic functional groups (predators and trophic competitors of larvae) on three experimental station Bouzigues, Listel and Marseillan.

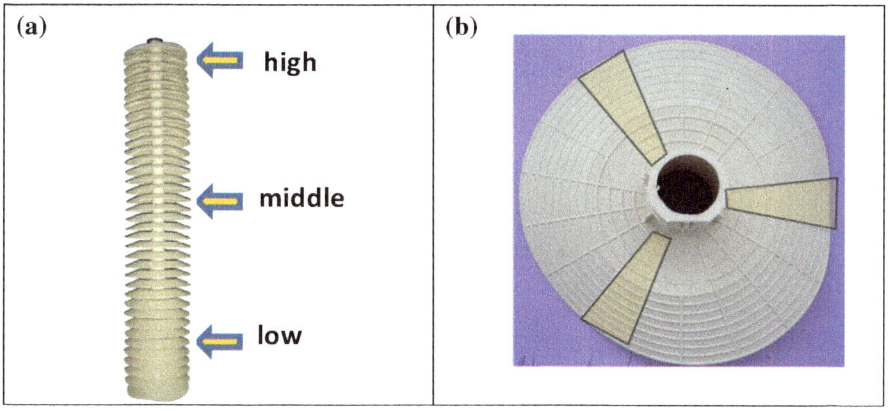

Fig. 27.3 **a** Photography of a complete cup collector showing the levels sampled. **b** Photography of a cup seen from above, with triplicated counting subunits in yellow (counting subunit = 1 basic subunit in this example). Figure recreated from Lagarde et al (2015), http://archimer.ifremer.fr/doc/00279/39054

27.3 Results

The target variables were the pediveligers, postlarvae and spat abundances by cup. The "pediveliger" variable was used as a first step to describe spat settlement on collectors (larval supply and settlement period). The relationship between pediveliger and spat abundances showed that the recruitment across the Thau lagoon was highly variable and that fixation on cups did not lead, in many cases, to settlement and seed formation. The metamorphosis stage sometimes seems to be a biological lock while the phase between postlarvae and spat is not an obstacle to larval development. Our results showed that "postlarvae" numbers are strongly correlated with "spat" numbers ($\rho = 0.68$, p value = $2.2e^{-16}$) (Fig. 27.4).

Spat abundances according to the harvest date and the experimental station are shown in Fig. 27.5. These relationships characterised the spatial and temporal spat heterogeneity in Thau lagoon for 2012, 2013 and 2014. The experimental stations "Mèze OSFZ", "Listel OSFZ" and "Bouzigues OSFZ" revealed spatfall intensity from medium to high in August and September for 2013 and 2014. Hence, oyster spat fields in Thau lagoon were observed outside oyster farmed areas. The "Listel" station presented low spat abundance variability (126 ind. cup^{-1}, 91 ind. cup^{-1}, 113 ind. cup^{-1}) over the studied years. The "Meze OSFZ" station is defined as having the highest densities of spat (up to 1112 ind. cup^{-1}) and the largest number of high spatfall occurrences for the years 2013 and 2014.

A Principal Component Analysis (PCA) was used to study the spatiotemporal heterogeneity of the environmental data set acquired in the "Bouzigues", "Marseillan" and "Listel" sites. These analyzes were based on the integration of 65

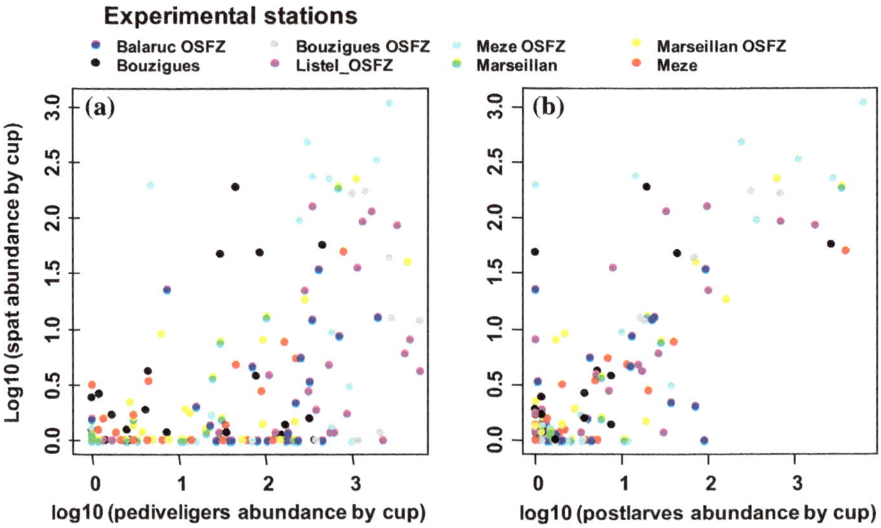

Fig. 27.4 Scatterplot between log10 abundances per cup of **a** "pediveliger" and "spat" and **b** "post larvae" and "spat" across all 8 experimental stations for the years 2012, 2013 and 2014

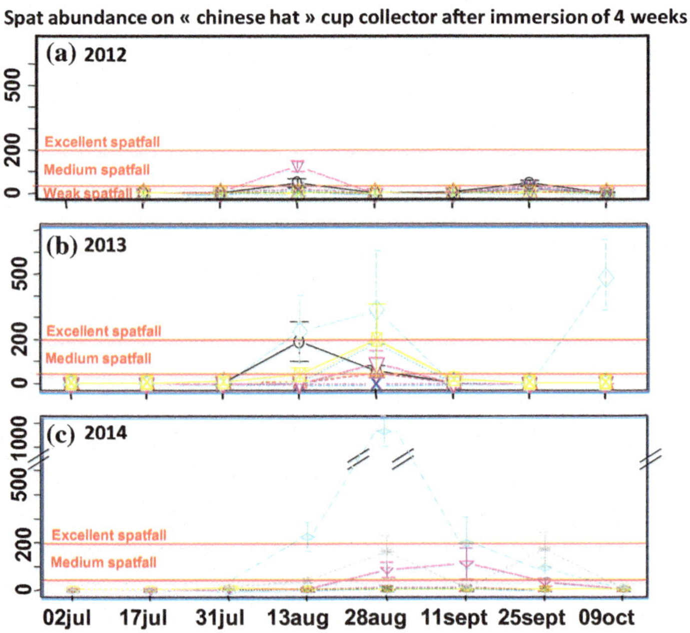

Fig. 27.5 Means (± SE) of spat abundance by cup in **a** 2012, **b** 2013 and **c** 2014 on the experimental stations

observations of spat collection related to 17 environmental variables including the pelagic larval and attachment period. The first axis (24% of the explained variance) was built by the contrast between the abundance of ciliates and dinoflagellates and the abundances of picoeukaryotes, cryptophytes, heterotrophic flagellates as well as the picophytoplankton biomass and salinity. This first axis expressed the temporal succession of the fauna, flora and the evolution of various hydrological factors during the experimental periods. The second axis (22% of the explained variance) contrasts ciliates, dinoflagellates, picoeukaryotes, cryptophytes to nanophytoplankton, microphytoplanktonic biomass, diatoms, cyanophytes, predators, trophic competitors and the water temperature: it reflects the lagoon's ecological heterogeneity, *i.e.* the impact of the effect of shellfish structure and lagoon confinement according to a East-West gradient (Guelorget and Perthuisot 1989; Guelorget et al. 1993) observed in hydrobiological cues with opposition of inside and outside shellfish farm zone biocoenosis (Fig. 27.6).

The overlaid factor representation also allowed to see that the first axis shows a temporal dynamic. It opposes the years 2013 and 2014 (Fig. 27.7) and expresses monthly changes. A spatial gradient is expressed in the second axis, spreading the sampling sites from top to bottom according to temperature, nanophytoplankton biomass, nano and microphytoplancton abundances, diatoms (*Chaetoceros sp.*) abundances for Marseillan, Bouzigues and Listel, respectively. In the same way, this second axis suggests a gradient in spat collection categories, reflecting a natural spat collection gradient according to nul, weak and medium categories. Note that within the three year monitoring program, no overabundant spat collection was recorded on any of the Bouzigues, Marseillan and Listel sites.

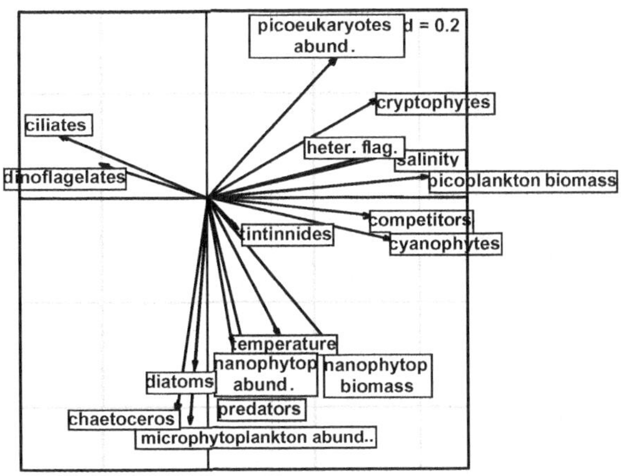

Fig. 27.6 PCA biplot of environmental data (n = 65, % PCA1 = 24%, % PCA2 = 22%). Figure recreated from Lagarde et al (2015), http://archimer.ifremer.fr/doc/00279/39054

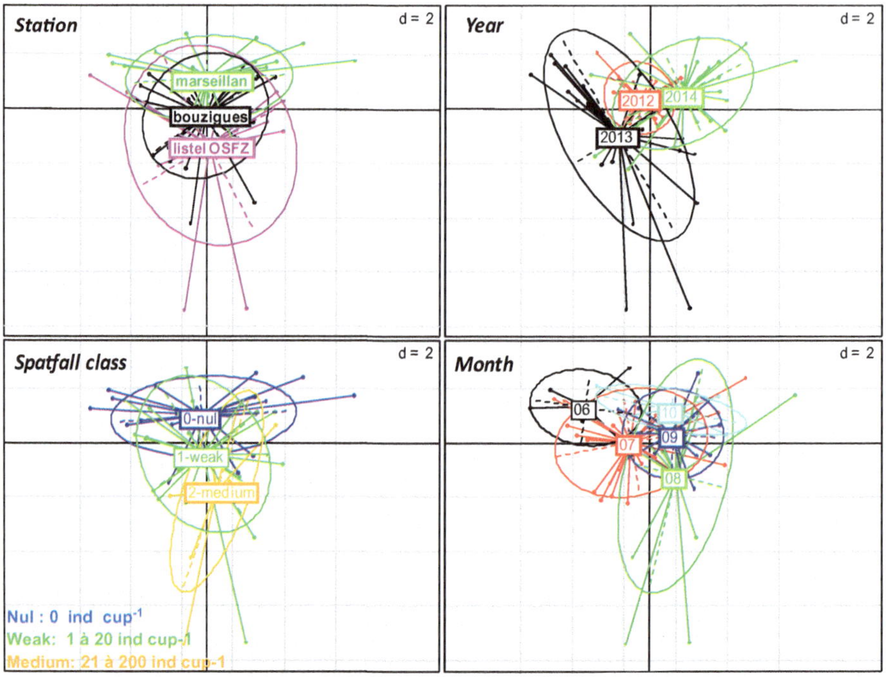

Fig. 27.7 PCA biplot of environmental data (n = 65, % PCA1 = 24%, % PCA2 = 22%) with overlaid 90% confidence ellipse for (above, left) experimental stations, (above, right) study year, (below, left) spatfall class, and (below, right) months of data acquisition. Figure recreated from Lagarde et al (2015), http://archimer.ifremer.fr/doc/00279/39054

27.4 Discussion

The present study is the first, to our knowledge, to document the recruitment of Japanese oysters in a Mediterranean lagoon. Observations showed a high variability of the spatfall success where certain sites would be highly profitable for the local industry in harvesting seed.

These results showed as well the effect of high nanoplankton, diatoms and *Chaetoceros sp.,* microphytoplankton abundances and water temperature on the success of the spat collection. Spat nutrition appears to play a limiting factor role for metamorphosis with the nanophytoplankton biomass being a key variable.

The intensity of the spatfall appears as an ecological gradient related firstly to the lagoon confinement (East -West direction between Bouzigues and Marseillan) and secondly, to the ecological differences between shellfish farmed biocoenosis (represented by stations "Bouzigues" & "Marseillan") and "out of shellfish area " biocoenosis as station "Listel".

These results should be taken into account while considering the study of the functioning of the Thau lagoon. This is particularly true since the lagoon is under a oligotrophication process since the 2000s (Collos et al. 2009; Bec et al. 2011; Leruste 2013). The successes and failures of larval cycles reflect the spatial and temporal heterogeneity (habitats and ecological niches) of natural oyster collection in a dynamic natural environment. This variability is particularly interesting to study firstly the characterization of the optimal habitats and define ecological niches of this species with its interactions in the Mediterranean lagoon and secondly to find the in situ tolerances of this introduced species during the larval critical phase. A collaboration between France and Japan could fit into this framework provided by the study of *Crassostrea gigas* in dynamic ecosystems nearing oligotrophication in France and Japan (Yanagi 2015).

Acknowledgements This study was funded by grants from Ifremer, Occitanie Région, France-Agrimer, Comité Régional de la Conchyliculture de Méditerranée, Cepralmar et Fond Européen pour la Pêche.

References

Bec B, Collos Y, Souchu P, Vaquer A, Lautier J, Fiandrino A, Benau L, Orsoni V, Laugier T (2011) Distribution of picophytoplankton and nanophytoplankton along an anthropogenic eutrophication gradient in French Mediterranean coastal lagoons. Aquat Microb Ecol 63:29–45

Collos Y, Bec B, Jauzein C, Abadie E, Laugier T, Lautier J, Pastoureaud A, Souchu P, Vaquer A (2009) Oligotrophication and emergence of picocyanobacteria and a toxic dinoflagellate in Thau lagoon, southern France. J Sea Res 61:68–75

Gangnery A, Bacher C, Buestel D (2001) Assessing the production and the impact of cultivated oysters in the Thau lagoon (Mediterranee, France) with a population dynamics model. Can J Fish Aquat Sci 58:1012–1020

Gangnery A, Bacher C, Buestel D (2004) Application of a population dynamics model to the Mediterranean mussel, *Mytilus galloprovincialis*, reared in Thau Lagoon (France). Aquaculture 229:289–313

Guelorget O, Perthuisot JP (1989) The Paralic Realm. Geological, biological and economic expressions of confinement. Rome FA0:142

Guelorget O, Perthuisot J-P, Lamy N, Lefebvre A (1993) Structure et organisation de l'étang de Thau d'après la faune benthique. Relations avec le confinement. Oceanol Acta 17:105–114

Leruste A (2013) First steps of ecological restoration in Mediterranean lagoons: shifts in phytoplankton communities. Estuar Coast Shelf Sci 180:190–203

Pernet F, Barret J, Le Gall P, Corporeau C, Dégremont L, Lagarde F, Pépin JF, Keck N (2012) Mass mortalities of Pacific oysters *Crassostrea gigas* reflect infectious diseases and vary with farming practices in the Mediterranean Thau lagoon, France. Aquac Environ Interact 2:215–237

Pernet F, Lagarde F, Le Gall P, D'Orbcastel ER (2014a) Associations between farming practices and disease mortality of Pacific oyster *Crassostrea gigas* in a Mediterranean lagoon. Aquac Environ Interact 5:99–106

Pernet F, Lagarde F, Jeanné N, Daigle G, Barret J, Gall P Le, Quere C, D'orbcastel ER (2014) Spatial and temporal dynamics of mass mortalities in oysters is influenced by energetic reserves and food quality. PLoS One 9(2):e88469

Pouvreau S, Bernard I, Le Souchu P, Queau I, et al (2012) Observer, Analyser et Gérer la variabilité de la reproduction et du recrutement de l'huître creuse en France: Le Réseau Velyger. Rapport annuel 2011. R.INT.BREST RBE/PFOM/PI 2012-1. https://archimer.ifremer.fr/doc/00135/24649/

Yanagi T (2015) Estuaries of the World, Eutrophication and Oligotrophication in Japanese Estuaries. The present and future tasks. Springer

Chapter 28
Suitable Oyster Culture Density in Oginohama Bay, Miyagi, Japan

Yutaka Okumura, Akatsuki Nawata, Hiroshi Ito, Akio Oshino and Motoyuki Hara

Abstract We investigated regular oyster dietary conditions, and shellfish and periphyton growth from 2013 to 2014 in Oginohama Bay, Miyagi Prefecture, Japan. We calculated phytoplankton biomass in the aquaculture areas minus the total phytoplankton filtration by oysters and periphyton. We estimated that the suitable oyster culture density is when that value is >0. The mass balance of phytoplankton biomass for food availability increased from January through April, and decreased from May to June. Although the mass balance of phytoplankton biomass fluctuated after June, no significant changes were seen until December. Therefore, the risk of a decrease in oyster food was thought to increase after June. The mass balance of phytoplankton biomass in both low (360 ropes × 25masters × 10ind./master) and high (400 ropes × 30 masters × 15ind./master) culture density (approximately 450 oyster longline facilities) was always >0 after the tsunami. The number of longline facilities before the 2011 tsunami (approximately 1,100) did not negatively impact the oyster food supply if culture density was low. However the mass balance of phytoplankton biomass under high culture density before the tsunami was calculated as negative from June to August and from

Y. Okumura (✉)
Tohoku National Fisheries Research Institute, Japan Fisheries Research and Education Agency, 3-27-5, Shinhama, Shiogama, Miyagi 985-0001, Japan
e-mail: okumura@affrc.go.jp

A. Nawata
Miyagi Prefectural Government, 3-8-1, Honmachi, Aoba-ku, Sendai, Miyagi 980-8570, Japan

H. Ito
Miyagi Prefecture Fisheries Technology Institute, 97-6, Sodenohama, Watanoha-aza 986-2135, Japan

A. Oshino
Kesennuma Fisheries Experiment Station, Miyagi Prefecture Fisheries Technology Institute, 107, Iwaisaki, Hajikami, Kesennuma, Miyagi 988-0241, Japan

M. Hara
Tohoku University, 468-1, Aobayama, Aramaki-aza, Aoba-ku, Sendai, Miyagi 981-0845, Japan

T. Komatsu et al. (eds.), *Oceanography Challenges to Future Earth*,
https://doi.org/10.1007/978-3-030-00138-4_28

October to November. We speculate that an increase in the number of longline facilities and high-density culture would result in a decrease in oyster food. Food availability under low culture density before the tsunami (1,118 longline facilities × 360 ropes × 25 masters × 10 ind./master) was the most suitable, because the mass balance of phytoplankton biomass was >0, and oyster density was the highest of the four conditions.

Keywords Oginohama bay · Japan · Oyster culture · Phytoplankton biomass · Oyster density

28.1 Introduction

The huge tsunami generated by the Great East Japan Earthquakeon 11th March, 2011, caused serious damage to the aquaculture in the Tohoku coastal area. For example, the oyster harvest decreased from 40,000–60,000 tons shell weight before the tsunami to 20,000 tons afterwards (2011–2014; Fig. 28.1. Ministry of Agriculture, Forestry and Fisheries 2016).

Although over 5 years have passed since the disaster, oyster production has not yet recovered to pre-earthquake levels. Additionally, high oyster mortalities occurred in a part of aquaculture areas, as a result of either high seawater temperatures during the summer in 2012 or a large periphyton outbreak in 2013. If unsuitable environmental events occur under the overcrowded culture condition of the oyster, it may cause high oyster mortality. Therefore, it is necessary to determine the suitable oyster culture density to effectively recover pre-earthquake conditions. Moreover, if the oyster culture density recovers to pre-earthquake conditions, overcrowding may be a problem in the future. The aim of this study was to calculate suitable oyster culture quantities to avoid overcrowding.

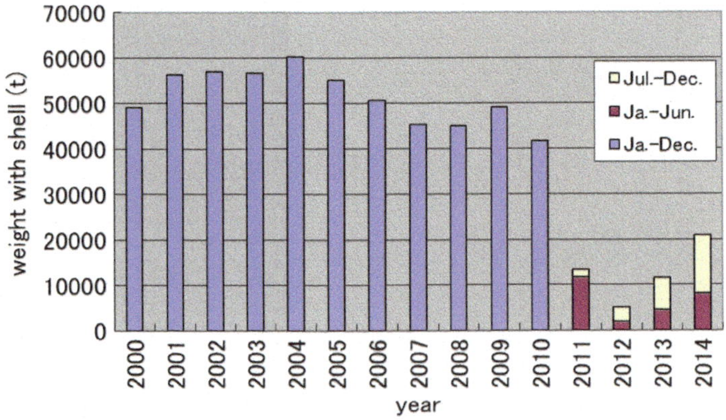

Fig. 28.1 Temporal trend of oyster production in Miyagi Prefecture (reproduced from M.A.F.F. 2016)

The suitable quantity has been calculated for various shellfish (Byron et al. 2013), such as oysters (Filgueira et al. 2014), scallops (Nagasawa et al. 2016), and pearl oysters (Abo et al. 2001). Shellfish aquaculture methods vary from place to place, even for the same species. For example, Japan and France use vertical (suspending method) and horizontal oyster growing techniques, respectively (Koike 2015). The culture period to shipment also varies (Hiroshima City Agric. For. Fish. Promo. Ctr. 2016), because shellfish growth and filtration are influenced by water temperature and food availability (Brown and Hartwick 1988; Akashige 2005; Comeau et al. 2008). Therefore, suitable oyster culture conditions need to be calculated for each area.

In this study, we periodically investigated total diet quantity of all shellfish and periphyton in the selected aquaculture grounds, and determined shellfish growth seasonally in Oginohama Bay, Miyagi Prefecture, Japan, where oysters are commonly cultured. We calculated the appropriate quantity of oysters by determining the difference between diet quantity and total shellfish filtration.

28.2 Materials and Methods

28.2.1 Continuous Chlorophyll a *and Water Temperature Observations*

To determine oyster feeding, the concentrations of chlorophyll *a* (Chl *a*), which is a phytoplankton pigment and indicator of water quality, were measured from seawater collected in Momonoura, Oginohama Bay (38.3889 N, 141.4122 E), a major aquaculture area, from January to December 2014 (Fig. 28.2).

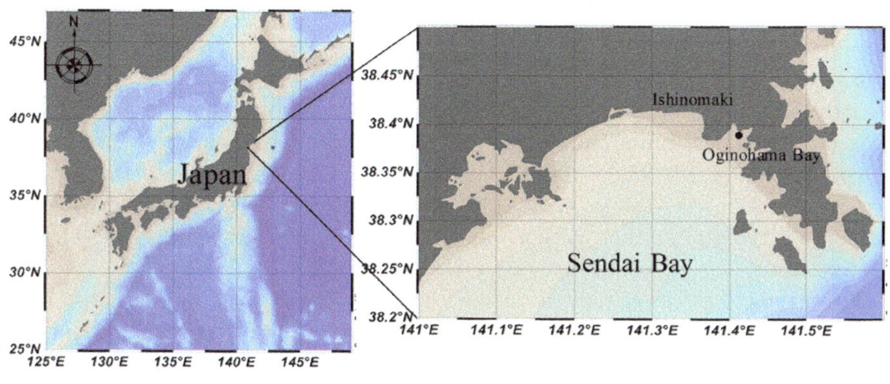

Fig. 28.2 Sampling site of Momonoura, Oginohama Bay, in Miyagi Prefecture, Japan

Chlorophyll (Chl) sensor (ACLW-USB, JEF Advantech CO., Ltd., Tokyo, Japan), and water temperature sensor (ACTW-USB, JEF Advantech CO., Ltd., Tokyo, Japan) were attached to oyster longline facilities in the aquaculture areas. Temporal changes in Chl *a* were determined from fluorescence values of the Chl sensors and Chl *a* values measured by high performance liquid chromatography (HPLC). After filtering the seawater by glass fiber filter (Whatmann GF/F), the phytoplankton pigments, which were extracted from the filter by methanol, were analyzed by HPLC (Zapata et al. 2000).

28.2.2 Specific Growth Rates of Phytoplankton Assemblages in Oginohama Bay

To calculate phytoplankton growth in the culture area, the specific phytoplankton growth rate in seawater was measured at about monthly intervals from January 2014 to January 2015. At two other sampling sites (38.3724 N, 141.4453 E and 38.3674 N, 141.4138 E) in Oginohama Bay, we collected seawater every 5 m depth (0, 5, 10, and/or 15 m). First, 150 ml seawater was filtered and Chl *a* was then analyzed by the above-mentioned method (2.1). Approximately 200 ml of residual seawater was filtered by plankton net (100 μm size), and the filtrates were then incubated (approximately 5000 l ×, room temperature) for one day. After incubation, Chl *a* concentration was analyzed by the above-mentioned method (2.1). The specific growth rate (μ) was calculated by the equation. $\mu = \frac{lnN1 - lnN0}{t}$ where, N0 is initial Chl *a* concentration (μg/L), N1 is Chl *a* concentration after *t* days (t = 1).

28.2.3 Shellfish and Major Periphyton Filtration in the Culture Area

To calculate filtration rate and growth rate, the weight of oysters in culture nets were measured over time. The oyster filtration rates were calculated by substituting either equation (Table 28.1), which was modified by Okumura (pers. comm.) from original equations (Akashige et al. 2005). The oyster growth rates were also calculated according to the method of Akashige and Takayama (2003), with modifications.

To calculate filtration rate of major periphyton, an oyster culture rope with mussels, and ascidians was purchased every three months, and the weights of organisms were also recorded for each taxa present. The filtration rates of mussels and ascidians were calculated by substituting either equation (Table 28.1) for water temperature and weight. We interviewed members of the fisheries cooperative association to ascertain the number of oyster longline facilities, culture ropes, oysters per longline facility, and culture methods (Fig. 28.3 and Table 28.2).

Table 28.1 Equations of filtration rates by oyster and the periphyton

Species	Calculations (F)	DW/WW	Conditions	References
Crassostrea gigas	$0.26e^{(0.1584xT)} \times WW$		T < 11.8	Okumura pers. Comm.
	(0.70 × T - 6.6) × WW		11.8 ≤ T < 18.5	Akashige et al. (2005)
	4.9 × WW		18.5 ≤ T	Akashige et al. (2005)
Mytilus galloprovincialis	$6.773 \times DW^{0.678}$	5(%)		Riisgard et al. (2014)
				Miura et al. (2013)
Ascidiella aspersa	$54.4 \times DW^{1.05}/1000 \times 60$	5(%)	conversion from ml/min	Randlov and Riisgard (1979)
Ciona intestinalis	$46.4 \times DW^{0.84}/1000 \times 60$	5(%)	conversion from ml/min	Randlov and Riisgard (1979)
Styela clava Herdman	$\exp(0.119T\text{-}4.54) \times DW^{0.6745}$	8.5(%)	T < 29	Kim and Moon (1998)
Halocynthia roretzi	$(0.1956 + 0.0182T) \times DW^{0.7978\exp(-0.0273T)}$		5 < T<25	Jeong and Cho (2013)
		12.2(%)		Yamamoto and Handa (2001)
Halocynthia aurantium	We substituted the calculation of *H. roretzi*.			

F: filtration (L/ind./h), DW: dry weight (g), T: temperature (C), WW: wet weight

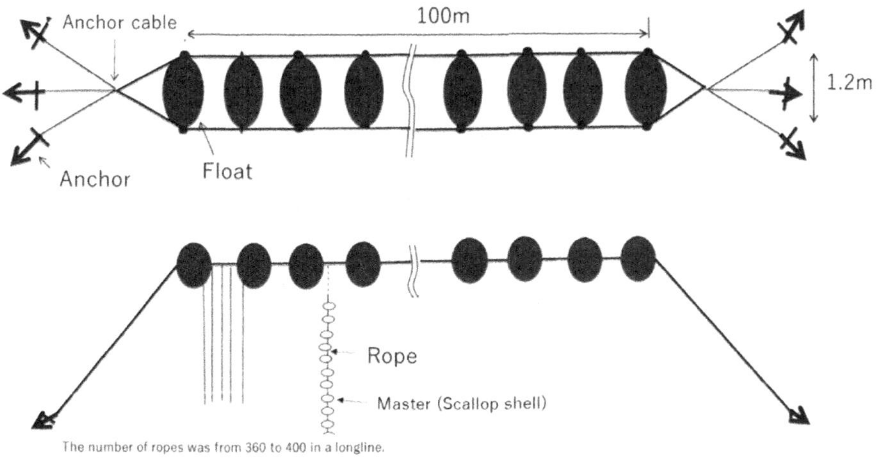

Fig. 28.3 Schematic drawings of longline facilities of oyster culture (from Japan fisheries cooperative document)

We also calculated the quantity of the oyster, which attached to the quay, by multiplying a literature value (Mizuo et al. 2008) by the length of the coast line (Table 28.3).

Table 28.2 Appropriate quantity of oyster culture in Oginohama Bay

	Number of longline facilities	Number of ropes/ longline facilities	Number of masters (Scallop shell)	Number of oysters/ master
High density culture post-tsunami	447	400	30	15
Low density culture post-tsunami	447	360	25	10
High density culture pre-tsunami	1118	400	30	15
Low density culture pre-tsunami	1118	360	25	10
Appropriate culture under high density	559	400	30	15

28.2.4 Calculation of Suitable Oyster Culture Density

We defined the suitable oyster culture quantity as phytoplankton quantity in the aquaculture areas minus the total filtration of phytoplankton by oysters and periphyton, such as mussel and ascidians (2.3), and that value is more than zero. Phytoplankton growth in the culture area was calculated by multiplying Chl *a* concentration (2.1) by average growth rate (Table 28.3). Furthermore, we added an inflow of phytoplankton from outside the culture area and subtracted an out flow from the culture area (Table 28.3). Seawater exchange rate, which is a parameter for checking the inflow and outflow of phytoplankton across aquaculture area boundaries, was substituted with the value from a box model of Sendai Bay (Okumura et al. 2013) in reference to a method by Unoki (1998). Finally, changes in phytoplankton flux on the aquaculture areas were calculated daily.

28.2.5 Coastline Map and Measurement

The map was created in ODV software (AWI, Denmark). The length of the coastline in Oginohama Bay was measured by Image J software (NIH, Maryland, USA) from a map.

Table 28.3 Other parameters used in this model

Item	Value	Unit	Remarks
(1) Area of Oginohama Bay	18 163 550	m^2	Calculated by Image J
(2) Water depth	10	m	
(3) Ratio of culture area (culture area/ Oginohama Bay)	0,74		Culture area was calculated by Image J
(4) Specific growth rate of phytoplankton in seawater	0,54	/day	The mean of the result of the study
(5) Growth rate of phytoplankton in seawater	0,72		exp(item(4))-1
(6) Outflow of seawater from culture area/ the capacity of Oginohama Bay	0.03–0.69		Calculated monthly from Box model (Okumura et al. 2013)
(7) Inflow of seawater from culture area/ the capacity of Oginohama Bay	0.03–0.69		Calculated monthly from Box model (Okumura et al. 2013)
(8) Phytoplankton quantity		μg/L	Chl *a* concentation in Oginohama Bay
(9) Length of the shoreline around the culture area	23	km	Calculated by Image J
(10) Number of oysters attached to seawall	80	ind./m^2	A literature value (Mizuo et al. 2008)
(11) The month of starting the culture	June		
(12) The shipment time of oysters	From October of the next year to Marchof two years later, sequentially		
(13) Blue mussel attached to the rope	After one year from starting oyster culture		
(14) Filtration rate of oyster	Table 28.1, Fig. 28.6		
(15) Filtration rate of periphyton	Table 28.1, Fig. 28.6		
(16) Total number of oysters in longline	Table 28.2		
(17) Total weight of periphyton in longline	Figure 28.7		
The capacity of culture area in Oginohama Bay	(1) × (2) × (3)		
The daily fluctuation of the phytoplankton quantity	(5) × {(8) + (8) × (7) - (8) × (6)} - {(14) × (16) + (14) × (9) × (10) + (15) × (17)} × (8)		

28.3 Results and Discussion

28.3.1 Temporal Chl a Trend and Specific Growth Rate

In 2014, the monthly average Chl *a* concentration ranged from 0.96 to 4.1 μg/L (see Fig. 28.4).

Chl *a* concentration in January, February, and December, when temperatures are typically low, were lower than in other months. The average Chl *a* concentration in April was 4.1 μg/L, which was the highest of the year. May was the next highest at 3.3 μg/L; a spring bloom occurred at this time. After June, Chl *a* concentration decreased gradually, and the average concentration from June to August was

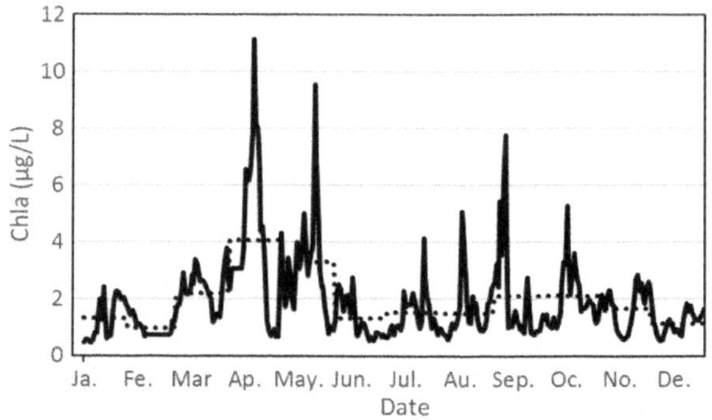

Fig. 28.4 Temporal change of Chl *a*. Solid line is daily Chl *a*. Dot line is Moving average of two weeks

1.4 µg/L. The Chl *a* concentration increased slightly from September to November at 1.9 µg/L. The annual average in 2013 and 2014 was 1.59 (unpublished data) and 1.92 µg/L, respectively. Chl *a* concentration in 2014 was slightly higher than that in 2013.

Chl *a* concentration has been measured in several shellfish culture areas (Gangnery et al. 2003; Chavez-Villalba et al. 2010). The concentrations recorded in Oginohama Bay were similar to those in the Mediterranean and off Mexico. The average phytoplankton specific growth rate was 0.53/day (Fig. 28.5).

This value was within the known limits for diatoms and dinoflagellates (Ministry of the Environment, Japan 2006).

28.3.2 Temporal Water Temperature Trend

The water temperature decreased from January to February, and was lowest at the beginning of February at 5.2 °C (Fig. 28.6). The water temperature gradually increased after February, and was highest in mid-August at 24.1 °C. Water

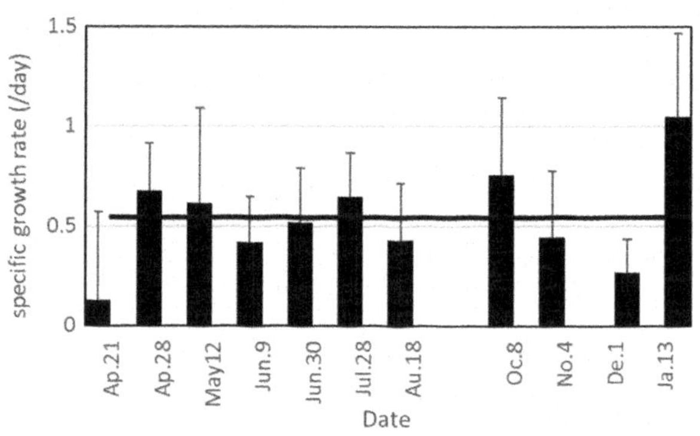

Fig. 28.5 The specific growth rates of phytoplankton in Oginohama Bay

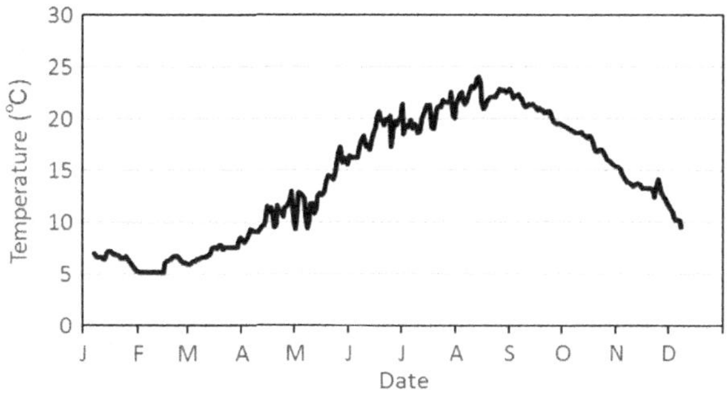

Fig. 28.6 Temporal change of water temperature in Momonoura, Oginohama Bay

temperature then gradually decreased to December. In 2014, the annual average water temperature in Oginohama Bay was 14.3 °C.

Water temperature has also been measured in several areas (Saràa et al. 1998; Hyun et al. 2001; Gangnery et al. 2003; Comeau et al. 2008; Chavez-Villalba et al. 2010) and that in Oginohama Bay was lower than the Mediterranean and off Mexico, slightly lower than off Korea, and slightly higher than off Canada.

28.3.3 Suitable Oyster Culture Density

Total oyster, mussel, and ascidian weight per oyster culture rope ranged from 29 to 99 kg in the following order, oysters > mussels > ascidians (Fig. 28.7). Mussels were the predominant periphyton, accounting for 7–34% of the total weight. While, filtration occurred in the following order, mussels > oysters > ascidians; mussels accounted for 41–87% of the total filtration (Fig. 28.8). As an oyster shell weight, which is unrelated to the filtration, is heavier than that of a mussel, mussel filtration was higher than that of oyster. Moreover, mussel filtration under low water temperature is higher than that of oysters (Comeau et al. 2008), thus mussel filtration in winter was higher than that of oyster.

The culture ropes were attached to the oyster longline facilities at the end of July, when the supply of mussel larvae from the aquaculture areas had decreased. Therefore, we assumed that mussels only attach to the culture ropes in 1 year of the 2 year oyster culture period. According to the calculation, total mussel filtration in the aquaculture areas was lower than that of the oysters from May onward. Moreover filtration of oysters attached to the quay did not even account for 10% of the total filtration of aquaculture (Fig. 28.9). We presumed that total cultured oyster filtration was higher than that of mussels and ascidians after spring in the culture grounds by schedules of oyster culture, and that oysters attached to the quay.

The mass balance of phytoplankton under four conditions (see Table 28.2) changed over time (Fig. 28.10).

The phytoplankton biomass for food availability increased in all conditions from January through April and decreased from May to June. Although phytoplankton

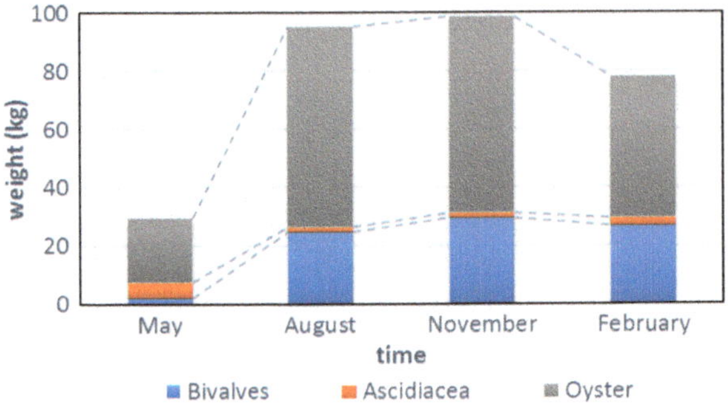

Fig. 28.7 Quantity of shellfish and periphyton per a rope

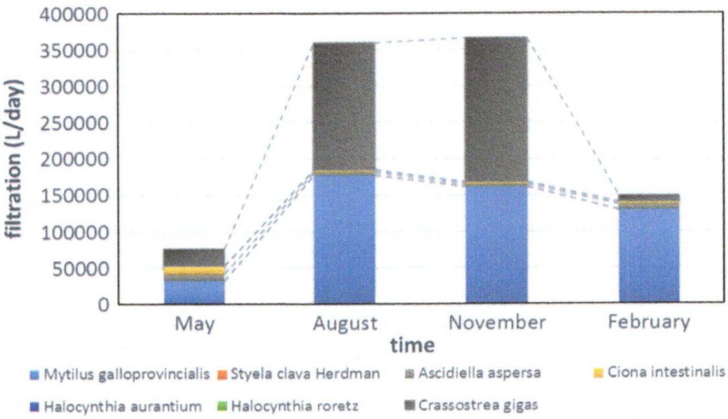

Fig. 28.8 Calculated filtration per rope

biomass fluctuated after June, a big change was not seen until December. Phytoplankton biomass remained high until May because a spring bloom occurred. However, phytoplankton biomass had decreased by the end of the spring bloom. Because oysters grow with rising water temperature, filtration also increased. Moreover, periphyton filtration also increased due to mussel's growth. These factors clearly had a role in the phytoplankton biomass decrease. Therefore, the risk of a decrease in oyster food availability was greater after June.

Phytoplankton biomass varied after June under the mass balance simulation conditions (see Fig. 28.10). Estimations of phytoplankton biomass as calculated for the culture density post-tsunami were always greater than zero (denoting sufficient food). Therefore, we estimated that the number of oyster longline facilities (approximately 450) after the tsunami did not result in reduced food availability, although the quantity of phytoplankton decreased with the cessation of blooming. Additionally, phytoplankton biomass in low-density conditions before the tsunami was calculated as always positive (indicating sufficient food). We believe that the number of longline facilities (approximately 1,100) before the tsunami did not

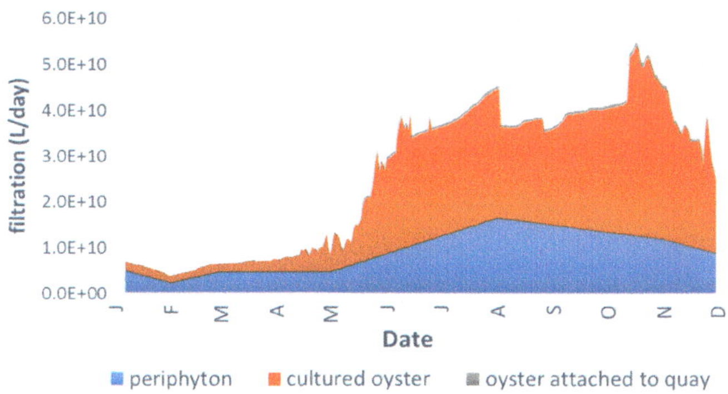

Fig. 28.9 Total filtration considered the culture conditions in Oginohama

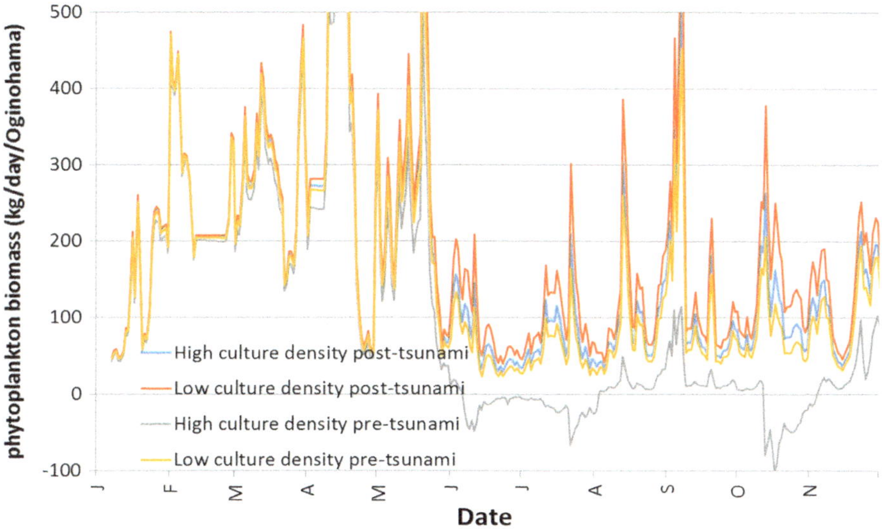

Fig. 28.10 Change of phytoplankton biomass in 2014 as calculated according to information in Table 28.3. The X axis shows the estimated values; we judged X value inferior to 0 to denote a shortage of food for oyster in the aquaculture area, and values higher than 0 as indicating the presence of sufficient food

negatively impact the oyster food supply, if the culture density was low. However, phytoplankton biomass in high density before the tsunami was calculated as negative (indicating food shortage) from June to August and from October to November. We speculate that oyster food availability would decrease, if we increased the number of longline facilities and carried out high-density culture.

Total oyster numbers in the four conditions was in the following order; low culture density post-tsunami < high culture density post-tsunami < low culture density pre-tsunami < high culture density pre-tsunami (see Table 28.2). Food availability in low culture density before the tsunami was the most suitable of the four conditions because oyster density was highest and phytoplankton biomass was calculated as negative (indicating food shortage, see Fig. 28.10). Total oyster density under low culture density before the tsunami (1,118 facilities × 360 ropes × 25 masters × 10ind./master) was equivalent to approximately 560 high culture density longline facilities used before the tsunami (400 ropes × 30 masters × 15ind./master). Although we thought that this culture condition was the most suitable at present, the phytoplankton biomass in the aquaculture area changes year by year. Therefore we intend to improve the accuracy of our results in the future by increasing our dataset.

Acknowledgements This work was mainly supported by a grant for the recovery from earthquake damage from the Fisheries Agency and the Reconstruction Agency, Japan. Thanks also to Dr. T. Kamiyama (TNFRI), and Dr. M. Yamasaki (TNFRI), for advice on this study.

References

Akashige S, Takayama K (2003) Technology development project of productivity improvements of Hiroshima's oyster. Ann Rep Hiroshima Fish Exp Stn 22:32–34 (in Japanese)

Akashige S, Hirata Y, Takayama K, Soramoto K (2005) Seasonal changes in oxygen consumption rates and filtration rates of the cultured Pacific oyster *Crasssotrea gigas*. Nippon Suisan Gakk 71:762–767 (in Japanese with English abstract)

Abo K, Toda S (2001) Evaluation model of farming density of Japanese pearl oyster, *Pinctada fucata martensii*, based on physiology and food environment. Bull Jpn Soc Fish Oceanogr 65:135–144 (in Japanese)

Brown JR, Hartwick EB (1988) Influences of temperature, salinity and available food upon suspended culture of the Pacific oyster, *Crassostrea gigas*: I Absolute and allometric growth. Aquaculture 70:231–251

Byron CJ, Costa-Pierce BA (2013) Carrying capacity tools for use in the implementation of an ecosystems approach to aquaculture. In: Ross LG, Telfer TC, Falconer L, Soto D, Aguilar-Manjarrez J (eds) Site selection and carrying capacities for inland and coastal aquaculture, pp 87–101

Chávez-Villalba J, Arreola-Lizárraga A, Burrola-Sánchez S, Hoyos-Chairez F (2010) Growth, condition, and survival of the Pacific oyster *Crassostrea gigas* cultivated within and outside a subtropical lagoon. Aquaculture 300:128–136

Comeau LA, Pernet F, Tremblay R, Bates SS, Leblanc A (2008) Comparison of eastern oyster (*Crassostrea virginica*) and blue mussel (*Mytilus edulis*) filtration rates at low temperatures. Can Tech Rep Fish Aquat Sci 2810:1–17

Filgueira R, Guyondet T, Comeaua LA, Grant J (2014) fully-spatial ecosystem-DEB model of oyster (*Crassostrea virginica*) carrying capacity in the Richibucto Estuary, Eastern Canada. J Mar Sys 136:42–54

Gangnery A, Chabirand J-M, Lagarde F, Gall PL, Oheix J, Bacher C, Buestel D (2003) Growth model of the Pacific oyster, *Crassostrea gigas*, cultured in Thau Lagoon (Me´diterrane´e, France). Aquaculture 215:267–290

Hiroshima City Agriculture, Forestry, and Fisheries Promotion Center (2016) Cultured method of the Hiroshima oyster. (in Japanese) http://www.haff.city.hiroshima.jp/suisansc/kaki_ikadasiki.html

Hyun K-H, Pang I-C, Klinck JM, Choi K-S, Lee J-B, Powell EN, Hofmann EE, Bochenek EA (2001) The effect of food composition on Pacific oyster *Crassostrea gigas* (Thunberg) growth in Korea: a modeling study. Aquaculture 199:41–62

Jeong WG, Cho SM (2013) Effects of Water Temperature and Body Weight on the Filtration Rate of Sea Squirt *Halocynthia roretzi*. Kor J Fish Aquat Sci 46:813–818 (In Korean with English abstract)

Kim YS, Moon TS (1998) Filtering Rate with Effect of Water Temperature and Size of two farming Ascidians *Styela clava* and *S. plicata*, and a farming Mussel *Mytilus edulis*. J Korean Fish Soc 31:272–277 (in Korean with English abstract)

Koike Y (2015) An Example of friendship and cooperation between France and Japan: oyster farming in Sanriku Area (Tohoku Region, Northern Japan) before and after tsunami—restoration and technical adaptation of culture systems. In: Ceccaldi HJ, Hénocque Y, Koike Y, Komatsu T, Stora G, Tusseau-Vuillemin M-H (eds) Proceedings of the 15th French-Japanese Oceanography Symposium, pp 3–14

Ministry of Agriculture, Forestry and Fisheries, Japan (2016) Marine aquaculture. 2. The yield of Oyster and Seaweed http://www.maff.go.jp/j/tokei/kouhyou/kaimen_gyosei/index.html (in Japanese)

Ministry of the Environment, Japan (2006) The adjustment and consideration of the knowledge regarding increase, a large scale of umber of instances of the red tide https://www.env.go.jp/council/20ari-yatsu/y200-23/mat02_2_2.pdf (in Japanese)

Miura H, Ito Y, Yoshida T (2013) Ecosystem Structure of a Fishing Port, and Presumption of a living Thing standing Stock. Coast Eng Comm B2(69):1211–1215

Mizuo H, Koichi Y, Shimomura K, Nishi E, Kimura T (2008) The purification of the Red-Tide seawater by oysters in the Nippon-Maru Dock in Yokohama. Ann Rep Yokohama Environ Sci Res Ins 32:56–61 (in Japanese)

Nagasawa K, Takahashi D, Itoh N, Takahashi KG, Osada M (2016) Growth of Japanese scallop *Mizuhopecten yessoensis* farmed by the ear-hanging method between different water layers and estimation of the productivity for farmed scallops in Ogatsu Bay (northeastern Japan) following the 2011 Great East Japan Earthquake. Nippon Suisan Gakk 82:321–329 (in Japanese with English abstract)

Okumura Y, Kakehi S, Yamashita Y (2013) Mass Balance of Dioxins Derived from Pesticides in Sendai Bay, Japan. JARQ 47:115–126

Randlov and Riisgard (1979) Efficiency of Particle Retention and Filtration Rate in Four Species of Ascidians. Mar Ecol Prog Ser 1:55–59

Riisgard HU, Larsen P, Pleissner D (2014) Allometric equations for maximum filtration rate in blue mussels *Mytilus edulis* and importance of condition index. Helgol Mar Res 68:193–198

Saràa G, Manganaro A, Cortese G, Pusceddu A, Mazzola A (1998) The relationship between food availability and growth in *Mytilus galloprovincialis* in the open sea (southern Mediterranean). Aquaculture 167:1–15

Unoki S (1998) Relation between the transport of gravitation circulation and the river discharge in bays. Oceanogr Jpn 7:192–283 (In Japanese with English abstract)

Yamamoto K, Handa T (2001) New Method for Direct Measurement of Ventilation Volume of the Ascidian, *Halocynthia roretzi*. J Natl Fish Univ 50:31–36 (in Japanese with English abstract)

Zapata M, Rodriguez F, Garrido JL (2000) Separation of chlorophylls and carotenoids from marine phytoplankton: a new HPLC method using a reversed phase C8 column and pyridine-conteining mobile phases. Mar Ecol Prog Ser 195:29–45

Chapter 29
Population Dynamics of the Manila Clam *Ruditapes Philippinarum* and Implications of the 2011 Tsunami Impact in Two Shallow, Semi-enclosed Bays in Northeastern Japan

Hirokazu Abe, Masami Hamaguchi, Naoto Kajihara, Yuichi Taniai, Akio Oshino, Akihiro Moriyama and Takashi Kamiyama

Abstract We investigated Manila clam population dynamics and evaluated the impact of the 2011 giant tsunami in Matsushima Bay and Mangoku-ura Inlet on the Pacific coast of north-eastern Japan. The clams have one spawning period: the planktonic larvae occurred from June to October with the peak period from mid-July to early-August. The larval density observed in the present study was relatively high compared with previous reports in other clam habitats around Japan, which was attributable to the abundance of large-sized adult clams that were mainly distributed around entrance area of the bays. Larval settlement (clams <1 mm shell length) and recruitment (clams >1 mm shell length) were mainly observed from July to November and September to January, respectively. The growth rate in shell length of the clam was 15 mm per year after settlement. In Matsushima Bay

H. Abe (✉)
Department of Biology, Center for Liberal Arts & Sciences, Iwate Medical University, 2-1-1 Nishitokuta Yahaba-Cho, Shiwa-Gun, Iwate 028-3694, Japan
e-mail: habe@iwate-med.ac.jp

H. Abe · T. Kamiyama
Tohoku National Fisheries Research Institute, Fisheries Research and Education Agency, 3-27-5 Shinhama-Cho, Shiogama, Miyagi 985-0001, Japan

H. Abe
Faculty of Science, Toho University, 2-2-1 Miyama, Funabashi, Chiba 274-8510, Japan

M. Hamaguchi · N. Kajihara
National Research Institute of Fisheries and Environment of Inland Sea, Fisheries Research and Education Agency, 2-17-5 Maruishi, Hatsukaichi, Hiroshima 739-0452, Japan

Y. Taniai · A. Oshino · A. Moriyama
Miyagi Prefecture Fisheries Technology Institute, 97-6 Watanoha Aza Sodenohama, Ishinomaki, Miyagi 986-2135, Japan

T. Komatsu et al. (eds.), *Oceanography Challenges to Future Earth*,
https://doi.org/10.1007/978-3-030-00138-4_29

and Mangoku-ura Inlet, the 2011 tsunami inundation heights were decreased by the presence of natural geographical barriers and the clam populations in both bays were less tsunami affected within the clam populations in north-eastern Japan, which suffered catastrophic impacts from the 2011 tsunami.

Keywords *Ruditapes philippinarum* · Artificial tidal flat · 2011 off the pacific coast of tohoku earthquake · Tsunami impact

29.1 Introduction

The Manila clam *Ruditapes philippinarum* is native along western pacific coastlines from Russia to China. In recent years, the clam is one of the most important commercial shellfish in the world. In Japan, clam production started to decrease since mid-1980s, and the 2010s levels (<30,000 tons) had decreased to less than 20% of the peak (160,000 tons) recorded in 1983 (Ministry of Agriculture, Forestry and Fisheries, Japan 2015). The reasons for the drastic decline in Japanese clam resources have long been discussed, and various factors have been considered as possible candidates for the decline, e.g., coastal reclamation (Toba 2004), over-fishing (Matsukawa et al. 2008), water quality deterioration (Uzaki et al. 2003; Toba 2004; Tsutsumi 2006), predation (Toba 2004; Tezuka et al. 2014), parasites (Waki et al. 2012), heavy-metal pollution (Tsutsumi et al. 2003; Tsutsumi 2006), decrease in the amount of sediment supply (Tsutsumi 2006), and decrease in primary production (Hamaguchi 2011). Furthermore, transplantation of juvenile clams from other areas to seed supply has led to many unintentionally introduced species, and one of these species, the moon snail *Laguncula pulchella*, is now a serious Manila clam predator in northeastern Japan. Many other unintentionally introduced species along with the clam transplantation are also recognized as a major problem in Japan (Okoshi 2004). Understanding of the clam population dynamics in northeastern Japan, which remains poorly understood, is necessary to establish a self-reliant production system that does not require clam seed transplantation in this region.

The 2011 Tohoku Earthquake (Mw 9.0) that occurred off the Pacific coast of northeastern Japan on March 11, 2011 (Ide et al. 2011; Ozawa et al. 2011; Sato et al. 2011; Simons et al. 2011) generated tsunami waves up to 43.3-m run-up height. The tsunami struck a 2000-km stretch of eastern Japan's Pacific coastline (Goto et al. 2011, 2012; Hinata et al. 2011; Kawai et al. 2011; Mori et al. 2011, 2012), and subsequently, markedly affected many marine organisms. Coseismic land subsidence up to approximately 1.2 m also generated by the earthquake in coastal areas (Ozawa et al. 2011) and altered the tidal flat landscape in many areas; some tidal flats were completely lost and became sub-littoral. These topographic alterations induced by land subsidence are thought to have changed habitat conditions and have significantly impacted many intertidal species.

The Manila clam is a representative species of tidal flat areas (Williams 1980; Miyawaki and Sekiguchi 1999; Ishii et al. 2001; Toba et al. 2007), and are considered one of the most tsunami affected species in several tidal flats along the Pacific coastline of the Tohoku region (Kanaya et al. 2015; Matsumasa et al. 2015; Abe et al. 2017). Various scale of the tsunami among bays according to the angle of incidence, seabed topography, bay shape, and the presence of either natural (e.g., sand bars and wind break forests) or artificial (concrete blocks) barriers leading to localized variation in the impacts of the tsunami on marine organisms (Miura et al. 2012; Urabe et al. 2013); this indicates that the tsunami impacts were different among clam populations in northeastern Japan.

Two shallow, semi-enclosed bays, Matsushima Bay and Mangoku-ura Inlet, the main clam production areas in northern Japan (Fig. 29.1), are on the Pacific coast of northeastern Japan near the epicenter of the 2011 Tohoku earthquake and were affected by both the earthquake and the resulting tsunami. In Mangoku-ura Inlet, tidal flats completely disappeared after the earthquake as a result of coseismic land subsidence (Okoshi 2012). As a countermeasure against such severe land subsidence, construction of three artificial tidal flats, which comprising a total area of 0.08 km^2, began in October 2013 for clam stock enhancement (Okoshi 2015): that is to say, Mangoku-ura Inlet is a unique field in which to assess the natural clam recruitment process.

The purpose of the present study is to investigate the population dynamics of the Manila clam in Matsushima Bay and Mangoku-ura Inlet, especially focused on the newly created artificial tidal flats, on the Pacific coast of northeastern Japan following the 2011 giant tsunami and to evaluate the impact of the tsunami on the clam populations in both bays.

29.2 Materials and Methods

29.2.1 Study Sites

Matsushima Bay (Fig. 29.1c: 38°21′N, 141°5′E) is a small, shallow, semi-enclosed embayment with an area of 35.3 km^2 (<4 m depth). The bay entrance is 1.7 km wide, and the bay is connected to the Pacific Ocean by six narrow channels. Wave action in the bay is weak, and the bottom is mainly comprised of muddy sediments with high silt content. There are over 200 small islands mainly around the bay entrance area and the most part of clam habitats are located around these islands (Sakai and Takahashi 1992).

Mangoku-ura Inlet (Fig. 29.1d: 38°25′N, 141°23′E) is a small, shallow, sac-like lagoon with an area of 7.4 km^2 (<4 m depth) located at the base of the Oshika Peninsula. The bay is connected to the north part of Sendai Bay by a narrow channel (500 m long and 100 m wide). Prior to the 2011 Tohoku Earthquake, there were 1 km^2 of tidal flats, including 0.7 km^2 of the artificially constructed tidal flats for clam stock enhancement. However, these tidal flats completely disappeared after

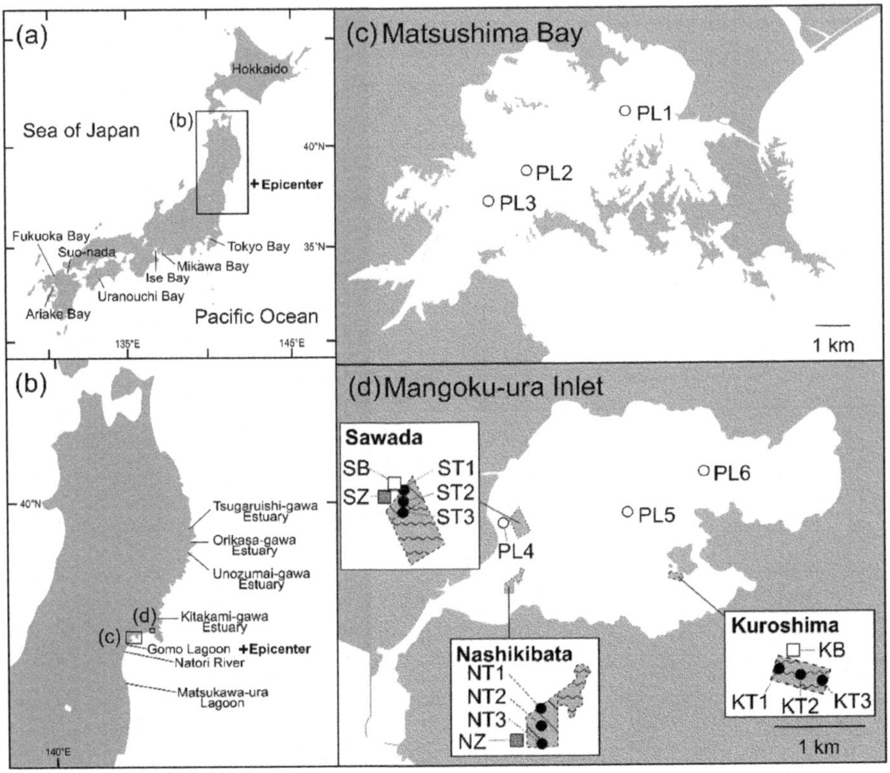

Fig. 29.1 Maps of sampling locations in Japan (**a**) and northeastern Japan (**b**). **c, d** Three planktonic larvae sampling stations in Matsushima Bay (PL1–3) and Mangoku-ura Inlet (PL4–6). **d** Three artificial tidal flats (Sawada, Nashikibata, and Kuroshima tidal flats) and four subtidal stations in *Zostera* beds (SZ and NZ: shaded squares) and either sandy or muddy bottoms (SB and KB: open squares) adjacent to the three artificial tidal flats in Mangoku-ura Inlet. Closed circles represent each three sampling stations in Sawada (ST1–3), Nashikibata (NT1–3), and Kuroshima (KT1–3) artificial tidal flats. Diagonal-lined and wavy-lined areas represent the artificial tidal flats which the constructions were completed in June 2014 and April 2015, respectively

the earthquake as a result of the estimated 0.8 m of coseismic land subsidence caused by the earthquake (Okoshi 2012, 2015). As a countermeasure against such severe land subsidence, three artificial tidal flats (Sawada, Nashikibata, and Kuroshima) with 0.08 km^2 total area have been constructed from mountain sediment for clam stock enhancement in the inlet. Partial construction of these artificial tidal flats was completed in June 2014 (Nashikibata and north part of Sawada) and April 2015 (Kuroshima and south part of Sawada) (Fig. 29.1d).

29.2.2 Larval Investigations

To investigate temporal variation in Manila clam larval density, plankton sampling was conducted at three stations in Matsushima Bay (PL1–3) from 2012 to 2014, and three (PL4–6) and one (PL5) stations in Mangoku-ura Inlet in 2013 and 2014, respectively (Fig. 29.1c, d). Samples were collected twice a week from July to mid-September and once every week or two in June and from mid-September through October. Horizontal larval distributions were also investigated at 21 and 13 stations in Matsushima Bay and Mangoku-ura Inlet, respectively on July 24, August 7, and August 30, 2013 (Fig. 29.4). The clam larvae were collected in one vertical haul from 2.5-m depth to the surface using a Kitahara's Quantitative Plankton Net with a 22.5-cm diameter mouth and a mesh size of 75 μm. The water volume filtered by the plankton net was calculated to be 100 L from the distance towed and the opening area of the net (0.04 m^2), assuming no overflow (i.e. 100% filtration efficiency). Larval samples were stored at −80 °C until they were processed for microscopic observation. Water temperature and salinity were measured with a Multiparameter Water Quality Sonde (Hydrolab DS5, Hach, Loveland, CO) at each sampling station.

Prior to microscopic observation, thawed samples were filtered through a 40-μm mesh, and all bivalve larvae were sorted and transferred into 1.5-ml tubes under an Olympus SZX16 stereomicroscope. The sorted samples were divided into 1–10 aliquots depending on larval density. The clam larvae were identified by an indirect immunofluorescence method using the Manila clam larvae-specific mouse monoclonal antibody (Hamaguchi 1999) as described in previous studies (Matsumura et al. 2001; Kasuya et al. 2004; Toba et al. 2007). The clam larvae were counted under an Olympus IX70 inverted fluorescent microscope with 490-nm excitation light and weak transmitted light. The clam larvae shell length (SL) was measured with an ocular micrometer and separated into three developmental stages: D-shaped (<130 μm), umbo (130–180 μm), and fully grown (>180 μm) larvae.

29.2.3 Clam Recruitment in Artificial Tidal Flats

The Manila clam recruitment process was investigated at three stations (ST1–3, NT1–3, and KT1–3; nine stations in total) in each of the three artificial tidal flats (Sawada, Nashikibata, and Kuroshima) in Mangoku-ura Inlet (Fig. 29.1d) once every 2 months from July 2014 to January 2016. At each station, sediment samples were collected using either a 3.7-cm dia. core sampler (only July and September, 2014; three replicates) or a 20-cm quadrat method to a depth of 10 cm (one replicate) and sieved through a 1 mm mesh to determine Manila clam density. All of the clams collected were counted and the shell lengths of the clams were measured with digital calipers. Newly settled juveniles were also collected at each station by taking three replicates of surface sediment samples to a depth of 3 cm using a 3.7-cm dia. core sampler. Newly settled juveniles that passed through a 1-mm mesh were sieved in three steps with 125-, 250-, and 500-μm meshes. The juveniles that remained on the sieves of each mesh were counted. In the present study, newly settled juveniles (new settlers) and recruits were defined as individuals that either passed through or remained on a 1-mm mesh, respectively.

29.2.4 Clam Distribution

In Matsushima Bay, clam density and size distribution were investigated at 16 stations from May to October 2015 (Fig. 29.6a). Three to five replicate samples were collected in tidal flats (Fig. 29.6a; Sts. A, B, and C) by the 20-cm quadrat method, in a subtidal area (Fig. 29.6a; St. F) by SCUBA, and in subtidal areas (Fig. 29.6a; the residual stations) by using a Joren (a long-handled, clam-harvesting dredge traditionally used in Japan: Masaki and Onohara 2003), which has a stainless basket 20 cm wide, 12 cm high, and with a 1-cm mesh spacing at the tip of a 3-m long handle. The Joren sampling areas were calculated by the width (20 cm) and dredge distances. The sediment samples collected by quadrat were sieved through a 2-mm mesh. All of the clams collected were counted and the shell lengths of the clams were measured with digital calipers.

Hearing investigation performed from eight local fisheries cooperatives and their branches about catches and distribution of the clam in the bay. Clam density in each fishing ground was roughly classified into three categories: i.e., abundant (>250 ind. m^{-2}), average (50–250 ind. m^{-2}), and scarce (<50 ind. m^{-2}).

In Mangoku-ura Inlet, clam density and size distribution were investigated on sandy mud and muddy bottoms (SB and KB), and the seagrass *Zostera marina* beds (SZ and NZ) adjacent to the three artificial tidal flats in October 2015 (Fig. 29.1d). Five replicate samples were collected by the 20-cm quadrat method at each station by SCUBA. The samples were sieved through a 2-mm mesh and all of the clams collected were counted and the shell lengths of the clams were measured with digital calipers.

29.3 Results

29.3.1 Larval Dynamics

The planktonic larvae of the clam were observed from June to October (Fig. 29.2). The mean density of the clam larvae fluctuated from 15 to 5,980 ind. m^{-3}, 30 to 16,033 ind. m^{-3}, and 0 to 22,313 ind. m^{-3} in 2012, 2013, and 2014, respectively, in Matsushima Bay, and from 30 to 28,550 ind. m^{-3} and 110 to 13,280 ind. m^{-3} in 2013 and 2014, respectively, in Mangoku-ura Inlet. The highest densities were observed at PL2 in Matsushima Bay on July 19 in 2013 (28,200 ind. m^{-3}) and at

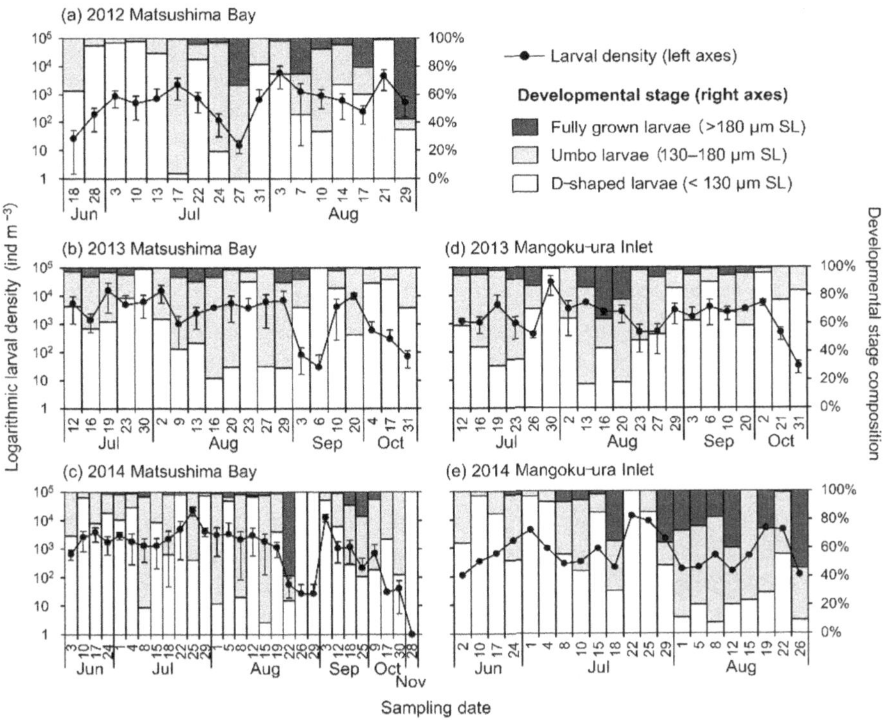

Fig. 29.2 Manila clam larval density (left axis) and the composition of each developmental stage (right axis) in Matsushima Bay (**a–c**) and Mangoku-ura Inlet (**d**, **e**) from 2012 to 2014. Error bars indicate standard deviations

PL6 in Mangoku-ura Inlet on July 30 in 2013 (49,800 ind. m^{-3}). Peak larval densities were usually observed from mid-July to early-August and a second peak occurred occasionally in September.

D-shaped larvae were the most abundant in each developmental stage (Fig. 29.2). The maximum mean densities of D-shaped larvae were 4,482 (Aug. 21), 9,953 (July 19), and 12,093 (Sept. 3) ind. m^{-3} in 2012, 2013, and 2014, respectively, in Matsushima Bay, and 28,250 (July 30) and 13,280 (July 22) ind. m^{-3} in 2013, and 2014, respectively, in Mangoku-ura Inlet. Umbo larvae density sometimes increased and exceeded half of the total larval density. The maximum mean umbo larvae densities were 2,165 (July 17), 5,587 (July 19), and 10,643 (July 25) ind. m^{-3} in 2012, 2013, and 2014, respectively, in Matsushima Bay, and 3,770 (Aug. 13) and 2,160 (Aug. 19) ind. m^{-3} in 2013, and 2014, respectively, in Mangoku-ura Inlet. Fully grown larvae densities were relatively low during the entire sampling period. The maximum mean fully grown larvae densities were 314 (Aug. 7), 730 (Aug. 29), and 123 (Aug. 5) ind. m^{-3} in 2012, 2013, and 2014, respectively, in Matsushima Bay, and 950 (Aug. 16) and 1,280 (Aug. 19) ind. m^{-3} in 2013, and 2014, respectively, in Mangoku-ura Inlet.

The mean surface water temperature at the three sampling stations ranged from 9.8 to 29.7 °C and 15.9 to 27.9 °C in Matsushima Bay and Mangoku-ura Inlet, respectively, during the sampling period (Fig. 29.3). The highest water temperature was observed in August. Temperature trends were similar in these two bays, except that it tended to be lower in Mangoku-ura Inlet in July. The mean surface salinity ranged from 18.5 to 32.1 and from 28.2 to 31.4 in Matsushima Bay and Mangoku-ura Inlet, respectively (Fig. 29.3). Salinity occasionally decreased in Matsushima Bay in July and August because of heavy rain fall but not in Mangoku-ura Inlet. The clam larvae usually occurred above 20 °C water temperature (Fig. 29.3). D-shaped larvae were observed in high densities at 20–26 °C, but the density tended to be lower above 27 °C (Fig. 29.3a). In contrast, umbo and fully-grown larvae generally occurred in higher water temperatures (Fig. 29.3b, c). The density of D-shaped and umbo larvae also tended to be high at lower salinities (Fig. 29.3a, b).

In the horizontal larval investigation, clam larvae were widely distributed in Matsushima Bay (Fig. 29.4a–c). The highest larval densities were recorded in the eastern part of the bay on July 24 (54,200 ind. m^{-3}) and August 30 (10,820 ind. m^{-3}), and at the mouth of the bay on August 7 (6,100 ind. m^{-3}). High larval densities (>5,000 ind. m^{-3}) were also observed near the central part of the bay on July 24 (16,700 ind. m^{-3}) and at the mouth of the bay on August 30 (8,360 ind. m^{-3}). Horizontal larval density differed by up to three orders of magnitude in Matsushima Bay and larval distribution was notably non-homogeneous. The clam larvae tended to be distributed in higher densities at the mouth and eastern parts of the bay. In Mangoku-ura Inlet, the clam larvae were only observed at low densities in the horizontal larval investigation (Fig. 29.4d, e) and were almost evenly distributed in the inlet on July 24.

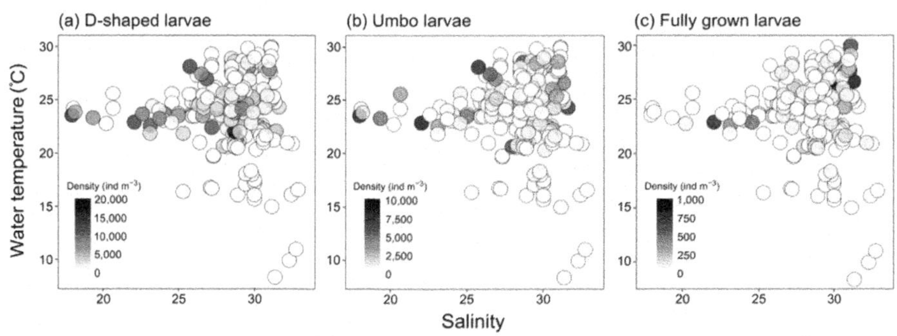

Fig. 29.3 Temperature-salinity diagrams of Manila clam D-shaped (**a**), umbo (**b**), and fully grown (**c**) larval density in Matsushima Bay and Mangoku-ura Inlet

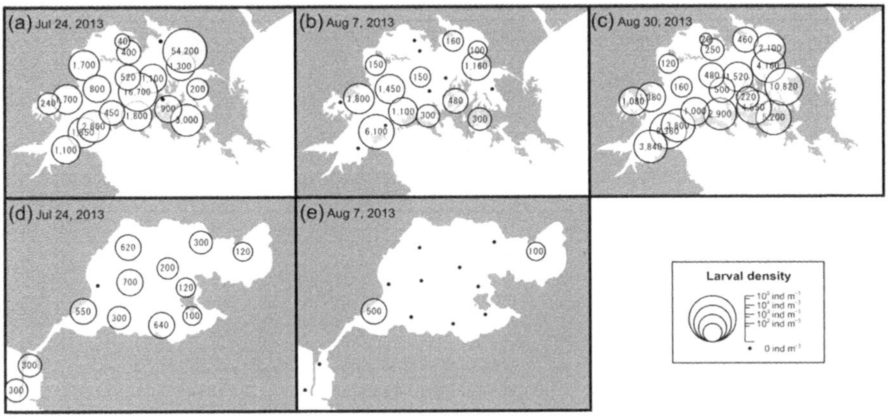

Fig. 29.4 Horizontal distributions of Manila clam larvae in Matsushima Bay (**a**–**c**) and Mangoku-ura Inlet (**d**, **e**) on July 24, August 7, and August 30, 2013. Circle diameter corresponds to logarithmic larval density. Black spots indicate no larvae. Numbers in the circles show larval densities (ind. m^{-3}) at each sampling station

29.3.2 Recruitment Process in Artificial Tidal Flats

No Manila clams remained on the 1-mm mesh in July 2014, when the first investigation was conducted after the construction of the artificial tidal flats (Sawada and Nashikibata) completed in June 2014 (Fig. 29.5d). In September 2014, both newly settled juveniles that passed through a 1-mm mesh and clams that remained on the 1-mm mesh were observed. The peak of new settler densities on the 250-μm (1,240 ind. m^{-2}) and 500-μm meshes (2,274 ind. m^{-2}) were observed in November 2014 (Fig. 29.5b, c), while the peak of new settler density on the 125-μm mesh was observed in July 2015 (1,860 ind. m^{-2}, Fig. 29.5a). The mean density of the clams that remained on the 1-mm mesh ranged from 800 to 2,967 ind. m^{-2} and from 708 to 1,908 ind. m^{-2} in the Sawada and Nashikibata artificial tidal flats, respectively, during the period from September 2014 to January 2016 (Fig. 29.5d). The shell length distribution of clams that remained on 1 mm mesh peaked at 2 mm from September 2014 to January 2015 (Fig. 29.5e). After that, the clam shell lengths increased and reached at 19 mm in January 2016. Juveniles from the 2015-year class that remained on the 1-mm mesh was observed from September 2015 and it was considerably more abundant in Sawada artificial tidal flat than in Nashikibata (Fig. 29.5e). Manila clam settlement and recruitment were not observed in Kuroshima artificial tidal flat, which was constructed in April 2015, during the period from May to September 2015 (Fig. 29.5a–d).

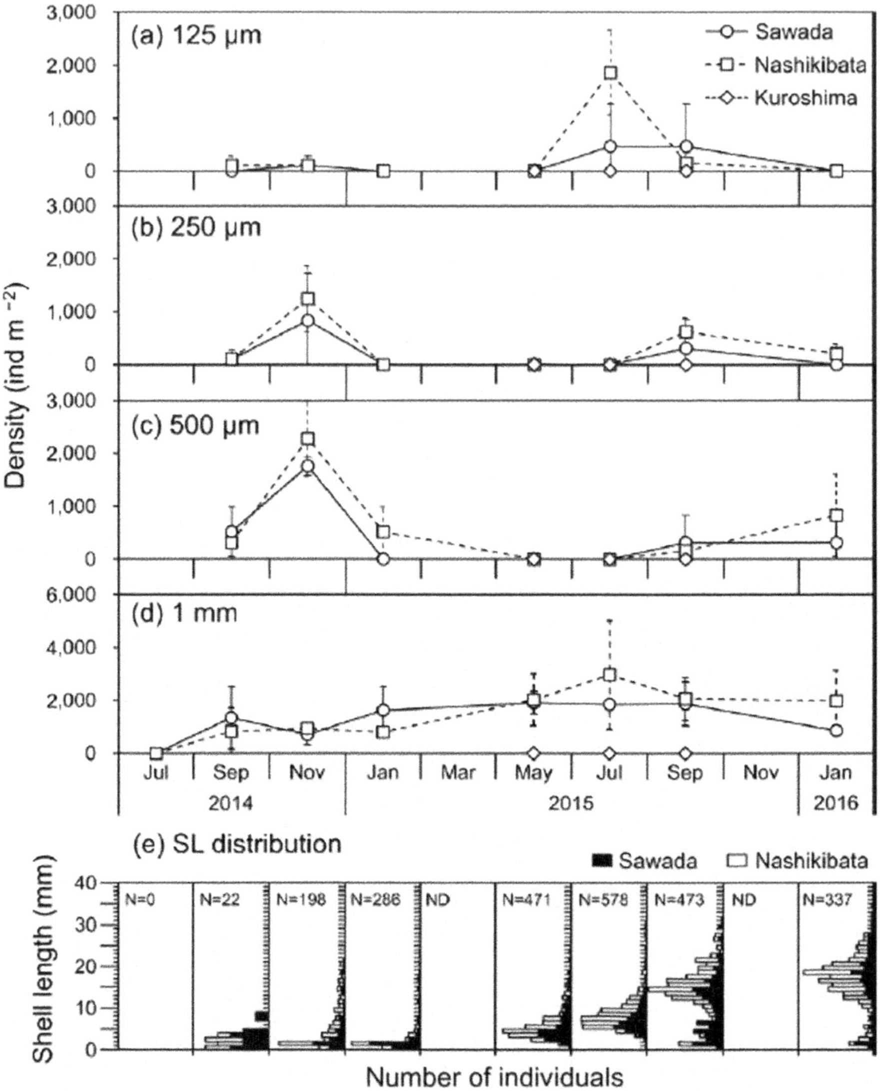

Fig. 29.5 Density of Manila clams that remained on the 125-μm (**a**), 250-μm (**b**), 500-μm (**c**), and 1-mm mesh (**d**), and shell length distribution of clams that remained on the 1-mm mesh (**e**) in three artificial tidal flats in Mangoku-ura Inlet during the period from July 2014 to January 2016. ND indicates no data

29.3.3 Clam Distribution

In Matsushima Bay, the Manila clam was collected at six out of 16 stations investigated (Fig. 29.6a). At these stations, clam density ranged from 42 to 575 ind. m^{-2} and shell length ranged from 2 to 49 mm (Fig. 29.6a, b). As a results of hearing investigation from local fisheries cooperatives and their branches, the clam habitats were mainly distributed around islands in the bay entrance, and that clam densities were also high at some sites in the inner part of the bay (Fig. 29.6c).

In Mangoku-ura Inlet, the Manila clam was collected at three stations adjacent to the two artificial tidal flats located near a channel (SB, SZ, and NZ), while none were collected at the muddy bottom station (KB) adjacent to Kuroshima tidal flat located in the inner part of the inlet (Fig. 29.7a). Clams were collected from both in a sandy bottom (SB) and among the seagrass *Z. marina* beds (SZ and NZ). Clam density ranged from 110 to 555 ind. m^{-2} and shell length from 2 to 48 mm at these stations (Fig. 29.7).

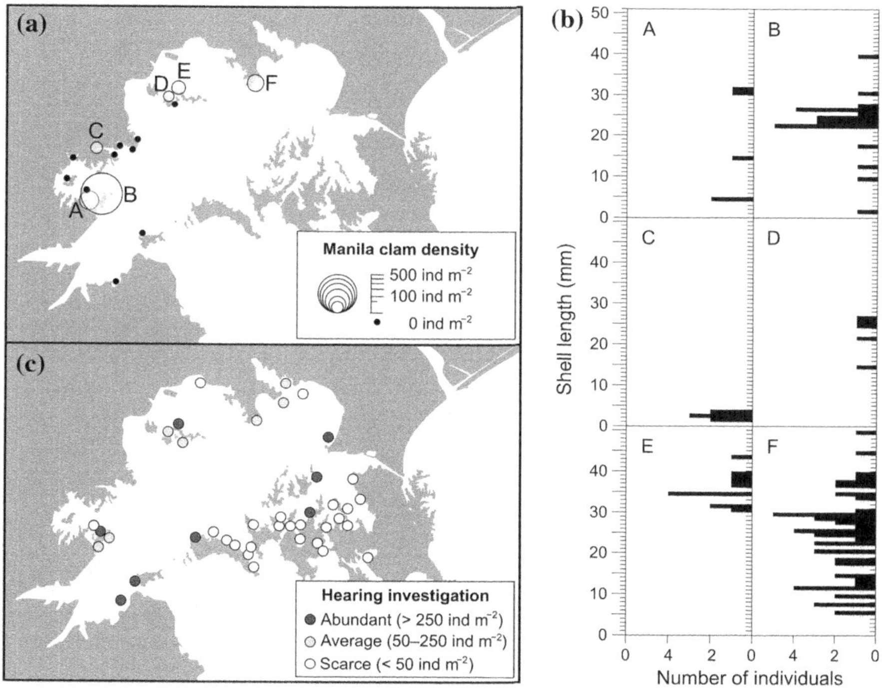

Fig. 29.6 **a** Clam density at the 16 Matsushima Bay stations investigated in the present study. Circle diameters correspond to clam densities. Black spots indicate no clams. **b** Clam shell length distribution collected at stations A to F in Fig. 29.6a. **c** Clam distribution information gathered from talking to the local fisheries cooperatives. Clam density at each site was roughly classified as abundant (>250 ind. m^{-2}), average (50–250 ind. m^{-2}) and scarce (<50 ind. m^{-2})

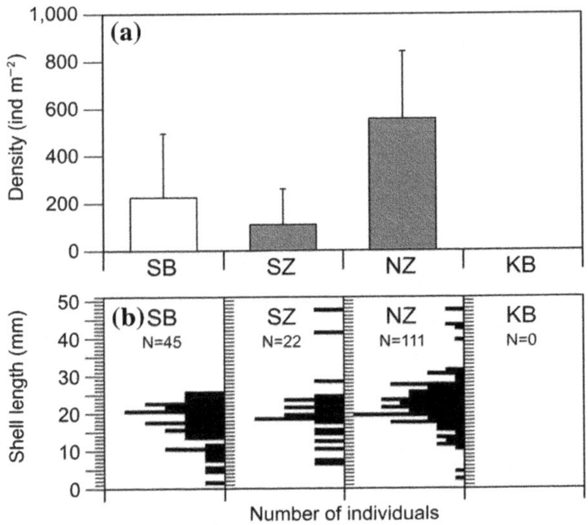

Fig. 29.7 Density (**a**) and shell length distribution (**b**) of the Manila clam at four subtidal stations in *Zostera* beds (SZ and NZ) and either sandy or muddy bottoms (SB and KB) adjacent to the three artificial tidal flats in Mangoku-ura Inlet

29.4 Discussion

29.4.1 Manila Clam Population Dynamics in Matsushima Bay and Mangoku-Ura Inlet

Planktonic larvae of the clam occurred from June to October with the peak period from mid-July to early-August in both study areas (Fig. 29.2). Previous studies (Shishido 1895; Miyagi Prefecture Fisheries Experimental Station 1916) also estimated that the spawning season of the clam in Sendai Bay was from mid-June to September by histological analysis. Peak Manila clam spawning occurs once a year in the summer in northern Japan and more than once a year in spring and fall in southern Japan (Toba and Miyama 1995; Matsumoto et al. 2014). Such geographic differences in spawning events are the result of differences in water temperature (Ponurovsky and Yakovlev 1992; Toba and Miyama 1995). In this study, no larval density peak was observed in autumn, suggesting that the clam populations spawn once a year in the summer. The period with high water temperature which is sufficient for their spawning was considered to be too short for biannual spawning of the species in Matsushima Bay and Mangoku-ura Inlet. D-shaped larvae were mainly observed at 20–26 °C (Fig. 29.3a), indicating that this temperature is suitable for spawning in the study areas; similar Manila clam spawning temperatures have been reported in other localities (Ponurovsky and Yakovlev 1992; Sladonja et al. 2011; Chung et al. 1994, 2001). Furthermore, larval peaks in Matsushima Bay coincided with low salinity (Fig. 29.3), suggesting that a reduction in salinity may stimulate spawning of this species as has been reported in other bivalve species (Southgate and Lee 1998).

The highest clam larval densities recorded in the present study were relatively high compared with those in the other Manila clam fishing ground in Japan (Table 29.1), even considering differences in study methods, e.g. identification methodology and plankton net mesh sizes. In the previous study, because of the low abundance of newly settled juveniles, it was concluded that the reproductive potential of the natural Manila clam population is low due to low adult clam density in Matsushima Bay (Miyagi Prefecture 1984). However, in the present study, the clam larvae occurred during the entire reproduction period (Fig. 29.2) and the larval density was comparable to that of other major clam habitats in Japan (e.g. Tokyo Bay and Mikawa Bay, Table 29.1). Therefore, we concluded that the reproductive potential of the clam populations in Matsushima Bay and Mangoku-ura Inlet is relatively high. The high larval density in these two bays is likely the result of the abundance of the large-sized adult clams (Figs. 29.6 and 29.7) as the clam fecundity is a function of size (Hasegawa et al. 2014).

Clam habitats tend to be distributed around the bay entrance area in Matsushima Bay (Fig. 29.6c). The bottom sediments in a large portion of Matsushima Bay are muddy with high silt content, particularly in the central bay area (Sakai and Takahashi 1992). The Manila clam basically favors sandy sediment habitats (Vincenzi et al. 2011), and coarse sandy sediment >0.5 mm is necessary for Manila clam larval settlement (Yamamoto 2005). For these reasons, clam populations were only found in tidal flats along the coastline and sandy mud areas around small islands in the bay (Fig. 29.6). The similar distribution patterns were also observed in Mangoku-ura Inlet (Fig. 29.7) and Matsukawa-ura Lagoon (Abe et al. 2017). This clam distribution pattern may be attributable to the presence of fast tidal flow at the bay entrance area compared with the inner bays, because fast currents usually increase food availability for the clams (Abe et al. 2015a) and create a sandy environment. The larval distribution pattern observed in the present study, i.e. higher density at the mouth and eastern parts of Matsushima Bay (Fig. 29.4a–c), was considered to reflect the adult distribution pattern. The clam habitats around islands located at bay entrance area are assumed to be a major source of larval supply in the bay.

In the artificial tidal flats in Mangoku-ura Inlet which were constructed in May 2014 (Sawada and Nashikibata), the Manila clam started to recruit in September 2014 (Fig. 29.5d). Since the clam recruitment was also observed from September in 2015 and small clams (<3 mm SL) were abundant until January (Fig. 29.5e) in 2015 and 2016, the clam recruitment season was considered to be from September to January. Additionally, the period from July to November, when the newly settled juveniles (<1 mm SL) were observed, was considered the larval settlement season (Fig. 29.5a–c). This settlement period corresponded well with the occurrence period of planktonic clam larvae (June to October, Fig. 29.2).

Shell length growth rate of the clam was 15 mm per year after settlement in the present study (Fig. 29.5e). The growth rate was higher than those reported in northern Japan (Hokkaido: Komorita et al. 2014; Tamura et al. 2014), lower than those in central Japan (Tokyo Bay: Nakamura et al. 2002; Toba et al. 2007), and comparable to those in other areas in north-eastern Japan (Matsukawa-ura Lagoon:

Table 29.1 The highest larval densities of the Manila clam recorded at various localities in Japan

Area	Highest density (ind m^{-3})	Year and month	Study period	Sampling interval	Number of stations	Mesh size (μm)[a]	Identification method	References
Matsushima Bay	10,720	Aug, 2012	Jun–Aug, 2012	Once–twice a week	3	75	Immunofluorescence	This study
Matsushima Bay	54,200	Jul, 2013	Jul–Oct, 2013	Once–twice a week	3–21	75	Immunofluorescence	This study
Matsushima Bay	29,400	Jul, 2014	Jun–Nov, 2014	Once–twice a week	3	75	Immunofluorescence	This study
Mangoku-ura Inlet	49,800	Jul, 2013	Jul–Oct, 2013	Once–twice a week	3–13	75	Immunofluorescence	This study
Mangoku-ura Inlet	13,280	Jul, 2014	Jun–Aug, 2014	Once–twice a week	1	75	Immunofluorescence	This study
Tokyo Bay	57,390	Sep, 2002	Apr, 2001–Dec, 2003	Once–twice a week	6	50	Immunofluorescence	Toba et al. (2007)
Tokyo Bay	186,830		2001–2003	Once two–four weeks	8	50	Immunofluorescence	Toba et al. (2012)
Mikawa Bay	>800	May, 1992	Apr, 1992–Mar, 1993	Once a week or month	1–28	100[b]	Morphological	Yamada et al. (1996)
Mikawa Ray	7,268	Aug, 1998	Apr, 1998–Mar, 1999	Once a month	6	100	Immunofluorescence	Matsumura et al. (2001)
Mikawa Bay	>25,000	Jul, 2000	May–Nov, 2000	Once a month	14	50	Immunofluorescence	Kuroda and Ochiai (2002)
Mikawa Bay	152,000	Jun, 1999	May–Nov, 1999 & 2001		3–4	100	Immunofluorescence	Kuroda (2005)
Ise Bay	ca. 300		2005–2006					Mizuno et al. (2009)
Uranouchi Bay	86,200	Nov, 2004	Jun, 2003–Mar, 2005	Once–twice a month	6	45	Immunofluorescence	Taino et al. (2006)
Uranouchi Bay	45,420	Nov, 2005	Apr–Dec, 2005	Twice a month	2–6	45	Immunofluorescence	Ueno et al. (2007)
Uranouchi Bay	6,035	Oct, 2006	Apr–Dec, 2006 & Nov–Dec, 2007	Twice a month	2	45	Immunofluorescence	Taino et al. (2009)

(continued)

Table 29.1 (continued)

Area	Highest density (ind m^{-3})	Year and month	Study period	Sampling interval	Number of stations	Mesh size (μm)[a]	Identification method	References
Uranouchi Bay	2,050	Dec, 2008	Sep–Dec, 2008	Once two weeks	2	45	Immunofluorescence	Taino and Hayashi (2010)
Uranouchi Bay	7,060	Oct, 2009	Sep-Dec, 2009	Once two weeks	2	45	Immunofluorescence	Taino and Ishikawa (2011)
Uranouchi Bay	6,550	Oct, 2010	Sep–Dec, 2010	Once two weeks	2	45	Immunofluorescence	Taino et al. (2012)
Uranouchi Bay	1,220	Dec, 2011	Sep–Dec, 2011	Once two weeks	2	45	Immunofluorescence	Kodama et al. (2013)
Suo-nada	1,940	Nov, 2006	May–Nov, 2005 & 2007, & Oct–Nov, 2006	ca. once a month	8–11	50	Immunofluorescence	Iwano et al. (2008)
Suo-nada	2,125	Jun, 2007	May–Nov, 2007 (except Jul & Sep)	Once a month	23	50	Immunofluorescence	Tezuka et al. (2008)
Suo-nada	255	Sep. 2006	May–Nov, 2005–2007	Once a month	10–11	50	Immunofluorescence	Yoshimatsu et al. (2008)
Fukuoka Bay	ca. 100	Jul, 2005	Apr, 2005–Mar, 2006	Once a month	2	100	Immunofluorescence	Ikeuchi and Satoh (2007)
Ariake Bay	4,750	Oct. 1997	Feb, 1997–Dec, 1998	Twice a week	2	40[c]	Morphological	Ishii et al. (2001)
Ariake Bay	1,119	Nov, 2004	Apr–Dec, 2004 & 2005	Once or twice a month	3–48	100	Immunofluorescence	Nishihama et al. (2011)

[a]Mesh size of plankton net used for collecting the clam larvae
[b]Count > 150 μm-larvae in diameter
[c]Count only umbo veliger larvae

Abe et al. 2017). These results indicate that the clam growth rate is mainly affected by water temperature although food availability is also important (Solidoro et al. 2000; Flye-Sainte-Marie et al. 2007).

In Kuroshima tidal flat, which was completed in April 2015, clams did not settle or recruit throughout study period (Fig. 29.5). No adult clams were collected from the subtidal muddy bottom sediment adjacent to Kuroshima tidal flat (Fig. 29.7); however, it seems unlikely to be because of a lack of larval supply, given the larval distribution results in this inlet (Fig. 29.4d). The sediment in Kuroshima tidal flat had a high broken oyster shell content, which was intended to promote Manila clam larval settlement and protect the clams from the predation by the moon snail (Sakai and Suto 2006). However, this might had a negative effect on clam recruitment. In this case, oyster shell paste (cracked oyster shell powder in seawater) was observed in the voids in the sediment and this inhibited seawater infiltration into the sediment. Because macrobenthos density was considerably low, particularly in infaunal species in Kuroshima tidal flat (Abe, unpublished data), Manila clam recruitment is thought to have been inhibited by the unfavorable sediment conditions. The sediment in the Sawada and Nashikibata tidal flats contained less cracked oyster shells and the similar phenomenon was not observed.

29.4.2 Impacts of the 2011 Tsunami on the Manila Clam Populations in Matsushima Bay and Mangoku-Ura Inlet

Matsushima Bay and Mangoku-ura Inlet, which are located near the 2011 Tohoku earthquake epicenter, were affected by the huge 2011 tsunami. However, the impact of the tsunami was relatively weak in both bays compared with other areas in northeastern Japan. The tsunami inundation heights were approximately 3 and <1 m in these two bays, respectively, which was relatively low compared with adjacent tsunami affected areas where recorded inundation heights ranged from 8 (Higashi-Matsushima) to 20 m (Onagawa Bay: Abe et al. 2015b). The low tsunami height was caused by the presence of many islands around bay entrance area in Matsushima Bay and the unique landlocked geography of Mangoku-ura Inlet (Fig. 29.1c, d). These geographical features acted as natural barriers and decreased the height of the 2011 tsunami.

Manila clam populations in Matsushima Bay and Mangoku-ura Inlet were subjected to relatively weak tsunami impacts compared with populations in other habitats in northeastern Japan. Only a small number of Manila clams survived the 2011 tsunami in Matsukawa-ura Lagoon (Fig. 29.1b: Abe et al. 2017), and the clam populations were almost completely wiped out by the tsunami in Tsugaruishi-gawa estuary, Orikasa-gawa estuary, Unozumai-gawa estuary, Kitakami-gawa estuary, and Gamo lagoon (Fig. 29.1b: Abe et al. 2014; Kanaya et al. 2014; Matsumasa et al. 2015). However, Manila clam populations in Matsushima Bay and

Mangoku-ura Inlet remained relatively unscathed. In Mangoku-ura Inlet, the largest tsunami impact was limited to the bay entrance area where the seagrass beds were also wiped out by the tsunami (Shoji and Morimoto 2016), while in the inner part of the inlet, seagrass beds and clams were less affected by the direct tsunami disturbance (Tamaki and Muraoka 2013; Muraoka et al. 2016). In the northeastern inner part of Matsushima Bay, the Manila clam and seagrass beds almost completely disappeared after the tsunami (Sato 2012; Muraoka et al. 2016). However, this is a peculiar case because the tsunami did not attack this area from the direction of bay entrance but attacked by crossed over the land from the east. Abundant recruitment of the Manila clam was observed during the period from July to November 2011 shortly after the tsunami in the previous studies (Okoshi 2012; Sato 2012). This indicates that many clams had survived to reproduce in Matsushima Bay. In the present study, planktonic larval density was high at the time of only one year after the 2011 tsunami (Table 29.1, Fig. 29.2). Furthermore, clam density and distribution patterns in Matsushima Bay after the tsunami (Fig. 29.6) were not notably different from that in 1982, before the tsunami (Miyagi Prefecture 1984). These results suggest that the clam populations in Matsushima Bay and Mangoku-ura Inlet were relatively unaffected by the earthquake and tsunami. The impact of the tsunami on Manila clam populations would be related to the inundation height, as previously reported in macrobenthic communities (Urabe et al. 2013).

Speed of clam population recovery following the tsunami seems to be closely related to degree of the tsunami impact. In Matsukawa-ura Lagoon (Fig. 29.1b), where only a small number of relatively large sized Manila clams survived the 2011 tsunami, the clam recruitment was observed in the same year of the tsunami and the population recovered rapidly (Abe et al. 2017). However, in Tsugaruishi-gawa estuary, Orikasa-gawa estuary, Unozumai-gawa estuary, Kitakami-gawa estuary, and Gamo lagoon (Fig. 29.1b), where clams were almost wiped out by the tsunami, the clam recruitment was observed in summer 2012, one year after the tsunami and population recovery delayed (Abe et al. 2014; Kanaya et al. 2014; Matsumasa et al. 2015). In Matsushima Bay and Mangoku-ura Inlet, the clam recruitment was reported to be observed from July to November 2011, shortly after the tsunami (Okoshi 2012; Sato 2012; Tamaki and Muraoka 2013). This may support the relatively low impact of the tsunami on the clam populations in both bays.

Habitat alteration due to topographic changes caused by the tsunami and seismic land subsidence was considered to have long-term effects on clam populations. In Mangoku-ura Inlet, the clams were affected by the almost complete disappearance of tidal flats. In 2004, before the tsunami, the clam densities were 1–472 ind. m^{-2} in the tidal flats which existed near the constructed site of Kuroshima tidal flat (Sakai and Suto 2006). However, in the present study, no clams were collected from the subtidal mud bottom around Kuroshima tidal flat (KB, Fig. 29.7). Silt content of the bottom sediment gradually increased to approximately 20% in this area as a result of reduced seawater velocity caused by increased water depth (Tamaki and Muraoka 2013). Increased silt content probably resulted in the disappearance of the

Manila clam in this area because high mud content in sediment is unsuitable for inhabitation of the clam (Miyagi Prefecture 1984). In addition to the land subsidence, topographic changes were induced also by tsunami deposition. In a tidal flat in Matsushima Bay, topographic instability due to the topographic changes by the 2011 tsunami deposition continuously lead to the decrease in density of macrobenthos including Manila clam (Abe et al., unpublished data). In the brackish environment of the Natori river, north-eastern Japan (Fig. 29.1b), the shape of the river mouth was completely altered by the 2011 earthquake and tsunami, and this topographic change led decrease in salinity and had long-term effects on benthic bivalves (Ito et al. 2016). Habitat alteration resulting from topographic changes by the earthquakes and tsunamis can cause longer-term impacts on benthic animals rather than direct tsunami impacts.

Acknowledgements We would like to express the deepest appreciation to Ritsuko Takahashi of Tohoku National Fisheries Research Institute for her help during sample treatment. We are grateful to Nobuhisa Abe, Shoichi Hanawa, Hiroshi Ito, Akira Kumagai, and Yuko Nishizawa of the Miyagi Prefecture Fisheries Research and Development Center, and Captain Kazuo Suzuki for their help with field sampling. We are also grateful to Matsushima branch, Miyato Seibu branch, Naruse branch, Shichigahama branch, Shiogama Dai-ichi branch, Urato branch, and Urato Tobu branch of Miyagi Prefecture Fisheries Cooperative and Shiogama City Fisheries Cooperative for their kind cooperation to our hearing investigation. This study was partially supported by a grant from the Fisheries Agency of Japan.

References

Abe H, Kondoh T, Kobayashi G, Sato-Okoshi W (2014) Environmental and macrobenthic faunal changes occurred in river month, tidal flat and bay after the 2011 tsunami—examples of Gamo Lagoon and Onagawa Bay. Kaiyo Monthly 46:48–55 (in Japanese)

Abe H, Hasegawa N, Yoon S, Kishi MJ (2015a) Evaluation of Manila clam (*Ruditapes philippinarum*) growth and microphytobenthos resuspension in a subarctic lagoon in Japan. Hydrobiologia 758: 87–98

Abe H, Kobayashi G, Sato-Okoshi W (2015b) Impacts of the 2011 tsunami on the subtidal polychaete assemblage and the following recolonization in Onagawa Bay, northeastern Japan. Mar Environ Res 112: 86–95

Abe H, Sato T, Iwasaki T, Wada T, Tomiyama T, Sato T, Hamaguchi M, Kajihara N, Kamiyama T (2017) Impact of the 2011 tsunami on the Manila clam *Ruditapes philippinarum* and subsequent population recovery in Matsukawa-ura Lagoon, Fukushima, northeastern Japan. Reg Stud Mar Sci 9:97–105

Chung E-Y, Ryou D-K, Lee J-H (1994) Gonadal development, age and growth of the shortnecked clam, *Ruditapes philippinarum* (Pelecypoda: Veneridae), on the coast of Kimje, Korea. Korean J Malacol 10:38–54

Chung E-Y, Hur SB, Hur Y-B, Lee JS (2001) Gonadal maturation and artificial spawning of the Manila clam *Ruditapes philippinarum* (Bivalvia: Veneridae), in Komso Bay, Korea. J Fish Sci Technol 4:208–218

Flye-Sainte-Marie J, Jean F, Paillard C, Ford S, Powell E, Hofmann E, Klinck J (2007) Ecophysiological dynamic model of individual growth of *Ruditapes philippinarum*. Aquaculture 266:130–143

Goto K, Chagué-Goff C, Fujino S, Goff J, Jaffe B, Nishimura Y, Richmond B, Sugawara D, Szczuciński W, Tappin DR, Witter RC, Yulianto E (2011) New insights of tsunami hazard from the 2011 Tohoku-oki event. Mar Geol 290:46–50
Goto K, Chagué-Goff C, Goff J, Jaffe B (2012) The future of tsunami research following the 2011 Tohoku-oki event. Sed Geol 282:1–13
Hamaguchi M (1999) *Ruditapes philippinarum* larvae-specific monoclonal antibody. Japan Pat Appl 1998–2913026
Hamaguchi M (2011) Relationship among primary production, marine environment and fisheries production of bivalves in the Seto Inland Sea. Bull Fish Res Agen 34:33–47 (in Japanese with English abstract)
Hasegawa N, Sawaguchi S, Unuma T, Onitsuka T, Hamaguchi M (2014) Variation in Manila clam (*Ruditapes philippinarum*) fecundity in eastern Hokkaido, Japan. J Shellfish Res 33:739–746
Hinata H, Fujii S, Furukawa K, Kataoka T, Miyata M, Kobayashi T, Mizutani M, Kokai T, Kanatsu N (2011) Propagating tsunami wave and subsequent resonant response signals detected by HF radar in the Kii Channel, Japan. Estuar Coast Shelf Sci 95:268–273
Ide S, Baltay A, Beroza GC (2011) Shallow dynamic overshoot and energetic deep rupture in the 2011 Mw 9.0 Tohoku-Oki Earthquake. Science 332:1426–1429
Ikeuchi H, Satoh H (2007) Appearance of larvae of Japanese littleneck clam in Fukuoka Bay. Bull Fukuoka Fish Mar Technol Res Cent 17:89–92 (in Japanese)
Ishii R, Sekiguchi H, Nakahara Y, Jinnai Y (2001) Larval recruitment of the Manila clam *Ruditapes philippinarum* in Ariake Sound, southern Japan. Fish Sci 67:579–591
Ito K, Katayama A, Shizuka K, Monna N (2016) Effects of the Great East Japan Tsunami on fish populations and ecosystem recovery. The Natori River; northeastern Japan. In: Santiago-Fandiño V, Tanaka H, Spiske M (eds) Tsunamis and earthquakes in coastal environments. Coastal Research Library, vol 14. Springer, Switzerland, pp 201–216
Iwano H, Fukuda Y, Egashira J, Hirakawa C (2008) Hydrographic survey for improvement and development of extensive fisheries grounds of Japanese littleneck clam. Report of 2007 commissioned survey projects for the improvement of fisheries industry. Fisheries Agency, Tokyo, 17 pp (in Japanese)
Kanaya G, Maki H, Suzuki T, Sato-Okoshi W, Kikuchi E (2014) Tsunami-induced changes in a shallow brackish lagoon ecosystem (Gamo Lagoon) in Sendai Bay, Japan. Global Environ Res 18:35–46
Kanaya G, Suzuki T, Kikuchi E (2015) Impacts of the 2011 tsunami on sediment characteristics and macrozoobenthic assemblages in a shallow eutrophic lagoon, Sendai Bay, Japan. PLoS One 10:e0135125
Kasuya T, Hamaguchi M, Furukawa K (2004) Detailed observation of spatial abundance of clam larva *Ruditapes philippinarum* in Tokyo Bay, central Japan. J Oceanogr 60:631–636
Kawai H, Satoh M, Kawaguchi K, Seki K (2011) The 2011 off Pacific coast of Tohoku earthquake tsunami observed by GPS buoys. J Jpn Soc Civ Eng Ser B2 67:1291–1295
Kodama O, Taino S, Suzuki R (2013) Guidance for the clam fishery. Annu Rep Kochi Pref Fish Exp Stn 109:177–194 (in Japanese)
Komorita T, Kajihara R, Tsutsumi H, Shibanuma S, Yamada T, Montani S (2014) Food sources for *Ruditapes philippinarum* in a coastal lagoon determined by mass balance and stable isotope approaches. PLoS One 9:e86732
Kuroda N (2005) Larval transportation and settlement mechanism to tidal flat in Japanese littleneck clam *Ruditapes philippinarum*. Bull Fish Res Agen Suppl 3:67–77 (in Japanese with English abstract)
Kuroda N, Ochiai M (2002) Distribution of D-shaped larvae of Japanese littleneck clam *Ruditapes philippinarum* in Mikawa Bay. Bull Aichi Fish Res Inst 9:19–26 (in Japanese with English abstract)
Masaki K, Onohara T (2003) Condition of fisheries and distribution of ark shell, *Scapharca sativa* in the Saga portion of the Ariake Sound. Bull Saga Pref Ariake Fish Res Dev Cent 21:29–36 (in Japanese)

Matsukawa Y, Cho N, Katayama S, Kamio K (2008) Factors responsible for the drastic catch decline of the Manila clam *Ruditapes philippinarum* in Japan. Nippon Suisan Gakkaishi 74:137–143 (in Japanese with English abstract)

Matsumasa M, Kinoshita K, Ito H, Kojima S (2015) Between tides on Sanriku coast: the vulnerability and robustness to catastrophic natural disturbances. DNA Polymorphism 23:9–16 (in Japanese)

Matsumoto T, Awaji M, Higano J, Hasegawa N, Yamamoto T, Shibata R, Hada Y, Sakurai I, Miyawaki D, Hirai A, Hodokawa K, Hanyu K, Ikushima N, Uchikawa J, Chow S (2014) Reproductive cycle of the Manila clam *Ruditapes philippinarum* in Japan. Nippon Suisan Gakkaishi 80:548–560 (in Japanese with English abstract)

Matsumura T, Okamoto S, Kuroda N, Hamaguchi M (2001) Temporal and spatial distributions of planktonic larvae of the clam *Ruditapes philippinarum* in Mikawa Bay; Application of an immunofluorescence identification method. Jpn J Benthol 56:1–8 (in Japanese with English abstract)

Ministry of Agriculture, Forestry and Fisheries, Japan (2015) Annual report of statistics on fisheries and aquaculture production in 2014. Ministry of Agriculture, Forestry and Fisheries, Japan, Tokyo, 29 pp (in Japanese)

Miura O, Sasaki Y, Chiba S (2012) Destruction of population of *Batillaria attramentaria* (Caenogastropoda: Batillaridae) by tsunami waves of the 2011 Tohoku earthquake. J Molluscan Stud 78:377–380

Miyagi Prefectural Fisheries Experimental Station (1916) Miyagiken-san Gyokairui Sanranki Chousa 3, Asari [Spawning period survey of fish and shellfish in Miyagi Prefecture III, Japanese littleneck clam]. Miyagiken Gyogyou Kihon Chousa Houkokusyo [Miyagi Prefectural Report of the Basic Fisheries Survey] 6:123–127 (in Japanese)

Miyagi Prefecture (1984) Matsushima Bay area. In: Fisheries Agency (ed) Daikibo Zousyokujyou Zousei Jigyou Chousa Sougou Houkokusyo [Report on the survey of large-scale development of fishing ground project]. Fisheries Agency, Japan, Tokyo, pp 1–64 (in Japanese)

Miyawaki D, Sekiguchi H (1999) Interannual variation of bivalve populations on temperate tidal flats. Fish Sci 65:817–829

Mizuno T, Maruyama T, Higano J (2009) Transition and prospect of Asari clam (*Ruditapes philippinarum*) fisheries in Ise Bay, Mie Prefecture. Bull Mie Pref Fish Res Inst 17:1–21 (in Japanese)

Mori N, Takahashi T, The 2011 Tohoku Earthquake Tsunami Joint Survey Group (2012) Nation wide post event survey and analysis of the 2011 Tohoku Earthquake Tsunami. Coast Eng J 54:1250001

Mori N, Takahashi T, Yasuda T, Yanagisawa H (2011) Survey of 2011 Tohoku earthquake tsunami inundation and run-up. Geophys Res Lett 38:L00G14

Muraoka D, Shimizu D, Shirafuji N, Tamaki H, Noda T, Hamaguchi M, Fujinami Y, Matsumoto Y (2016) Impact and recovery process of the Great East Japan Earthquake and the following tsunami on *Zostera* meadows. Nippon Suisan Gakkaishi 85:142 (in Japanese)

Nakamura Y, Hagimo M, Hiwatari T, Iijima A, Kohata K, Furota T (2002) Growth of the Manila clam *Ruditapes philippinarum* in Sanbanse, the shallow coastal area in Tokyo Bay. Fish Sci 68:1309–1316

Nishihama S, Yurimoto T, Naito T, Mori Y, Fujii A, Nasu H, Kimoto K, Maeno Y (2011) Trends in population density and shell length of the planktonic larva *Ruditapes philippinarum* in Ariake Bay. Aquaculture Sci 59:255–264 (in Japanese with English abstract)

Okoshi K (2004) Alien species introduced with imported clams: the clam-eating moon snail *Euspira fortunei* and other unintentionally introduced species. Jpn J Benthol 59:74–82 (in Japanese with English abstract)

Okoshi K (2012) How do the 3.11 tsunami affect the intertidal animals? Jpn J Benthol 66:117–119 (in Japanese)

Okoshi K (2015) Impact of repeating massive earthquakes on intertidal mollusk community in Japan. In: Ceccaldi H-J et al (eds) Marine productivity: disturbance and resilience of coastal socio-ecosystems. Springer, Switzerland, pp 55–62

Ozawa S, Nishimura T, Suito H, Kobayashi T, Tobita M, Imakiire T (2011) Coseismic and postseismic slip of the 2011 magnitude-9 Tohoku-Oki earthquake. Nature 475:373–376

Ponurovsky SK, Yakovlev YM (1992) The reproductive biology of the Japanese littleneck, *Tapes philippinarum* (A. Adams and Reeve, 1850) (Bivalvia: Veneridae). J Shellfish Res 11:265–277

Sakai K, Suto A (2006) Improvement of sand bed of Manila-clam *Ruditapes philippinarum* for prevention of predator; Moon snail *Neverita didyma*. Miyagi Pref Rep Fish Sci 6:83–86 (in Japanese)

Sakai K, Takahashi K (1992) Reclamation of seabed for the Japanese littleneck *Tapes philippinarum* in Matsushima Bay. Fish Eng 29:41–45 (in Japanese)

Sato S (2012) Faunal changes of mollusks around the Tona coast, Miyagi Prefecture, northern Japan before and after the Great East Japan Earthquake. Jpn J Benthol 66:122–123 (in Japanese)

Sato M, Ishikawa T, Ujihara N, Yoshida S, Fujita M, Mochizuki M, Asada A (2011) Displacement above the hypocenter of the 2011 Tohoku-Oki Earthquake. Science 332:1395

Shishido I (1895) Sendai-chihou Kairui Sanranki [Spawning season of shellfishes in Sendai area]. Suisan Chousa Houkoku [Fisheries Survey Reports] 3:27–116 (in Japanese)

Shoji J, Morimoto M (2016) Changes in fish community in seagrass beds in Mangoku-ura Bay from 2009 to 2014, the period before and after the tsunami following the 2011 off the Pacific coast of Tohoku earthquake. J Oceanogr 72:91–98

Simons M, Minson SE, Sladen A, Ortega F, Jiang J, Owen SE, Meng L, Ampuero J-P, Wei S, Chu R, Helmberger DV, Kanamori H, Hetland E, Moore AW, Webb FH (2011) The 2011 magnitude 9.0 Tohoku-Oki Earthquake: mosaicking the megathrust from seconds to centuries. Science 332:1421–1425

Sladonja B, Bettoso N, Zentilin A, Tamberlich F, Acquavita A (2011) Manila clam (*Tapes philippinarum* Adams & Reeve, 1852) in the Lagoon of Marano and Grado (Northern Adriatic Sea, Italy): socio-economic and environmental pathway of a shell farm. In: Sladonja B (ed) Aquaculture and the environment—A shared destiny. In Tech, Rijeka, pp 51–78

Solidoro C, Pastres R, Melaku Canu D, Pellizzato M, Rossi R (2000) Modelling the growth of *Tapes philippinarum* in northern Adriatic lagoons. Mar Ecol Prog Ser 199:137–148

Southgate PC, Lee PS (1998) Hatchery rearing of the tropical blacklip oyster *Saccostrea echinata* (Quoy and Gaimard). Aquaculture 169:275–281

Taino S, Hayashi Y (2010) Occurrence situation of Japanese littleneck clam planktonic larvae in Uranouchi Inlet in 2008. Annu Rep Kochi Pref Fish Exp Stn 106:177–180 (in Japanese)

Taino S, Ishikawa T (2011) Fishery guidance I. Occurrence situation of Japanese littleneck clam planktonic larvae in Uranouchi Inlet in 2009. Annu Rep Kochi Pref Fish Exp Stn 107:221–225 (in Japanese)

Taino S, Ura Y, Hayashi Y, Ohkawa T, Ando H (2006) Investigation of the cause of Japanese littleneck clam poor catch. Seasonal changes of Japanese littleneck clam planktonic larvae in Uranouchi Inlet. Annu Rep Kochi Pref Fish Exp Stn 102:81–83 (in Japanese)

Taino S, Ueno Y, Ando H, Hayashi Y, Ohkawa T (2009) Occurrence situation of Japanese littleneck clam planktonic larvae in Uranouchi Inlet in 2006 and 2007. Annu Rep Kochi Pref Fish Exp Stn 105:187–190 (in Japanese)

Taino S, Sugimoto M, Suzuki R (2012) Fishery guidance I. Occurrence situation of Japanese littleneck clam planktonic larvae in Uranouchi Inlet in 2010. Annu Rep Kochi Pref Fish Exp Stn 108:219–223 (in Japanese)

Tamaki H, Muraoka D (2013) Disturbances of tidal flat, *Zostera* and *Eisenia bicyclis* habitats by the Great East Japan Earthquake and their recovery in Miyagi Prefecture. J Environ Conservat Eng 42:46–54 (in Japanese)

Tamura R, Nakagawa C, Goshima S (2014) Growth of short-necked clam *Ruditapes philippinarum* in Hakodate Bay, Hokkaido. Bull Fish Sci Hokkaido Univ 64:37–44 (in Japanese with English abstract)

Tezuka N, Hamaguchi M, Tarutani K, Kuwahara H, Saito H, Shimizu M, Takeoka H, Iseki K (2008) Hydrographic survey for improvement and development of extensive fisheries grounds of Japanese littleneck clam. Report of 2007 commissioned survey projects for the improvement of fisheries industry. Fisheries Agency, Tokyo, 10 pp (in Japanese)

Tezuka N, Kanematsu M, Asami K, Nakagawa N, Shigeta T, Uchida M, Usuki H (2014) *Ruditapes philippinarum* mortality and growth under netting treatments in a population-collapsed habitat. Coast Ecosyst 1:1–13

Toba M (2004) The decline of Manila clam stock in Tokyo Bay. Bull Fish Res Agen 1:13–18

Toba M, Miyama Y (1995) Influence of temperature on the sexual maturation in Manila clam, *Ruditapes philippinarum*. Aquac Sci 43:305–314

Toba M, Yamakawa H, Kobayashi Y, Sugiura Y, Honma K, Yamada H (2007) Observations on the maintenance mechanisms of metapopulations, with special reference to the early reproductive process of the Manila clam *Ruditapes philippinarum* (Adams & Reeve) in Tokyo Bay. J Shellfish Res 26:121–130

Toba M, Yamakawa H, Shoji N, Kobayashi Y (2012) Vertical distribution of Manila clam *Ruditapes philippinarum* larvae characterized through year-round observations in Tokyo Bay, Japan. Nippon Suisan Gakkaishi 76:1135–1148 (in Japanese with English abstract)

Tsutsumi H (2006) Critical events in the Ariake Bay ecosystem: clam population collapse, red tides, and hypoxic bottom water. Plankton Benthos Res 1:3–25

Tsutsumi H, Tsukuda M, Yoshioka M, Koga M, Shinohara R, Nomura Y, Choi K-S, Cho H-S, Hong J-S (2003) Heavy metal contamination in the sediment and its effect on the occurrence of the most dominant bivalve, *Ruditapes philippinarum* on the tidal flats of Ariake Bay in Kumamato prefecture, the west coast of Kyusyu, Japan. Benthos Res 58:121–130

Ueno Y, Ando H, Hayashi Y, Taino S, Ohkawa T (2007) Investigation of the cause of Japanese littleneck clam poor catch. Annu Rep Kochi Pref Fish Exp Stn 103:129–147 (in Japanese)

Urabe J, Suzuki T, Nishita T, Makino W (2013) Immediate ecological impacts of the 2011 Tohoku Earthquake Tsunami on intertidal flat communities. PLoS One 8:e62779

Uzaki N, Kai M, Aoyama H, Suzuki T (2003) Change in mortality rate and glycogen content of the Manila clam *Ruditapes philippinarum* during the development of oxygen-deficient waters. Fish Sci 69:936–943

Vincenzi S, Zucchetta M, Franzoi P, Pellizzato M, Pranovi F, De Leo GA, P. Torricelli GA (2011) Application of a random forest algorithm to predict spatial distribution of the potential yield of *Ruditapes philippinarum* in the Venice lagoon, Italy. Ecol Model 222:1471–1478

Waki T, Shimokawa J, Watanabe S, Yoshinaga T, Ogawa K (2012) Experimental challenges of wild Manila clams with *Perkinsus* species isolated from naturally infected wild Manila clams. J Invertebr Pathol 111:50–55

Williams JG (1980) Growth and survival in newly settled spat of the Manila clam, *Tapes japonica*. Fish Bull 77:891–900

Yamada S, Iwata Y, Yanagisawa T (1996) Spatial distribution of Manila clam larvae in Mikawa Bay—transporting, dispersing, recruiting processes. Kaiyo Monthly 28:150–156 (in Japanese)

Yamamoto M (2005) Sediment environment and the characteristics on the fishing ground of *Venerupis philippinarum*. Bull Fish Res Agen Suppl 3:17–25 (in Japanese with English abstract)

Yoshimatsu T, Watanabe S, Taga S, Matsuno S, Hatama T, Wanishi A (2008) Hydrographic survey for improvement and development of extensive fisheries grounds of Japanese littleneck clam. Report of 2007 commissioned survey projects for the improvement of fisheries industry. Fisheries Agency, Tokyo, 7 pp (in Japanese)

Chapter 30
Feed and Feeding in Certification Schemes of Sustainable Aquaculture

Catherine Mariojouls, Raphaëla Le Gouvello and François Simard

Abstract Certification programs for sustainable or responsible aquaculture have been widely developed. In the framework of IUCN project about sustainable feed in aquaculture, we considered the main existing certification schemes for sustainable aquaculture and analysed the present criteria taken into account for aquafeeds and their ingredients. As certification of sustainable aquaculture is a highly dynamic sector, we tried to identify the main trends in the organization of the certification schemes, and in the way the certification of sustainability for feed and raw materials is defined by the main actors. The growing substitution rate in fish feeds of fish meal and fish oil by other raw materials has widened the issue of fish feed sustainability. Therefore, certification of sustainability for aquafeed will increasingly rely on other certification schemes, issued by agriculture and feed mill sectors, where aquaculture has a limited weight. The present situation and the evolution of certification for sustainable aquaculture raise numerous questions, as their weight on small producers, and the need for a better governance for defining and applying sustainability to aquaculture, with a more balanced weight between public and private sectors.

Keywords Certification · Aquaculture · Feed · Sustainability

C. Mariojouls (✉)
AgroParisTech, 16 Rue Claude Bernard, 75231 Paris Cedex 05, France
e-mail: catherine.mariojouls@agroparistech.fr

R. Le Gouvello
IUCN/RespectOcean/STERMOR, 40 Rue de l'Eglise, 56760 Pénestin, France

F. Simard
IUCN, International Union of Conservation Nature, Gland, Switzerland

T. Komatsu et al. (eds.), *Oceanography Challenges to Future Earth*,
https://doi.org/10.1007/978-3-030-00138-4_30

30.1 Introduction: The Project of IUCN About Feed in Sustainable Aquaculture

The IUCN (International Union for Conservation of Nature) is involved *inter alia* in the reflexion about the development of aquaculture. The IUCN has been in charge in 2014–2015 of a project about "Feed for sustainable aquaculture", funded by the French Directorate for Fisheries & Aquaculture. The project is a multi-stakeholder project, as two workshops gathered numerous experts from private and public sectors. The project has been managed by a steering committee including CIPA (interprofessional body for fish culture) and SNIA/SPPA (syndicates of feed producers).

The output of the project is a report presenting the conclusions of the workshops, including chapters on peculiar points (by different authors), and proposing recommendations. One chapter presents the analysis of the certification schemes aiming at sustainable aquaculture, in the specifications they define for feed and feeding in aquaculture farms, and we present here the main findings.

30.1.1 Certification in Aquaculture

Globally, the certification programs concerning fishery and aquaculture products have grown considerably in the past two decades, either by using existing devices in agrifood sector, either, and widely, by inventing some new devices. The goals are to provide a guarantee on compliance with specifications on production (sometimes extended in the value chain, upstream to feed, or downstream to processing and retail), to improve product promotion, and to create a market segmentation.

The certification is defined by ISO[1] as a "the provision by an independent body of written assurance (a certificate) that the product, service or system in question meets specific requirements". Audits performed by an independent, third party and accredited certification body provide the highest guarantee. But note that we also find in aquaculture audits by a "first party", so self-checking its own specifications drawn up by a company or group of companies, and audits by a "second party" of specifications prepared by a customer or NGOs, done by their controllers, and some of these programs based on audits by first or second parties are included in reports on "labels and certifications" in aquaculture (Ababouch Washington 2011). Nevertheless, the third party certification brings the highest level of guarantee.

An additional difficulty in this still new and changing field is a lack of stabilized vocabulary and translations, which is source of errors or misunderstandings. For the remainder of this writing, the word "label" will be considered as "expression of a certification on the product packaging."

[1]ISO: International Organization for Standardization.

In aquaculture, certification schemes (or "quality-initiatives", "quality schemes", "labels") have various goals and a variety of settings: certification of origin (e.g. Protected Designation of Origin -PDO-, Protected Geographical Indication -PGI-, following 1992 EU regulation), certification of a particular quality (e.g."Label Rouge" i.e. the French label for a product of superior quality, "Irish quality Schemes" for several aquaculture species), certification of sustainable or responsible aquaculture (Aquaculture Stewardship Council -ASC-, Best aquaculture Practices of Global Aquaculture Alliance -BAP GAA-, Friend of the Sea -FoS-), certification of organic aquaculture (standards issued by government, as the EU standard, or standards issued by private organizations as Krav, Debio, Naturland, etc.), certification providing guarantees about sanitary quality (specifications of large retailers' groups such as Global-GAP, International Feature Standard -IFS-, British Retail Consortium -BRC-), certification of a national standard by the government or by professional organizations ("Aquaculture of our regions" of the French professional organization CIPA, Thai Quality Shrimp…). However, the specifications can combine various objectives and the creation of a typology by goal is not obvious.

Another typology can be made according to the type of organization initiating the program: a professional aquaculture sector (e.g. FEAP, CIPA, APROMAR…), the public authorities (e.g. organic aquaculture in the EU), some retailers (individually or as a group, e.g. GlobalGAP), some NGOs (e.g. ASC).

This set is difficult to understand because of its abounding and very changing nature. For more information, the interested reader can refer to the bibliography, including reports and information from international organizations (FAO 2011; IUCN 2009; Globefish news and reports). From a producer perspective, particularly in developing countries, the reports underline the difficulty of access to certification programs in the case of highly demanding specifications or because of the cost of audits for producers. So the complexity of the certification process acts as a barrier for small producers, especially in developing countries, except through groupings of small producers, when that organization is accepted by the certification program.

Analysis of this set also highlights the wide variety of certification programs and the difficulty of comparing, with a risk of confusion for the consumer. Interestingly, initiatives are underway to propose a framework for comparison, such as the Global Seafood Sustainability Initiative (GSSI) reported by Globefish (2015).

30.1.2 Certification of Sustainable Aquaculture

The area of certification programs to be considered is not obvious, for the reasons outlined above. In addition, as for sustainability, this criterion may appear secondarily in a program that did not explicitly focus on it, such is the case today of some Label Rouge for salmon.

This review focuses on four certification programs with international scope, which explicitly aim at sustainable or responsible aquaculture: BAP- GAA,

GlobalGAP, FoS and ASC. The analysis focuses on how the food sustainability is taken into account in these four major programs, and provides a brief overview of the specifications for organic aquaculture and the Red Label.

30.1.3 Organic Aquaculture, Label Rouge and Feeding

In the EU, the regulation for organic aquaculture[2] specifies that feed must cover the nutritional needs according to each aquaculture species and according to their stages of development. Marine raw materials must come from sustainable exploitation of fish stocks.[3] Other raw materials must come from organic farming or, for non-agricultural products, be natural products allowed (mentioned in a positive list). Growth promoters and synthetic amino acids are prohibited. Subsequent regulations specify certain points, introducing adjustments to the needs of species or groups.[4]

This brief review shows the spirit of the inclusion of feed in the definition of organic aquaculture in the EU. It also illustrates the existence of periodic changes in this context. Organic aquaculture ("organic aquaculture") globally represents a very diverse set of private and public programs, which analysis is currently underway in the European program ORAQUA (http://www.OrAqua.eu/).

Another important certification for aquaculture in France, the Label Rouge (i.e. Red Label) deserves a mention. It is part of the official French device of Identifying Signs of Quality and Origin, managed by the INAO.[5] This device combines a third party certification (by an accredited certification body) and a guarantee by a public device. The Label Rouge is based on a stndard defining the production conditions for a product of superior quality in its market. There is therefore no question of a priori guarantee of sustainability. However, we note that some Label Rouge in salmon farming mention a sustainability goal on their websites presenting the main points of the specifications (not publicly available in full version). For the Label Rouge Scottish Salmon LA No 33/90, it is mentioned that producers are engaged in a process of sustainability and environmental preservation. The specifications precise that the feed includes exclusively marine raw materials, vegetable

---

[2]Council Regulation (EC) N° 834/2007 of 28 June 2007 on organic production and labelling or organic products and repealing regulation (EEC) N° 292/91.

[3]As defined in Article 3 of COUNCIL REGULATION (EC) No 2371/2002 of 20 December 2002 on the conservation and sustainable exploitation of fisheries resources under the Common Fisheries Policy.

[4]The COMMISSION REGULATION (EC) No 889/2008 of 5 September 2008 laying down detailed rules for the implementation of Council Regulation (EC) No 834/2007 on organic production and labelling of organic products with regard to organic production, labelling and control. Le Règlement d'exécution (UE) n° 1358/2014 de la Commission du 18 décembre 2014; and COMMISSION IMPLEMENTING REGULATION (EU) No 836/2014.

[5]INAO: Institut National de la Qualité et de l'Origine = National Institute of origin and quality.

ingredients, vitamins, minerals and carotenoids. For Atlantic salmon Label Rouge LA No 31/05, it is emphasized that the farming is done in a "selected marine farm respecting environment" and the specifications stipulates that feed does not contain any products from land animal or GMOs (in the maximum limit of accidental contamination of 0.9%).

30.2 Presentation of the Four Certification Programs for Sustainable or Responsible Aquaculture, with International Scale

Created since 2000 or even very recently, these programs are based on a third-party certification, and differ by several aspects (see Table 30.1). BAP-GAA and ASC aim explicitly at responsible aquaculture, FoS at sustainable aquaculture, and GlobalGAP has a set of goals that include environmental and social components of sustainability.

It may be noted that these certification programs have different genesis:

- BAP-GAA is a program launched in 2000 by GAA (non-profit organization, or NGO) by stakeholders of the aquaculture sector (scientists, technicians, producers),
- GlobalGAP was created by a consortium of European distributors (GlobalGAP, formerly EurepGAP) for all agriculture and since 2004 includes an application to aquaculture,
- FoS is an NGO that has developed a certification program for sustainable fisheries, then in 2006, for sustainable aquaculture,
- ASC is a non-profit organization or NGO, created in 2010 by WWF and IDH (Dutch Sustainable Trade Initiative) to manage the standards issued from the multi-stakeholders Aquaculture Dialogues launched by WWF.

The procedures of these certifications are also different: ASC, BAP/GAA and FoS are certifications of the type B to C[6] with visible logo by the consumer, while GlobalGAP is of type B to B,[7] not visible to the consumer. They work on their own standards, developed either by species or generically for a set of species or any species, and standards are evolving in content and structure (for example, ASC launched in December 2014 a project to create a common standard, instead of 8 existing standards).

These certifications have different economic importance, but we see very strong growth currently for the number of farms and concerned products, notably for ASC and BAP–GAA. So today there is a strong competition to find farms to be certified, internationally (Table 30.1).

[6]B to C: Business to Consumer.

[7]B to B: Business to Business.

Table 30.1 Main characteristics of four certification programs for responsible or sustainable aquaculture, with international scope

Name and category	Creation year and type	Objectives	Standards for farming and other activities in supply chain	Economic weight (Dec 2015)
BAP of GAA *(NGO)*	(1997) 2000 B to C (logo)	**Responsible** aquaculture (environment, social)	Multi-species Crustacean and Finfish (incl. specific items for shrimp and tilapia), Salmon, Mussel, + Feed + Processing	1028 certified units: 92 hatcheries; 620 farms (Dec 2013: 352; 751 KT); 293 processors (1,.86 MT); 56 feed mills
ASC *(NGO)*	2010 B to C (logo)	**Responsible** aquaculture (environment, social)	Tilapia, Pangasius, Bivalves, Abalone, Salmon, Trout, Shrimp + Chain of Custody (CoC) for processors and retailers	Farms: 207 (Dec. 2014: 123; # 440 KT); CoC: 445 companies; Market: 4462 products (54 countries)
FoS *(NGO)*	2006 B to C (logo)	**Sustainable** aquaculture (environment, social)	Freshwater fishfarming, Marine fishfarming on shore, Shrimp, Mussel + Feed	Farms: # 100
Global G. A. P. *(NGO based on retailers group)*	(1997) 2004 for aquaculture B to B	Food safety, traceability, environment, animal welfare, social	Multi-species + Feed	2 MT, 20 countries (August 2012)

It is also interesting to note that these programs are changing, and tend to harmonize since 2013, with the signing of successive agreements of collaboration and exchange. As for feed in aquaculture, the "ASC Responsible Feed Project" launched by ASC from 2013 to 2015 is a comprehensive work on sustainable aquaculture feed, with a steering committee multi- agencies, and 5 technical working groups involving GAA and GlobalGAP.

30.3 Consideration of Feed in the Standards of Four Certification Programs for Sustainable or Responsible Aquaculture

The analysis of standards of the four certification programs reveals a convergence in objectives: supplies traceable and responsible/sustainable for raw materials, reduction of the use of fish meal and fish oil from dedicated fisheries. The main available features about feed are summarized in Table 30.2.

Table 30.2 Feed and feeding in four certification programs for responsible or sustainable aquaculture, with international scope—analysis of public standards (as available on internet by December 2014)

	BAP/GAA	ASC	FoS	GLOBALG. A.P.
Certification scheme for feed mills, or for feed	Certified according to BAP standard for feed mill, or equivalent	Responsible sourcing of raw materials, complying with international moratories for raw materials and regulation	• Feed certified by FoS, if available for concerned species • Or feed mill certified by IFFO RS	GLOBALG. A. P. Compound Feed Standard
All raw materials	• BAP standards or documents proving responsible sourcing • List of ingredients > 10% obtained from the supplier	Traceability, transparency, responsible sourcing for ingredients		Documents proving sustainability, traceability, transparency
Raw materials from marine origin	• List of ingredients > 1% • Sustainability of wild fisheries producing FM&FO • From June 2015, > 50% FM&FO issued from certified fisheries or by-products from certified fisheries	• Traceability by species & origin, Exclusion: species of IUCN Red List • FM&FO from fisheries certified by program accredited ISEAL (exp MSC): 10% at 3 years, 100% at 5 years. By 5 years: comply with FishSource score	• Trimmings • Anchovy and Pacific mackerel (Peru) • Menhaden (USA) • Sardines and mackerel (Marocco)	Exclusion of species of IUCN Red List, or from IUU fisheries
GMO ingredients	Control of GMO concentration (> 1%) in compound feed	Mention for GMO ingredient > 0,9% in feed	• In FishFeed 2009-3: exclusion of GMO use (limit 0,9% PCR) • Dec. 2014: no mention	

(continued)

Table 30.2 (continued)

	BAP/GAA	ASC	FoS	GLOBALG. A.P.
Soya		Certification RTRS (100% after 5 years)		
Palm oil		Certification RSPO		
Monitoring and indicators	Consumed feed, harvest volumes FCR FI-FO ratio	FFDRm (Fishmeal Forage Fish Dependency Ratio) $\leq$ 1.5 FFDRo (Fish Oil Forage Fish Dependency Ratio) $\leq$ 2.95 or EPA + DHA $\leq$ 9% FCR	FCR	

In the manufacture of feed and for traceability on raw materials, these organizations rely heavily on additional certification programs: specific programs for BAP-GAA and GlobalGAP (program CFM for *Compound Feed Manufacturers*); and for FoS, an own feed certification program or IFFO RS certification program.[8]

For fishmeal and fish oils, ASC ruled out the use of fish or by-products issued from species on the IUCN Red List or from illegal, unreported and unregulated fisheries (IUUF); and requires, within five years after the publication of the ASC concerned standard, that marine raw materials are issued from sources certified under a standard member of the ISEAL.[9] Alliance (example MSC.[10]). Before reaching this goal, supplies of marine raw materials must meet the conditions of sustainability indicators (FishSource scores).[11]

For vegetable raw materials, ASC requires soybeans certified by RTRS (RoundTable on Responsible Soy) and palm oil certified by RSPO (Roundtable for Sustainable Palm Oil).

Regarding the use of potentially GM raw materials, there is no announced exclusion (except for FoS in a 2009 document). It was noted in ASC standards a compulsory declaration of presence for GMO raw materials at a concentration greater than 0.9%, and in GlobalGAP standards, a necessary verification of the presence of GMO material in a concentration greater than 1% in the final feed.

In terms of feeding practices, feed efficiency is sought by all, both for the savings of raw materials and for reducing environmental impact. This concerns in particular

[8]IFFO RS: Global Standard and Certification Program for the Responsible Supply of Fish Meal and Fish Oil.

[9]ISEAL: International Social and Environmental Accreditation and Labelling Alliance.

[10]MSC: Marine Stewardship Council.

[11]FishSource Score: ratio from 1 to 10 characterizing the sustainability of a fishery—see https://www.fishsource.com/.

fish meal and fish oil. Thus, farms are required to do follow-ups of amounts of feed used, and to calculate indicators on the use of fish meal and fish oil or on feed efficiency (Table 30.2).

30.4 Discussion and Perspectives

The analysis of available standards shows the diversity of the means employed by several organizations for displayed similar objectives, but does not allow to fully apprehending it. The opening since 2013 of collaboration between GlobalGAP, ASC and BAP - GAA for a partial pooling on the audit phase, provides details for this comparison. The diagram (Fig. 30.1) of the initial inventory of the criteria for selection of fish meal and fish oil, for the three organizations, shows limited common share (traceability, absence of endangered species from the IUCN Red List, and preferably suppliers with responsible sourcing), but shows also important differences, ASC having the highest requirements.

It is noted that fish meal and fish oil are the only ingredients for which was conducted an important work about sustainability. Conversely, it must question how the raw materials other than fish meal and fish oil, increasingly used in fish

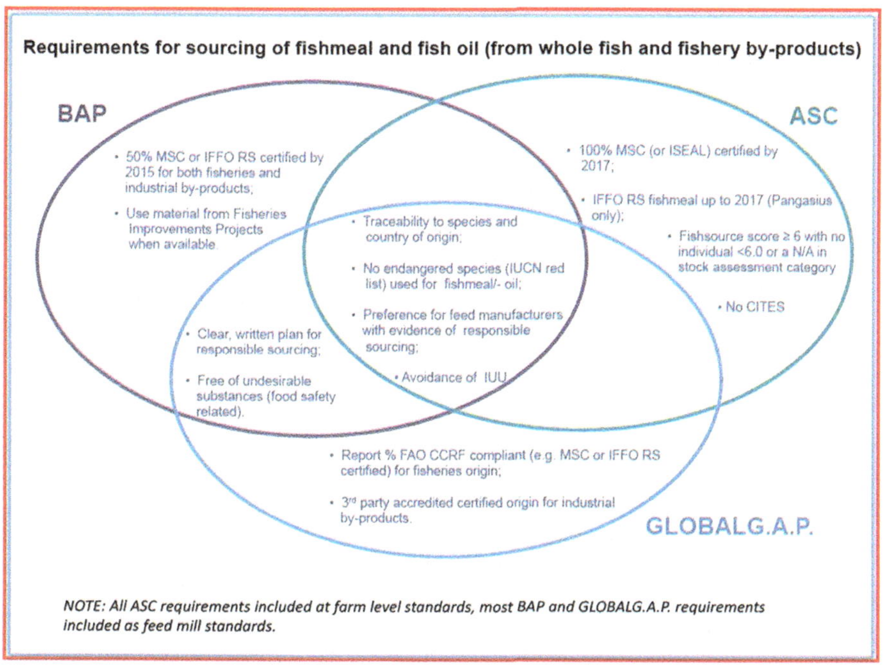

Fig. 30.1 Requirements for sourcing of fishmeal and fish oil for BAP/GAA, ASC & GlobalG.A.P. *Source* Anon (2013)

feed, will be evaluated from the perspective of sustainability. Beyond the need for responsible/sustainable supply, the plant ingredients are not specifically addressed by ASC, which for soybeans and palm oil, uses existing certification schemes in these important sectors. This movement could be developed, including other raw materials to the extent that certification of sustainable sources do exist, which is not the case for most raw materials used in animal feed.

30.5 Conclusion

The rise of certification programs for sustainable or responsible aquaculture is an important phenomenon. As for feed and feeding, the criteria included in the certifications at first focused sustainability of fisheries for fish meal and fish oils. However the increasing use of other raw materials in aquafeed, particularly plant raw materials, greatly expands the range of sectors whose sustainability must be assessed, through the use of existing specific certification programs, when they exist. It thus draws a system of certification programs used for the certification of sustainable aquafeed, bringing aquaculture in a much larger universe, that of animal feed industry and the connected agriculture supply chains, where aquaculture has a minor weight until today.

Through this paper about feed in certification programs for sustainable aquaculture is coming out a highly difficult question: where are the poles of decision for the conditions of aquaculture production, and definition of sustainability?

In addition to the regulatory framework, it appears that certification programs create an additional framework, even if it is voluntary. Internationally, the growing weight of private certification programs coming from Western organizations is widely criticized, particularly for two reasons (Bush et al. 2013):

- even if it is a response to a weak government involvement in the definition and regulation of sustainable aquaculture, today there is an imbalance of power between private and public sectors in the definition of sustainability,
- for Developing Countries, there is a form of interference and the risk of abuse of power, from organizations from Developed Countries.

Moreover, the development of aquaculture certification brings the aquaculture sector to a situation already well-known in the agricultural sector: a growing framing of the act of production by external devices. In particular, we see how the feed for sustainable aquaculture, due to an increasing use of new raw materials, will be dependent on pre-established certifications for these ingredients.

This situation rises also question of governance, at national and international levels, for defining what is feed for sustainable aquaculture, and how standards could be defined in a more balanced way, involving not only private organizations, but actors from aquaculture sector, and public authorities.

References

Anon (2013) Joint statement on requirements for the responsible sourcing of fishmeal and fish oil issued by ASC, GAA and GLOBALG.A.P. pp 2. http://www.asc-aqua.org/index.cfm?act=update.detail&uid=176&lng=1. Accessed 14 Nov 2013

Bush SR, Belton B, Hall D, Vandergeest P, Murray FJ, Ponte S, Oosterveer P, Islam MS, Mol AP, Hatanaka M, Kruijssen F, Ha TT, Little DC, Kusumawati R (2013) Certify sustainable aquaculture? Science 341:1067–1068 (6 Sep 2013)

FAO (2011) Technical guidelines on aquaculture certification. Rome/Roma, FAO, p 122

IUCN (2009). Guide for the Sustainable Development of Mediterranean Aquaculture 3. Aquaculture Responsible Practices and Certification. Gland, Switzerland and Malaga, Spain: IUCN, VI+ pp 70

Washington S, Ababouch L (2011) Private standards and certification in fisheries and aquaculture: current practice and emerging issues. FAO Fish Aquac Tech Paper 553:203 p

http://www.asc-aqua.org/index.cfm?lng=1

http://bap.gaalliance.org/

http://www.globalgap.org/uk_en/for-producers/aquaculture/

http://www.friendofthesea.org/FR/

http://www.saumonecossais.com/fr/saumon-ecossais-label-rouge/une-qualite-exceptionnelle

http://www.paq.fr/galerie/filiere-mer-aquaculture/saumon-atlantique/

http://www.bim.ie/our-services/grow-your-business/farmedfishqualitylabelling/

http://www.globefish.org/homepage.html

Part VII
Short and Preliminary Communications

Chapter 31
French Bluefin Tuna Longline Fishery Bycatch Programme

François Poisson, Sophie Arnaud-Haond, Hervé Demarcq, Luisa Métral, Blandine Brisset, Delphine Cornella and Bertrand Wendling

Abstract During the last decade, particular attention has been paid worldwide to the problem of bycatch and discards in fisheries. Collaborative research between fishermen and scientists is important to fisheries management. Partnerships with commercial longline fishermen were developed to enable them to participate in two research projects in order to integrate their information, experience and expertise. These programmes, financed by the fishing industry and regional councils were designed to describe the activity of the fisheries, to assess the scale of fishery effects on the various taxa, to study the ecology and explore spatial population genetic structure in the western part of the Mediterranean Sea of the blue shark (*Prionace glauca*) and sting rays (*Pteroplatytrygon violacea*) and finally to propose mitigation measures to reduce impacts on elasmobranchs, sea birds and sea turtles. Communication, education, post-implementation monitoring and long-standing collaboration are the key factors to success. This presentation shows the progress realized to date.

Keywords Bycatch · Longline · Shark · Megafauna · Mitigation · Mediterranean

Tuna and tuna-like species are of great importance as a global food resource and are of major economic value. Their reported global landings have increased continuously from less than 0.6 mt in 1950 to 7 mt in 2010 (FAO 2014). By-catch in tuna fisheries is the primary source of fishing mortality of some marine species (Bellido et al. 2011), and tuna longline fisheries have been associated as one of the major

F. Poisson (✉) · S. Arnaud-Haond · H. Demarcq · L. Métral · B. Brisset
MARBEC, Univ Montpellier, CNRS, Ifremer, IRD, Ville, France (Sète), Avenue Jean Monnet CS 30171, 34203 Sète, France
e-mail: francois.poisson@ifremer.fr

D. Cornella
IRD, Unité Halieutique Méditerranée (HM), UMR 212 - Ecosystème Marin Exploité (EME), B.P. 171, Av. Jean Monnet, 34203 Sète Cedex, France

B. Wendling
Association Méditerranéenne des Organisations de Producteurs - AMOP 29, Promenade J-B Marty, 34200 Sète, France

T. Komatsu et al. (eds.), *Oceanography Challenges to Future Earth*,
https://doi.org/10.1007/978-3-030-00138-4_31

threat to populations of various marine vertebrates, including marine turtles (Baez et al. 2013; Lewison et al. 2004; Roe et al. 2014), sharks (Gilman et al. 2007; Huang 2011), seabirds (Croxall et al. 2012; Lewison et al. 2012), marine mammals (Forney et al. 2011; Macías et al. 2012).

The pelagic longline is defined as a series of baited hooks regularly attached to a mainline suspended by buoys. This passive fishing method targets tunas (bluefin, bigeye, yellowfin, albacore) and swordfish. The impact of pelagic longline fisheries on noncommercial and protected species varies with many factors including season and location of fishing, fishing gear and method, abundance and behavior of potential bycatch species. Therefore, it is important not to generalize across fisheries when considering bycatch issues.

The French artisanal Bluefin tuna (*Thunnus thynnus*) fisheries in the Mediterranean Sea have changed over the past decade following the mandatory measures imposed by the Regional Fisheries Management Organizations (ICCAT and GFCM) and the environmental non-governmental organizations. While two main gears were used (rod and reel and large-scale pelagic driftnet) in the Gulf of Lions, the French fisheries components have deeply changed with a shift to longline as main gear. Fishers were interested in participating in cooperative research aimed at improving the sustainability of their fishery. The bycatch issue is well known to stakeholders and the fishing industry which accepted that the problem needed to be addressed. The reduction of bycatch mortality is an objective of the ecosystem approach to fisheries and a strong request made by consumers.

In 2013, a three-year research program was launched in collaboration with the domestic bluefin tuna longline fishery. This program is financed by the fishing sector (France FilièrePêche), by Ifremer and by three regional councils (Hérault, Pyrénées-Orientales and Languedoc Roussillon). The SELPAL programme conducted in collaboration with the fishing industry is designed to describe the activity of the fishery, to assess the scale of fishery effects on the various taxa, to study the ecology of the species which are or could be potentially impacted.

In 2016, 101 vessels are engaged in the domestic longline fleet which can be categorized by two vessel sizes (classes): a small boat class (vessels < 12 m), and a large boat class (vessels $\geq$ 12m). The domestic longline fishery is a surface fishery. Onboard small vessels, the line are cast manually and a hydraulic line hauler is used to retrieve the fishing gear, hooks are stored in baskets. While larger boats, the mainline is deployed from a hydraulically powered spool over the stern a monofilament reel system uses a hydraulic reel to haul and store the mainline.

The fishing season occurs from April to December. Logbooks were distributed to all domestic vessels, and fishers received subsidies to complete them on a regular basis. Logbook datamade up the major source of information used to assess the composition of the catches and catch per unit of effort (CPUE). The composition by number was 56% of pelagic sting ray (*Pteroplatytrygon violacea*), 35% of bluefin tuna (*Thunnusthynnus*), 8% of Blue shark (*Prionace glauca*) and 1% of swordfish (*Xiphias gladius*).

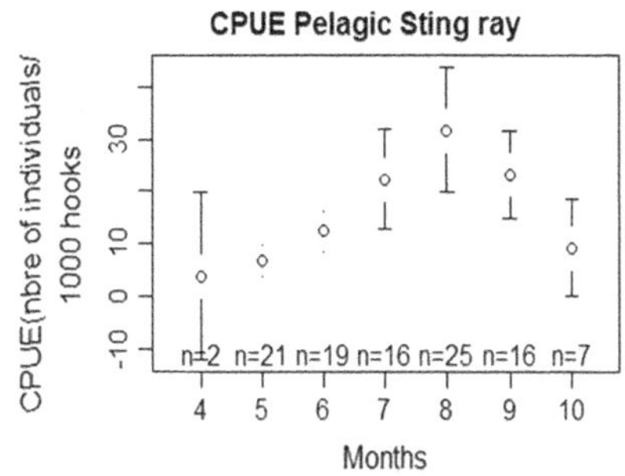

Fig. 31.1 Cpues of pelagic sting ray (*Pteroplatytrygonviolacea*) in the domestic blue fin tuna fishery (2013–2015)

Cpues for the pelagic sting ray, the first component species increases steadily from May to reach a peak in August then decrease and only few stingrays are caught after October (Fig. 31.1).

The project aims also at investigating on habitat occupancy, residency times and migratory pathways as well as providing behavioural data on temperature experience and swimming depth of the blue shark and pelagic sting ray. To date, 32 blue sharks have been tagged in the French coast of Mediterranean with different types of "smart tags", in the Alboran Sea and the eastern coast of Corsica. The preliminary analysis from the deployed tags, show that specimens tagged moved moves along the coast to shelf break of the continental shelf north of the Tyrrhenian Sea, the Ligurian Sea, and the Balearic Sea until Alboran Sea. Some live tracks have been displayed on two dedicated websites.[1] Tags deployed have provided both horizontal and vertical movement data for several individuals (e.g. Figure 31.2). These data have been yet fully analyzed (Fig. 31.3).

21 pelagic stingrays have been tagged mainly in the Gulf of Lions. Preliminary analyses of pelagic sting ray behavioral data indicated significant diel differences in vertical movements with thermal ranges of 3–10 °C experienced across a 24 h period.

At-vessel mortality and post-release survival rates are also investigated for these two species.

The preliminary results seem to indicate that at-vessel mortality are very high for both species (>95%). Several studies have highlighted the fact that handling practices is a key factor which can increase the survival of discarded fish (Abascal et al. 2011; Campana et al. 2009; Poisson et al. 2014). Dissemination of good handling and release practices and appropriate equipment have been identified as good methods to reduce mortality and injury of vulnerable species. Therefore, a good practices manual is under development. This manual will provide appropriate handling practices to ensure crew safety and increase the odds of survival for released animals, based on scientific observations and empirical knowledge from skippers and crew (Fig. 31.2).

[1](http://amop-selpal.com/index.php/suivi-des-especes-marquees and http://www.stellaris-asso.org/suivi-des-requins).

Fig. 31.2 Drawing showing how to release a sting ray with a line cutter (from the good practices manual to reduce the mortality of the bycatch in the longline fishery targeting bluefin tuna)

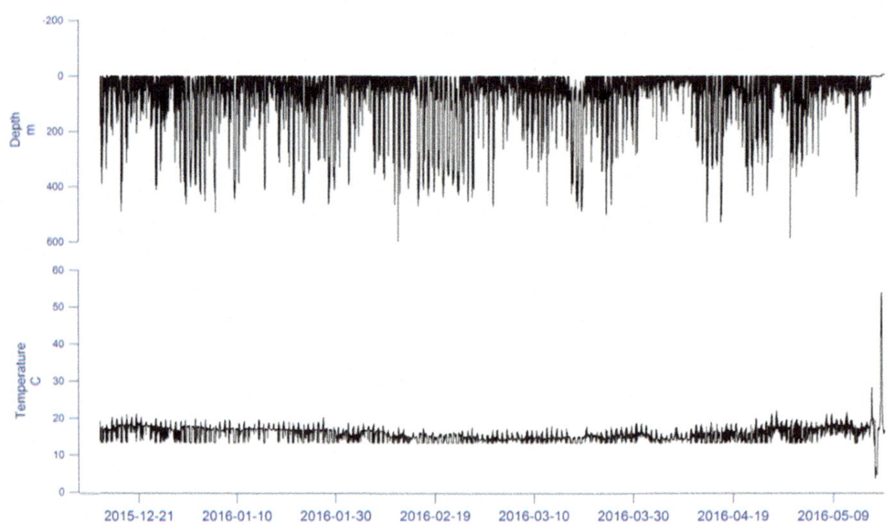

Fig. 31.3 Depth and temperature data for a period of 153 days recording from a splash tag (Wildlife Computers, Redmond, Washington, USA) deployed on a female blue shark (fork length = 1.72 m)

References

Abascal FJ, Quintans M, Ramos-Cartelle A, Mejuto J (2011) Movements and environmental preferences of the shortfin mako, Isurus oxyrinchus, in the southeastern Pacific Ocean. Mar Biol 158(5):1175–1184

Baez JC, Macias D, Caminas JA, de Urbina JMO, Garcia-Barcelona S, Bellido JJ, Real R (2013) By-catch frequency and size differentiation in loggerhead turtles as a function of surface longline gear type in the western Mediterranean Sea. J Mar Biol Assoc U K 93(5):1423–1427

Bellido J, Santos MB, Pennino MG, Valeiras X, Pierce G (2011) Fishery discards and bycatch: solutions for an ecosystem approach to fisheries management? Hydrobiologia 670(1):317–333

Campana SE, Joyce W, Manning MJ (2009) Bycatch and discard mortality in commercially caught blue sharks Prionace glauca assessed using archival satellite pop-up tags. Mar Ecol Prog Ser 387:241–253

Croxall JP, Butchart SHM, Lascelles B, Stattersfield AJ, Sullivan B, Symes A, Taylor P (2012) Seabird conservation status, threats and priority actions: a global assessment. Bird Conserv Int 22(01):1–34

FAO (2014) The State of World Fisheries and Aquaculture (2014) ISBN 978-92-5-108275-1. Food and Agriculture Organization of the United Nations, Rome

Forney KA, Kobayashi DR, Johnston DW, Marchetti JA, Marsik MG (2011) What's the catch? Patterns of cetacean bycatch and depredation in Hawaii-based pelagic longline fisheries. Mar Ecol 32(3):380–391

Gilman EL, Clarke S, Brothers NJ, A-S, Mandelman J, Mangel J, Petersen S, Piovano S, Thomson N, Dalzell P, Donoso M, Goren M, Werner T (2007) Shark depredation and unwanted bycatch in pelagic longline fisheries: Industry practices and attitudes, and shark avoidance strategies. Western Pacific Regional Fishery Management Council. Honolulu, HI

Huang HW (2011) Bycatch of high sea longline fisheries and measures taken by Taiwan: actions and challenges. Mar Policy 35(5):712–720

Lewison RL, Freeman SA, Crowder LB (2004) Quantifying the effects of fisheries on threatened species: the impact of pelagic longlines on loggerhead and leatherback sea turtles. Ecol Lett 7 (3):221–231

Lewison RL, Oro D, Godley B, Underhill L, Bearhop S, Wilson RP, Ainley D, Arcos JM, Boersma PD, Borboroglu PG, Boulinier T, Frederiksen M, Genovart M, González-Solís J, Green JA, Grémillet D, Hamer KC, Hilton GM, Hyrenbach KD, Martínez-Abraín A, Montevecchi WA, Phillips RA, Ryan PG, Sagar P, Sydeman WJ, Wanless S, Watanuki Y, Weimerskirch H, Yorio P (2012) Research priorities for seabirds: improving conservation and management in the 21st century. Endanger Species Res 17(2):93–121

Macías LD, Barcelona SG, Báez JC, De la Serna JM, Ortiz de Urbina, JM (2012) Marine mammal bycatch in Spanish Mediterranean large pelagic longline fisheries, with a focus on Risso's dolphin (Grampus griseus). Aquatic Living Resources 25(04):321–331

Poisson F, Filmalter JD, Vernet A-L, Dagorn L (2014) Mortality rate of silky sharks (Carcharhinus falciformis) caught in the tropical tuna purse seine fishery in the Indian Ocean. Can J Fish Aquat Sci 71(6):1-4. https://doi.org/10.1139/cjfas-2013-0561

Roe JH, Morreale SJ, Paladino FV, Shillinger GL, Benson SR, Eckert SA, Bailey H, Tomillo PS, Bograd SJ, Eguchi T, Dutton PH, Seminoff JA, Block BA, Spotila JR (2014) Predicting bycatch hotspots for endangered leatherback turtles on longlines in the Pacific Ocean. Proc Royal Soc B: Biol Sci 281(1777)

Chapter 32
^{137}Cs and Tritium Concentrations in Seawater off the Fukushima Prefecture: Results from the SOSO 5 Rivers Cruise (October 2014)

Michio Aoyama, Hervé Thébault, Y. Hamajima, Sabine Charmasson, Mireille Arnaud and Céline Duffa

32.1 Introduction

The Fukushima Dai-ichi Nuclear Power Plant (FDNPP1) accident, which occurred in Japan on March 11, 2011, resulted in the releases of large amount of various radionuclides especially caesium isotopes (^{137}Cs half-life 30.2y and ^{134}Cs half-life 2.06y) both in the atmosphere and in the oceans. Various authors have observed that caesium have been transported from contaminated watershed to rivers (Nagao et al. 2013). Three and a half years after the accident this study has been conducted in order to investigate the river influence on coastal areas.

32.2 Methods

In October 2014, the cruise SOSO 5 Rivers took place off the coast of the Fukushima prefecture. The sampling was targeted off the mouths of 5 coastal rivers located North of the FDNPP and whose watershed were strongly contaminated by accident fallout i.e. Mano, Nitta, Ota, Odaka and Ukedo rivers.

M. Aoyama
Institute of Environmental Radioactivity, Fukushima University, 1 Kanayagawa, Fukushima 960-1296, Japan

H. Thébault (✉) · S. Charmasson · M. Arnaud · C. Duffa
Institut de Radioprotection et de Sûreté Nucléaire (IRSN), CS 20330, La Seyne sur mer cedex 83507, France
e-mail: herve.thebault@irsn.fr

S. Charmasson
e-mail: sabine.charmasson@irsn.fr

Y. Hamajima
Low Level Radioactivity Laboratory, Kanazawa University, Wake, Nomi, Ishikawa 923-1224, Japan

T. Komatsu et al. (eds.), *Oceanography Challenges to Future Earth*,
https://doi.org/10.1007/978-3-030-00138-4_32

Seawater was sampled along radials comprising 5 stations in front of each of these 5 rivers. Seawater was sampled on surface and 1 m above bottom. In addition one water sample was realised in 4 estuaries (the Ukedo estuary was not accessible at that time). All the samples were filtered on 0.45 μm. An improved ammonium phosphomolybdate (AMP) procedure (Aoyama and Hirose 2008) was used to extract caesium from the samples. The weight yield of AMP/Cs compound generally exceeds 99% for 2 litre samples as well as radiochemical yield of radiocaesium. The activities of AMP/Cs compound were measured at the Ogoya Underground Facility of the Low Level Radioactivity Laboratory of Kanazawa University using high-efficiency, well-type ultra-low background Ge-detectors. One liter of seawater underwent tritium analysis through electrolytic enrichment at the Tritium Laboratory (Miami, USA).

32.3 Results and Discussion

Both ^{134}Cs and ^{137}Cs were detected in all the samples demonstrating a contamination from the FDNPP accident releases. Generally the concentrations were higher at coastal sites and decreased with distance from the coast, and were higher in the surface layer compared to the bottom layer. The relationship between ^{137}Cs and ^{134}Cs decay corrected to the date of the accident showed that pre-Fukushima ^{137}Cs activities due to global fallout from nuclear bomb testings (Aoyama et al. 2008) were close to 1.5 Bq.m^{-3}. The ^{137}Cs/^{134}Cs activity ratio was close to 1, a value consistent with the ratio observed nearby the release point after the accident (Buesseler et al. 2011).

Regarding tritium, only a part of all samples (11) were analysed (Table 32.1). The tritium levels are quite low but higher levels always characterized estuarine waters. In seawater they ranged from 1.48 $\pm$ 0.09 in front of the Mano river to 0.57 $\pm$ 0.09 TU at 6.7 nautic miles off the Nita mouth (Nitta 5). These results underlined the fact that away of rivers influence, seawater tritium content is less than 1 TU which corresponds to background level in the marine environment. In rivers tritium contents ranged from 4.10 $\pm$ 0.14 in the Ota estuary to 1.81 $\pm$ 0.09 in the Odaka estuary. These levels are in agreement with what is known from continental freshwater, the continental tritium reservoir being higher than the marine one. In Japan, levels up to 6TU are conservative estimate of the pre-Fukushima background level (Matsumoto et al. 2013). Therefore our dataset does not show any influence of the tritium released by the Fukushima accident on these rivers or on coastal waters north FDNPP in October 2014.

Table 32.1 Tritium content (±1 sigma) in rivers and coastal rivers (SoSo5 rivers project-October 2014)

Location	Date (yymmdd)	TU	eTU
Mano Estuary	14/10/11	3.28	0.11
Mano 1 surface	14/10/09	1.48	0.09
Nitta Estuary	14/10/11	3.89	0.13
Nitta 1 surface	14/10/19	0.89	0.09
Nitta 5 surface	14/10/19	0.57	0.09
Ota Estuary	14/10/11	4.10	0.14
Odaka Estuary	14/10/11	1.81	0.09
Odaka 1 surface	14/10/18	0.99	0.09
Odaka 5 surface	14/10/18	0.68	0.10
Ukedo 1 surface	14/10/17	0.92	0.09
Ukedo 5 surface	14/10/17	0.70	0.09

Acknowledgements This work was part of the AMORAD project (French National Agency for Research ANR-11-RSNR-0002). The authors thank the KANSO team for their helpful and expert assistance in field work.

References

Aoyama M, Hirose K (2008) In: Pavel PP (ed) Radioactivity in the environment. Elsevier

Aoyama M, Hirose K, Nemoto K, Takatsuki Y, Tsumune D (2008) Geophys Res Lett 35(1)

Buesseler K, Aoyama M, Fukasawa M (2011) Environ Sci Technol 45(23):9931–9935

Matsumoto T, Maruoka T, Shimoda G, Obata H, Kagi H, Suzuki K, Yamamoto K, Mitsuguchi T, Hagino K, Tomioka N, Sambandam C, Brummer D, Martin Klaus P Aggarwal (2013) Sci Total Environ 445–446, 365–370

Nagao S, Kanamori M, Ochiai S, Tomihara S, Fukushi K, Yamamoto M (2013) Biogeosciences 10:6215–6223

Author Index

A
Abe, Hirokazu, 365
Amano, Yosuke, 235
Amilhat, Elsa, 69
Aoki, Shigeru, 123
Aoyama, Michio, 407
Arakawa, Hisayuki, 21, 29, 195, 203
Arnaud-Haond, Sophie, 401
Arnaud, Mireille, 407
Asada, Minami, 95

B
Baffreau, Alexandrine, 105
Bec, B., 341
Bernard, I., 341
Boisneau, C., 69
Boisneau, Ph., 69
Bonnet, D., 341
Brisset, Blandine, 401

C
Ceccaldi, Hubert-Jean, 3
Charmasson, Sabine, 407
Chiantella, C., 341
Cochet, H., 341
Cornella, Delphine, 401

D
Dauvin, Jean-Claude, 105, 285
Demarcq, Hervé, 401
Devlin, Robert H., 223
Duffa, Céline, 407
Dufour, Florence, 295

F
Ferrou, Nelly, 295
Feunteun, Eric, 69
Fiandrino, A., 341
Fourneau, Gérard, 295
Foveau, Aurélie, 285
Fujio, Shinzou, 171
Fukuyo, Yasuwo, 305
Furuya, Ken, 305

G
Galgani, François, 55
Gervasoni, E., 341
Girault, Mathias, 181
Guillou, J. L., 341

H
Hamaguchi, Masami, 277, 365
Hamajima, Y., 407
Hara, Motoyuki, 351
Hasegawa, Kohei, 29
Hattori, Akihiro, 181
Hattori, Teruhisa, 161
Hayashi, Mitsuru, 45
Henocque, Yves, 325
Hirakawa, Naoto, 29
Hirota, Masahito, 277
Hori, Masakazu, 277

I
Igeta, Yosuke, 137
Ikeda, Yoshichika, 195, 203
Ishibashi, Kenichi, 217

T. Komatsu et al. (eds.), *Oceanography Challenges to Future Earth*,
https://doi.org/10.1007/978-3-030-00138-4

Ishizu, Miho, 171
Ito, Hiroshi, 351
Izumi, Mitsuru, 195

K
Kajihara, Naoto, 365
Kakehi, Shigeho, 161
Kamiyama, Takashi, 365
Kim, Hyonchol, 181
Kimura, Ryo, 277
Kitade, Yujiro, 123, 137
Kobayashi, Eiichi, 45
Kohama, Keiyu, 305
Koshimura, Shunich, 45
Kumaki, Yutaka, 137
Kunisato, Ritsuki, 161
Kurita, Yutaka, 235
Kurokura, Hisahi, 305

L
Laborde, Aurore, 295
Lagarde, Franck, 277, 341
Le Gall, P., 341
Le Gouvello, Raphaëla, 387
Le Pape, Olivier, 313
Leurion, A., 341

M
Magome, Shinya, 161
Makino, Mitsutaku, 277
Mariojouls, Catherine, 387
Masuda, Takako, 305
Matsumoto, Akira, 203
Matsuura, Kenji, 181
Meddah, S., 341
Messiaen, G., 341
Métral, Luisa, 401
Michelet, Nicolas, 69, 249
Miki, Motohiro, 195
Miron, G., 341
Miyamoto, Yoshinori, 29
Moriyama, Akihiro, 365
Mortreux, S., 341
Murata, Hiroki, 95

N
Nakada, Satoshi, 45
Nakane, Yukinori, 235
Nakano, Toshiki, 223
Nawata, Akatsuki, 351
Nishigaki, Hajime, 171

O
Odaka, Masao, 181
Ohshima, Kay I., 123
Oishi, Taro, 305
Okabe, Katsuaki, 161
Okumura, Yutaka, 351
Oshino, Akio, 351, 365

P
Penne, Aurélie, 295
Perignon, A., 341
Pezy, Jean-Philippe, 105
Poisson, François, 401
Pouvreau, S., 341
Prouzet, Patrick, 69, 249

R
Richard, M., 341
Roque D'orbcastel, E., 341
Roques, C., 341

S
Sasa, Shuji, 95
Sawayama, Shuhei, 95
Scourzic, Thomas, 295
Shen, Zhonghua, 305
Shibata, Rena, 217
Shimada, Keishi, 123
Shirai, Hidekazu, 161
Shirakawa, Hitoshi, 223
Simard, François, 387
Soga, Tomoyoshi, 223

T
Takagi, Takamasa, 161
Takayanagi, Kazufumi, 161
Tanaka, Kiyoshi, 171
Taniai, Yuichi, 365
Terauchi, Genki, 95
Terazono, Hideyuki, 181
Teruhisa, Komatsu, 15, 95
Tezuka, Naoaki, 277
Thébault, Hervé, 407
Togashi, Hiroyuki, 235
Toguchi, Kazuki, 203
Tsujimoto, Ryo, 95

U
Ubertini, M., 341
Uchida, Keiichi, 29
Uto, Yasuyuki, 217

V
Vaz, Sandrine, 277, 313

W
Wakita, Kazumi, 305
Watanabe, Tatsuro, 137
Wendling, Bertrand, 401

Y
Yada, Takashi, 217
Yagi, Nobuyuki, 305
Yamazaki, Keiichi, 137
Yanagimoto, Daigo, 171
Yanagi, Tetsuo, 95
Yasuda, Kenji, 181
Yeo, Giles, 223
Yoshida, Katsumi, 305

Z
Zaccari, Elodie, 295

Printed by Printforce, the Netherlands